Diversity of Bacterial Respiratory Systems

Volume I

Editor

Christopher J. Knowles, Ph.D.
Senior Lecturer in Biochemistry
Biological Laboratory
University of Kent
Canterbury, Kent
United Kingdom

CRC Press
Taylor & Francis Group
Boca Raton London New York

CRC Press is an imprint of the
Taylor & Francis Group, an **informa** business

First published 1980 by CRC Press
Taylor & Francis Group
6000 Broken Sound Parkway NW, Suite 300
Boca Raton, FL 33487-2742

Reissued 2018 by CRC Press

Library of Congress Cataloging in Publication Data

Main entry under title:

The Diversity of bacterial respiratory systems.
Bibliography: p.
Includes index.
I . Microbial respiration. 2. Bacteria—
Physiology. I. Knowles, C. J . [DNLM:
I. Bacteria—Physiology. 2. Cell membrane—
Physiology. 3. Respiration. QW52.3 D618]
QR89.D58 589.9'01 '2 79-17010
ISBN 0-8493-5399-8 (Volume I)
ISBN 0-8493-5400-5 (Volume II)

A Library of Congress record exists under LC control number: 79017010

Publisher's Note
The publisher has gone to great lengths to ensure the quality of this reprint but points out that some imperfections in the original copies may be apparent.

ISBN 13: 978-1-315-89239-9 (hbk)
ISBN 13: 978-1-351-07149-9 (ebk)

Visit the Taylor & Francis Web site at http://www.taylorandfrancis.com and the
CRC Press Web site at http://www.crcpress.com

PREFACE

Although wide differences occur in the composition and function of mitochondrial respiratory systems, there is also a distinct and fundamental similarity between them, whether they originate from plants, animals, or microorganisms. However, in bacteria, respiratory systems vary enormously in both composition and function from the very simple to complex mitochondrial-like systems, depending on the degree of evolutionary sophistication of the organism and the type of habitat in which they exist. For example, there are bacteria that respire with oxygen, nitrate, fumarate, or sulfur compounds as electron acceptors, whereas mitochondrial systems respire only to oxygen. Some bacteria even require reversal of electron transfer against the normal electrochemical gradient in order to grow.

It is the aim of this book to present reviews on a wide range of aspects of bacterial respiratory systems. Because of the on-going publication elsewhere of reviews on bacterial respiration, a "blanket" coverage of the field has not been attempted. Rather, a range of topics have been selected, either because they are of special current interest, they have not been reviewed recently, or they have never been reviewed.

C. J. Knowles

THE EDITOR

Christopher J. Knowles, Ph.D., is Senior Lecturer in Biochemistry in the Biological Laboratory of the University of Kent, Canterbury, England. Dr. Knowles received his B.Sc. in chemistry from the University of Leicester in 1964 and his Ph.D. in biochemistry in 1967. From September 1967 to September 1969 he was a Postdoctoral Fellow of the American Heart Foundation at Dartmouth Medical School, Hanover, New Hampshire, U.S.A. In 1969 he returned to Britain as a Science Research Council Postdoctoral Fellow for one year at the University of Warwick. In October 1970 he was appointed Lecturer in Biochemistry at the University of Kent and promoted to Senior Lecturer in October 1977.

CONTRIBUTORS

Assunta Baccarini-Melandri, Ph.D.
Assistant Professor of Plant Physiology
University of Bologna
Bologna, Italy

Werner Badziong, Dr. rer. nat.
Research Associate
Fachbereich Biologie
Philipps-Universität Marburg
Auf den Lahnbergen
Marburg/Lahn
Federal Republic of Germany

Philip D. Bragg, Ph.D.
Professor of Biochemistry
University of British Columbia
Vancouver, British Columbia
Canada

Jan William Drozd, Ph.D.
Fermentation and Microbiology Division
Shell Research Limited
Shell Biosciences Laboratory
Sittingbourne Research Center
Sittingbourne, Kent
United Kingdom

I. John Higgins, Ph.D.
Senior Lecturer in Biochemistry and Microbiology
Biological Laboratory
University of Kent
Canterbury, Kent
United Kingdom

Peter Jurtshuk, Jr., Ph.D.
Professor of Biology
University of Houston
Houston, Texas

Christopher J. Knowles, Ph.D.
Senior Lecturer in Biochemistry
Biological Laboratory
University of Kent
Canterbury, Kent
United Kingdom

Wil N. Konings, Ph.D.
Associate Professor of Microbiology
Department of Microbiology
Biological Center
University of Groningen
Groningen
The Netherlands

Achim Kröger, Dr. phil.
Akademischer Rat
Institut für Physiologische Chemie
Universität München
München
Federal Republic of Germany

Paul A. M. Michels, Ph.D.
Research Fellow
Department of Microbiology
Biological Center
University of Groningen
Groningen
The Netherlands

Oense M. Neijssel, Ph.D.
Lecturer
Laboratorium voor Microbiologie
Universiteit van Amsterdam
Amsterdam
The Netherlands

Jae Key Oh, Ph.D.
Research Associate
Department of Microbiology
University of Manitoba
Winnipeg, Manitoba
Canada

L. F. Oltmann, Ph.D.
Research Fellow
Biological Laboratory
Free University
Amsterdam
The Netherlands

Robert K. Poole, Ph.D.
Lecturer
Department of Microbiology
Queen Elizabeth College
University of London
Campden Hill, London
United Kingdom

Irmelin Probst, Ph.D.
Research Associate
Institut für Mikrobiologie
Universität Göttingen
Göttingen
Federal Republic of Germany

Belinda Seto, Ph.D.
Senior Staff Fellow
Laboratory of Biochemistry
National Heart, Lung, and Blood Institute
National Institutes of Health
Bethesda, Maryland

Adrian H. Stouthamer, Ph.D.
Professor of Microbiology
Biological Laboratory
Free University
Amsterdam
The Netherlands

Isamu Suzuki, Ph.D.
Professor and Head
Department of Microbiology
University of Manitoba
Winnipeg, Manitoba
Canada

David W. Tempest, D.Sc.
Professor
Laboratorium voor Microbiologie
Universiteit van Amsterdam
Amsterdam
The Netherlands

Rudolf K. Thauer, Dr. rer. nat.
Professor of Microbiology
Fachbereich Biologie
Philips-Universität Marburg
Auf den Lahnbergen
Marburg/Lahn
Federal Republic of Germany

Jan van't Riet, Ph.D.
Senior Lecturer in Biochemistry
Biochemical Laboratory
Free University
Amsterdam
The Netherlands

Ralph S. Wolfe, Ph.D.
Professor of Microbiology
Department of Microbiology
University of Illinois at Urbana-Champaign
Urbana, Illinois

Tsan-yen Yang, Ph.D.
Postdoctoral Fellow
Johnson Research Foundation
University of Pennsylvania
School of Medicine
Philadelphia, Pennsylvania

Davide Zannoni, Ph.D.
Assistant Professor of Plant Biochemistry
University of Bologna
Bologna, Italy

TABLE OF CONTENTS

Volume I

TABLE OF CONTENTS

Volume II

Chapter 1

GROWTH YIELD VALUES IN RELATION TO RESPIRATION

David W. Tempest and Oense M. Neijssel

TABLE OF CONTENTS

I. INTRODUCTION

Aerobic bacteria, and facultative anaerobes growing aerobically and utilizing oxygen, effect the bulk of their ATP synthesis through electron transport chain activity in which oxygen serves as the terminal electron acceptor. Hence, the specific rate of respiration may be taken as a measure of the specific rate of ATP synthesis. Further, since microbial growth demands, in addition to essential nutrient substances, a supply of energy in the form of ATP, one might anticipate that the specific rate at which aerobic organisms grow will be a function of the specific rate of oxygen consumption. Clearly, then, if this is the case, the amount of new material synthesized per unit of oxygen consumed should be a precise indicator of the ATP requirement of cell synthesis. Alternatively, assuming some common value for the ATP requirement of cell synthesis, the amount of biomass synthesized per unit of oxygen consumed may be used as a measure of the efficiency with which oxidative phosphorylation proceeds in different organisms.

In agreement with the above hypothesis, it has generally been found with cultures growing aerobically in either batch or continuous culture that the specific rates of carbon-substrate and oxygen consumption

$$q = -1/x \cdot ds/dt$$

correlate closely with the specific rate of cell synthesis

$$\mu = 1/x \cdot dx/dt$$

That is

$$\mu = Yq \qquad (1)$$

where Y is the proportionality factor or *yield value*. However, whereas this relationship holds true with many carbon-substrates and with many organisms growing at a constant rate, nevertheless the actual yield value expressed does vary to a greater or lesser extent with growth rate, and also with changes in many environmental parameters. In some cases a rational explanation has been forthcoming, but in others this is not so, and a unifying hypothesis that allows for all yield data and their variations to be interpreted in precise physiological and/or bioenergetic terms is still lacking. Fundamental to this problem is a paucity of knowledge regarding the stoichiometries of energy generation and of its consumption in cell synthesis and uncertainties as to the degree to which these two are coupled. This chapter, therefore, attempts to identify and analyze the main components of the problem and to draw broad conclusions regarding the relationships between energy-generating and energy-consuming reactions extant in aerobically growing heterotrophic organisms. However, the principles developed herein apply equally, though in an even simpler form, to phototrophs and chemolithotrophs since in these organisms the energy-yielding reactions are separate from those effecting intermediary metabolite synthesis.

II. GROWTH AND ASSOCIATED PROCESSES

A. Measurable Parameters

All living organisms seemingly contain carbon, hydrogen, oxygen, nitrogen, sulphur, and phosphorus, as well as potassium, magnesium, and a number of so-called "trace" elements. Therefore, each of these cellular components must be present in the environment in a suitably utilizable form, along with water and a source of energy. Then, if other conditions are propitious, microbial growth can occur.

Growth involves the uptake and chemical transformation of a small number of substances, the energy for this process being derived, in heterotrophic organisms, from the oxidation of the carbon substrate. Thus, simplifying the problem of energetic interrelationships, one can consider yield values primarily in the context of carbon substrate metabolism and pose the question: "What processes are involved in the conversion of carbon substrate to cells?" The least complex scheme is as follows:

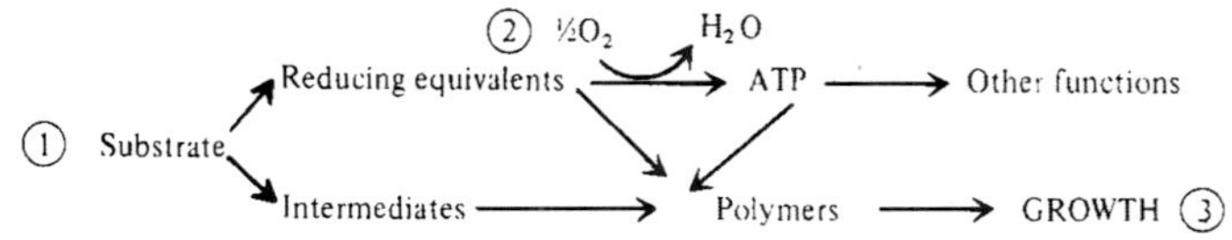

Here, substrate catabolism is depicted as generating intermediary metabolites and reducing equivalents. A portion of the latter is oxidized to generate ATP which, together with the other portion of reducing equivalents, is consumed in polymer synthesis and growth. It is necessary to emphasize here that ATP also would be synthesized by substrate-level phosphorylation reactions implicated in the formation of some intermediary metabolites, but that the amount of ATP so formed would be small in comparison to that generated by respiration-linked processes. Therefore, in the first instance, this might be ignored. Further, it is assumed that, though ATP is a substrate (that is, an intermediary metabolite) in nucleic acid synthesis, it is largely hydrolyzed to ADP and AMP in reactions leading to the synthesis of the other polymeric constituents of the cell and in the so-called "other functions" depicted in the above scheme.

Further consideration of this scheme reveals that three component processes (and only three) can be measured directly. These are (1) the rate of substrate uptake, (2) the rate of oxygen consumption, and (3) the rate of cell synthesis. For although the rates of synthesis of some polymers might be directly measurable, the rate of formation of intermediates, reducing equivalents, and particularly, ATP can only be derived indirectly. Therefore, for the moment, it is appropriate to concentrate on the three parameters that can be directly quantified and to analyze their relationships each to the other two.

1. Relationships Between Measurable Parameters

Firstly, it is unquestionably true that growth (as measured by a net increase in biomass) cannot occur in the absence of carbon substrate. In other words, for there to be a finite rate of biomass synthesis there must be a corresponding rate of carbon substrate uptake and assimilation. Similarly, growth either cannot occur, or else occurs with a markedly decreased efficiency in terms of $Y_{substrate}$ in the absence of oxygen. Yet again, oxygen uptake rate is markedly influenced by (and related to) the rate of substrate uptake. However, it still can occur to some extent in the absence of added substrate, as happens with washed suspensions of organisms that express an "endogenous" respiration. These three relationships, then, one might define as "strict", but there are a number of others that are not so clearly interdependent.

With facultatively anaerobic bacteria, for example, the substrate uptake rate is not dependent on the oxygen uptake rate. Indeed, the substrate uptake rate may increase when the oxygen uptake rate is deliberately impeded, the well known "Pasteur Effect" (see Reference 5). Moreover, under other circumstances (e.g., with washed suspensions of bacteria), neither substrate uptake rate nor oxygen consumption rate is dependent on growth rate. Thus, substrate can be oxidized at a high rate under conditions where growth is highly constrained, if not totally prevented.

It is clear, then, that in terms of Equation 1 μ is a function of q, but q is not a function of μ. This applies equally to the specific rate of carbon substrate uptake (q_{sub}) and the specific rate of oxygen consumption (q_{O_2}). Hence, to analyze this situation further, one must now consider the other components of the scheme depicted previously.

2. The Requirement for Reducing Equivalents

With aerobic bacteria, reducing equivalents (mainly in the form of NADH) provide substrate for oxidative phosphorylation reactions. Thus, they are fundamental to the synthesis of ATP, as mentioned previously. Equally, they are necessary for reducing intermediary metabolites to the level of cell substance, for in very many cases cell substance is vastly more reduced than the substrates from which it is synthesized. For example, the approximate empirical formula of *Klebsiella aerogenes* is $C_4H_7O_2N$ (see Herbert[6]). If these cells were to be synthesized from glucose and ammonia with a 100% conversion of substrate carbon to cell carbon, then no excess reducing equivalents would be available for oxidative phosphorylation reactions to proceed:

$$2/3\ C_6H_{12}O_6 + NH_3 \longrightarrow C_4H_7O_2N + 2\ H_2O$$

Moreover, if nitrate was supplied as the nitrogen source in place of ammonia, then a substantial deficiency of reductant would be manifest. Thus:

$$2/3\ C_6H_{12}O_6 + HNO_3 \longrightarrow C_4H_7O_2N + H_2O + 4(O)$$

Hence, for cell synthesis to occur, even with ammonia as the nitrogen source, extra reducing equivalents must be delivered up to supply the demands of respiratory ATP synthesis. It follows, therefore, that some portion of the intermediary metabolites must be further oxidized to CO_2 with a concomitant decrease in the carbon conversion efficiency. This, then, raises the important question of regulation. What controls the further oxidation of intermediary metabolites? How is the flow of reducing equivalents partitioned between respiration and biosynthesis? What part does the ATP/ADP ratio play in these regulatory processes?

Here, the situation has been somewhat oversimplified since a cursory study of the pathways by which intermediary metabolites are synthesized reveals that several decarboxylation reactions are involved in the synthesis of some key compounds and that these simultaneously generate reducing equivalents. In particular, 6-phosphogluconate dehydrogenase and isocitrate dehydrogenase usually generate NADPH, whereas pyruvate dehydrogenase and 2-oxoglutarate dehydrogenase generate NADH in reactions leading to the synthesis of key intermediates of nucleic acid and protein synthesis. Thus, it is possible that sufficient reducing equivalents may be made available by these decarboxylating reactions to meet the cells' requirement for both ATP and polymer precursor synthesis, and that the sole losses in carbon conversion efficiency stem from intermediary metabolite synthesis. That such is not the case, at least with cultures growing in a chemostat, can be deduced from the fact that the rate of CO_2 generation in carbon-substrate-limited chemostat cultures of bacteria, though a linear function of the growth rate, does not extrapolate to zero at zero growth rate. This is clearly evident from the data of Herbert[1,6] (Figure 1). At low growth rates, a large proportion of the carbon substrate is fully oxidized to CO_2 with a corresponding loss of carbon conversion efficiency. This situation is even more clearly revealed when the distribution of cell-carbon and CO_2-carbon is plotted as a function of the growth rate (Figure 2). Thus, at a dilution rate of 0.05/hr, almost two thirds of the substrate (glycerol) carbon was oxidized to CO_2 as compared with about one third with cultures growing at dilution rates higher than 0.3/hr. These findings are incompatible with the proposition that carbon conversion efficiency is specified solely by the requirements of intermediary metabolite synthesis.

Again, if organisms were not able to oxidize carbon substrates without concomitant biosynthesis, it is difficult to understand how washed suspensions manage to accomplish this in a Warburg apparatus, for example. For although they well may synthesize storage-type polymers such as glycogen (see Dawes and Senior[7]), the rates at which these are formed generally are insufficient to account for the high rate of substrate metabolism observed. Moreover, not all species that readily oxidize carbon substrates under growth-restricting conditions synthesize simultaneously storage-type polymers. Hence, since intermediary metabolites can be further oxidized under conditions where cells cannot grow, and since the reducing equivalents that are generated simultaneously can be totally disposed of by respiratory processes, it is pertinent to consider whether, and to what extent, these carbon and energy spilling reactions are suppressed in actively growing cells.

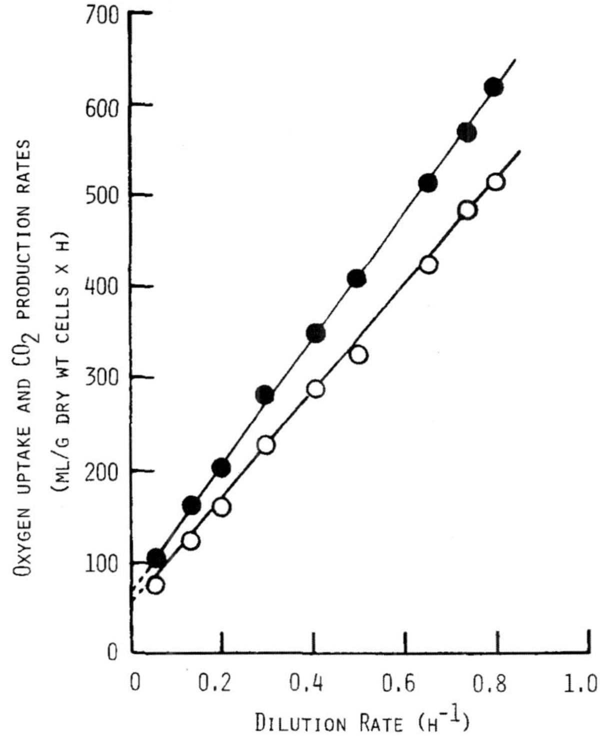

FIGURE 1. Specific rates of (●) oxygen consumption and (○) carbon dioxide production by a glycerol-limited culture of *Klebsiella aerogenes* growing in a chemostat culture at different dilution rates. Data of Herbert.[1]

3. Overflow Metabolism

The point has been made repeatedly that carbon substrate can be rapidly oxidized by nongrowing cells, but how do actively growing cells respond to carbon-sufficient (otherwise nutrient-limited) conditions? In this connection, it has been argued that organisms like *Escherichia coli* possess the capacity to regulate the uptake of carbon substrate such as to meet precisely their biosynthetic and bioenergetic needs. However, in contrast to this conclusion was the finding of Neijssel et al.[9] that glucose-limited

TABLE 1

Glucose Utilization Rates, and Rates of Product Formation, in Variously Limited Chemostat Cultures of *Klebsiella aerogenes* NCTC 418

	Limitation					
	Glucose	Sulphate	Ammonia	Phosphate	Magnesium	Potassium
Glucose used	36.8	98.7	107.4	112.8	124.6	175.0
Products formed						
Cells	20	20	20	20	20	20
CO_2	15.6	20.8	20.2	20.4	31.4	56.3
Pyruvate	—	21.9	5.2	—	9.6	10.1
2-Oxoglutarate	—	1.8	22.5	—	9.6	3.0
Acetate	—	9.6	4.0	4.5	17.7	13.3
Gluconate	—	—	—	9.5	0.1	31.9
2-Ketogluconate	—	—	—	39.9	11.9	20.5
Succinate	—	2.3	—	—	—	—
D-Lactate	—	—	—	—	10.9	—
Protein (exocellular)	—	1.8	—	—	2.5	8.0
Polysaccharide (exocellular)	—	7.0	36.0	15.1	—	—
Carbon recovery (%)	97	91	102	97	91	93
q_{O_2} (mmol/g × hr)	4.1	7.4	7.4	9.8	11.2	16.3
$q_{glucose}$ (mmol/g × hr)	2.1	5.0	5.9	6.3	7.0	9.9

Note: D = 0.17/hr, 35°C, and pH 6.8. All values expressed as milliatoms carbon per hour and normalized to a cell production rate of 20 matoms carbon per hour.

chemostat cultures of *Klebsiella aerogenes* possessed the capacity to oxidize glucose at a considerably greater rate than they could express in the glucose-limited culture. For if cell-saturating concentrations of glucose were pulsed into steady-state glucose-limited cultures, then there was an immediate and rapid increase in the oxygen consumption rate (Figure 3) and a concomitant excretion of partially oxidized products of glucose catabolism from the cells.

The fact that bacteria do not invariably regulate precisely the uptake of excess carbon substrate is best revealed by studies of the growth of *K. aerogenes* in glucose-sufficient (otherwise nutrient-limited) chemostat cultures. Here it was found[10] that there was a two- to five fold difference between the amount of glucose catabolized in the synthesis of a fixed amount of biomass by glucose-sufficient cultures as compared with those that were glucose-limited (Table 1). As can be seen, the excess glucose catabolized by the glucose-sufficient cultures was not completely oxidized to CO_2, and a range of intermediary metabolites accumulated in the culture extracellular fluids. Such behaviour has been termed "overflow metabolism".

It is obvious that whenever the rate of substrate uptake and catabolism is increased relative to the rate of cell synthesis there must initially accumulate within the cell intermediary metabolites and reducing equivalents. Anaerobically, certain intermediary metabolites can act as electron acceptors, thereafter being vented from the cell. Thus both a redox balance and a balance in the pool content of intermediary metabolites is re-established. Aerobically, this seemingly does not happen (Table 1). Instead, selected intermediary metabolites are excreted, and a redox balance is re-established by virtue of an increased terminal respiration rate. However, if respiration is tightly coupled to ATP synthesis, then solving a redox imbalance by means of an increased respiration rate must simultaneously create a severe energy overplus (that is, an imbalance in the ATP/ADP ratio) unless, of course, mechanisms exist for turning over the ATP pool at a high rate by growth-unassociated (energy spilling) reactions. On the other hand,

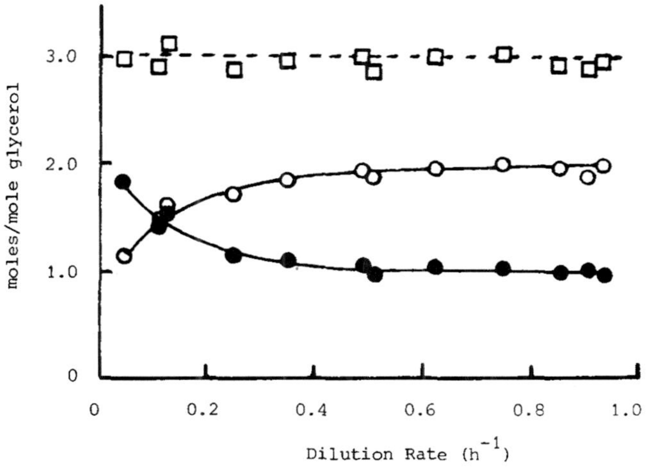

FIGURE 2. Distribution of substrate carbon between cells and carbon dioxide found with a glycerol-limited chemostat culture of *Klebsiella aerogenes* growing at different dilution rates. Cell carbon (○) and carbon dioxide (●) are plotted as moles per mole of glycerol consumed at each growth rate. A 100% carbon recovery is indicated by the broken line, and the actual values found are shown (□). Data of Herbert.[6]

it is still by no means certain that the transfer of electrons from NADH to oxygen *necessarily* requires concomitant ATP synthesis, or that the rate of ATP synthesis is related stoichiometrically (and invariantly) to the rate of oxygen reduction. With whole cells, it is not possible to demonstrate unequivocally the process of respiratory control to an extent that is characteristic of isolated mitochondria. However, certain observations do point to there being an increased rate of ATP synthesis associated with an increased respiration rate under conditions where the rate of cell synthesis is held constant. This evidence will be briefly considered in the next section.

B. Production and Utilization of ATP

It follows from the chemiosmotic hypothesis of Mitchell[11] that protons ejected from the cell during the the transfer of electrons along the respiratory chain re-enter the cell largely (though not entirely) through the membrane-bound ATPase. Thus, it is suggested that the plasma membrane *per se* is impermeable to protons except for specific sites associated with particular respiratory pigments and other proteins such as ATPase, permeases, and substrate transport mechanisms. If this is so, then an increased respiration rate must effect a simultaneous increase in either the rate of ATP synthesis or else the rate at which some transport processes run. The only other possibility demands one to accept the proposition that the transfer of electrons to oxygen does not necessarily require protons to traverse the cell membrane (i.e., that there can be invoked some kind of "short circuit"). However, such a proposition fails to accord with the experimental finding of a *stoichiometric* respiration-driven proton translocation in washed cell suspensions, a finding that forms the corner-stone of the Mitchell hypotehsis (see also Haddock and Jones[12]). Hence, it is logical to conclude that respiratory control in the bacterial cell is masked by the presence of cellular components that can turn over the ATP pool at a high rate. Corroborative evidence for this comes from the finding of John and Whatley[13] that membrane vesicles (prepared from *Paracoccus denitrificans)* do indeed exhibit respiratory control, whereas whole cells do not. Moreover, if in fact the rate of transfer of electrons to oxygen was being limited in carbon sufficient cells by the rate of influx of protons, then addition of uncouplers

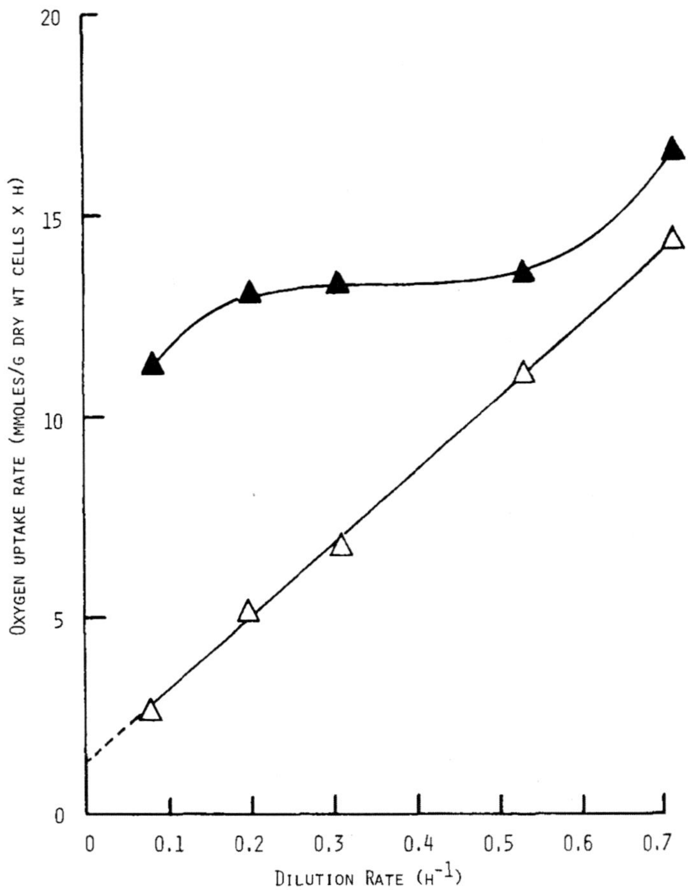

FIGURE 3. Plot of the changes in the "potential" rate of oxygen consumption (▲) and in situ rate (△) found with a glucose-limited culture of *Klebsiella aerogenes* growing in a chemostat at different dilution rates. The "potential" rate of respiration was determined by adding a cell-saturating pulse of glucose to the steady state culture and measuring the rate of oxygen consumption expressed after 2 min. Data of Neijssel and Tempest.[54]

that function as proton translocators to carbon-sufficient cultures ought to effect an immediate stimulation of respiration rate. Such was not observed when 2, 4-dinitrophenol (1 m*M* end concentration) was injected into variously limited glucose-sufficient cultures of *K. aerogenes.*[55]

The only evidence suggesting the existence of respiratory control processes in carbon-sufficient cultures of *K. aerogenes* was the finding of some decrease in respiration rate when the supply of medium to a phosphate-limited culture was interrupted (Figure 4[14]). What is also clearly evident in this figure is that the rate of decrease in respiration rate was markedly growth-rate dependent and, moreover, was such as to suggest that it was not regulated by some finely tuned quick-acting mechanism like that affecting respiratory control in mitochondria. In this connection, it might also be mentioned that these experiments were performed with cultures growing on glycerol. Since glycerol can readily penetrate the cell membrane, it is unlikely that changes in respiration rate were being provoked by regulation of carbon substrate uptake. On the other hand, the possibility that carbon substrate catabolism was being regulated, rather than respiration per se, cannot be excluded.

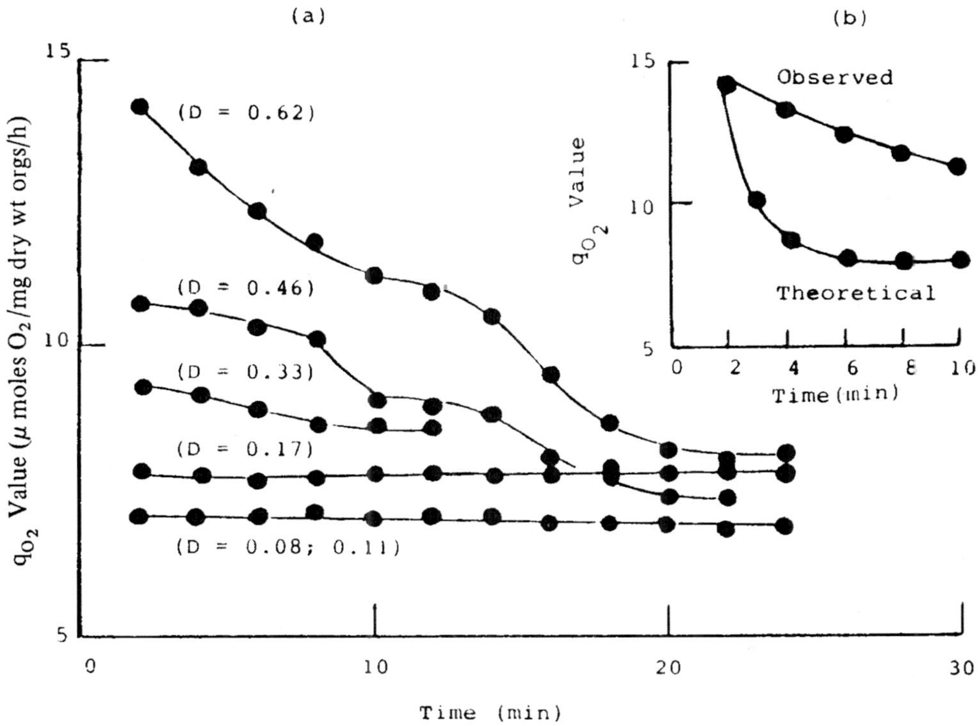

FIGURE 4. Influence of a "step-down to zero dilution rate on the specific rates of oxygen consumption of phosphate-limited *Klebsiella aerogenes*, cultures growing on glycerol at different rates. Data of Neijssel and Tempest.[14]

Although the addition of DNP to glucose-sufficient chemostat cultures of *K. aerogenes* did not effect an immediate stimulation of respiration rate. Nevertheless, if these cultures were allowed to grow for some extended period of time in the presence of this uncoupler, then steady state conditions ultimately became established in which the respiration rate was markedly elevated (Figure 5A). Similarly, the presence of 1 m*M* DNP in the feed medium caused a glucose-limited culture of *K. aerogenes* to express a greatly increased respiration rate at each growth rate value above 0.15/hr (Figure 5B). At low growth rate values, however, respiration was severely impeded, and much acetate and D-lactate were excreted (see Neijssel[15]). These experiments, then, show that although the addition of a proton translocator does not effect an immediate stimulation of respiration rate in growing organisms, they do respond to its continued presence by reorganizing their physiology to effect an overall increase in respiration rate. Significantly, when growing at dilution rates above 0.15/hr in the presence of 1 m*M* DNP, glucose-limited *K. aerogenes* cultures oxidized glucose almost completely to CO_2. That is, there was only a small excretion of acetate. Also, the increase in glucose uptake rate, though significant, was not sufficient to allow the conclusion that oxidative phosphorylation processes were totally prevented. Thus, it must be concluded that respiratory-coupled ATP synthesis could occur at a substantial rate even in the presence of high concentrations of this proton translocator.

There are yet other grounds for supposing that an elevated respiration rate, such as was found with carbon-sufficient cultures, is associated with an elevated rate of ATP synthesis. Thus, it has been found that glycogen synthesis in prokaryotic organisms is markedly stimulated by ATP and retarded by ADP and AMP (see Dawes and Senior[7]). In other words, it is potentiated by a high energy charge. In this connection, it has been routinely found that ammonia-limited cultures of *K. aerogenes* and *E. coli* syn-

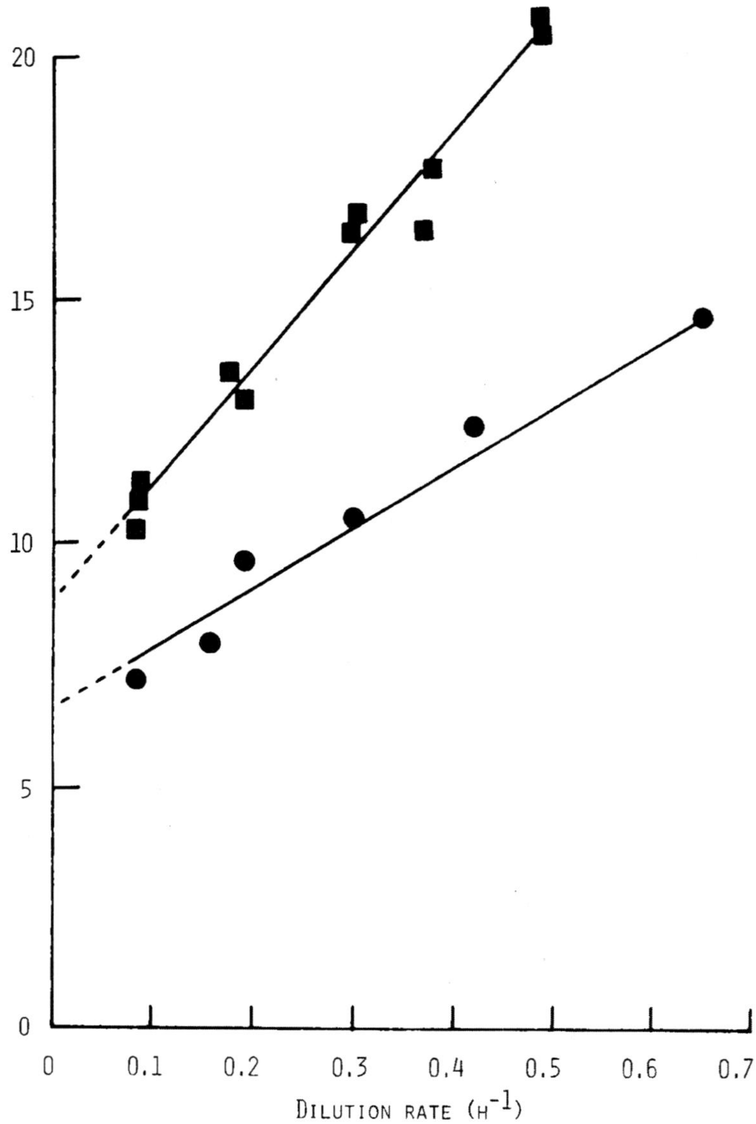

A

FIGURE 5. Influence of dinitrophenol (1 m*M* end concentration) on the steady state rate of oxygen consumption by *Klebsiella aerogenes* cultures growing at different rates. (A) an ammonia-limited culture growing in the presence (■) and absence (●) of DNP. (B) a glucose-limited culture growing in the presence (□) and absence (○) of DNP. Data of Neijssel.[15]

thesize much glycogen, particularly when growing at a slow rate. Moreover, species of *Klebsiella* may synthesize substantial amounts of exocellular polysaccharide. Again, this seemingly is provoked by a high carbon/nitrogen ratio in the feed medium. Not surprisingly, the synthesis of glycogen by *K. aerogenes* also was found to occur under conditions of phosphate and sulphate limitation. However, none was synthesized when the cultures were either magnesium or potassium limited,[16] even though such cultures expressed high oxygen uptake rates (Table 1).

1. Respiratory Chain Modification

Variations in the pattern of electron transport chain components can be induced,

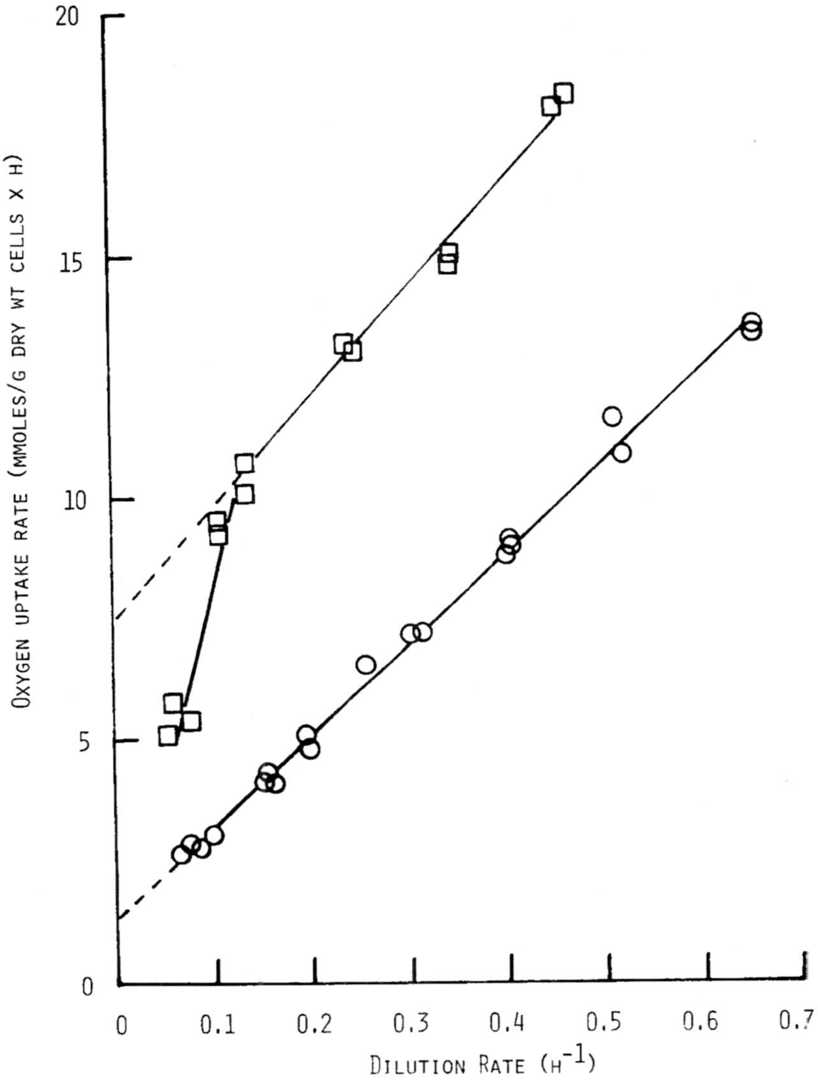

FIGURE 5B

within a single species, by altering the growth conditions (see Haddock and Jones[12]). This is particularly evident when growth is limited by the availability of oxygen where variations occur mainly among the terminal oxidases and quinones.[17,18] However, substantial quantitative changes in electron transport chain components, and in the efficiency with which oxidative phosphorylation proceeds, have been reported to follow growth (of *E. coli*) in media containing limiting concentrations of sulphate[19] or iron.[20] In these cases, it appeared to be the type *b* cytochromes, nonheme iron and iron-sulphur proteins whose levels were decreased. Similar observations have been made with iron- or sulphate-limited cultures of the yeast *Candida utilis*[21] where it could be shown that Site I energy conservation was lost. Similarly, potassium-limited growth of this yeast at a low dilution rate in glucose-containing chemostat culture effected loss of one site of energy conservation, though this seemingly was not Site I.[22]

It follows, therefore, that growth of bacteria in glucose-sufficient chemostat cultures also might induce modifications in the organization of the electron transport chain such as to allow respiration to proceed at an elevated rate without there being a concomitant increase in the rate of ATP synthesis. Nevertheless, the differences between the q_{O_2} expressed by glucose-sufficient cultures as compared with those which were glucose-limited (Table 1) is such as to suggest that, in the majority of cases, even the loss of a site of energy conservation would not prevent ATP being generated at a substantially higher rate than that at which it could be turned over by growth-associated processes.

2. ATP Turnover: Futile Cycles

It would seem prudent to accept, as a minimum hypothesis, the proposition that a high respiration rate has associated with it a high rate of ATP synthesis (from ADP and Pi). Hence, it is clear that there must be present within the cell a number of growth-unassociated functions whose activities can be extensively modulated and that serve to turn over the ATP pool at a high rate under conditions where the synthesis of cell material is grossly impeded. However, what is the nature of these energy-spilling reactions, and how are they invoked and regulated?

At the onset, one must admit that there is a paucity of precise information on the nature and mode of action of so-called "futile cycles" in prokaryotic microorganisms. All that can be definitely stated is that there are frequently found to be present within microbial cells a number of enzymes which, *if* they acted in concert, could effect the hydrolysis of ATP without there being produced a net change in any other component. For example, ammonia-limited *K. aerogenes* cells are rich in glutamine synthetase, but also possess an active glutaminase (see Brown and Stanley[23]). Together, these two enzymes could act as an ATPase system as follows:

$$\text{Glutamate} + \text{Ammonia} + \text{ATP} \longrightarrow \text{Glutamine} + \text{ADP} + \text{Pi}$$
$$\text{Glutamine} \longrightarrow \text{Glutamate} + \text{Ammonia}$$
$$\overline{\text{ATP} \longrightarrow \text{ADP} + \text{Pi}}$$

Of course, whether this potential futile cycle actually operates in actively growing ammonia-limited *K. aerogenes* organisms is unknown. Nor is it known how its activity is suppressed in the growing cell, if, in fact, it is suppressed. A clue to the functional significance of this potential futile cycle might be found by studying the changes in the pool free-ammonia level following an interruption in the supply of medium to an ammonia-limited culture. If under these conditions the pool ammonia level is maintained at some detectable concentration, then one might suppose that the futile cycle functions as a "sensor" for ammonia by keeping the assimilatory enzyme (glutamine synthetase) actively functioning. Thus, the spillage of ATP-associated energy may not, in reality, be the primary purpose of the futile cycle, but is the "price" that must be paid for its maintaining the cells in a state where they can respond rapidly to a sudden change in the supply of the limiting nutrient (ammonia). Moreover, if this is the function of this particular futile cycle, then one would anticipate that similar cycles might operate in the uptake of other possible growth-limiting nutrients such as phosphate and sulphate. Such evidence is, indeed, to be found in the literature (see Tempest[24]).

Although one would not expect futile cycles to operate in organisms that were limited in their growth by the availability of the carbon and energy source, nevertheless their presence is indicated by the fact that a rapid dissipation of energy is found to follow the sudden addition of cell-saturating concentrations of glucose to glucose-limited cultures. In this connection, *K. aerogenes* has recently been shown to possess both an NADH-linked pyruvate reductase that generates D-lactate and a D-lactate dehydrogenase that is a flavoprotein.[25] Together, these two enzymes are potentially capable of oxidizing NADH by a route that by-passes Site I of the respiratory chain, and significantly, the pyruvate reductase was found to be homotropic with respect to pyruvate. Thus, any transient rise in the intracellular pyruvate level (such as would occur in glucose-limited cultures that were pulsed with glucose) could, at least in theory, cause Site I to be circumvented. Here it is worth mentioning that sulphate-limited cultures of *K. aerogenes,* which reportedly lack a site of energy conservation[4], routinely are found to excrete much pyruvate into the medium when growing on glucose (Table 1). Further, carbon-sufficient chemostat cultures frequently are found to excrete much acetate which, it is insufficiently realized, can act as a potent uncoupler of oxidative phosphorylation, particularly if the culture pH is sufficiently low to permit significant amounts of acetic acid to be present in its undissociated form.[26]

III. MICROBIAL GROWTH YIELD VALUES

Having considered the relationships that are known, or thought to exist, between the different cellular components that contribute to the growth process, it is now appropriate to examine critically the theories that have been advanced to account for the variations in yield values that are routinely found with actively growing bacterial cultures. In particular, it is important to assess the physiological significance and utility of those mathematical relationships between the different measurable parameters that have been derived, and to identify possible errors of interpretation.

A. Theoretical Aspects

The starting point for most theoretical analyses of growth yields are data similar to those shown in Figure 1. That is, there is a linear relationship between the specific rate of oxygen consumption and growth rate (and between CO_2 production and growth rate) that does not extrapolate back through the origin. Thus, since by definition $Y = \mu/q$ (Equation 1), it follows that the actual yield value decreases towards zero as the growth rate is progressively lowered. In order to explain thisvariation in yield value with growth rate, it was proposed that a portion of the carbon substrate (and of oxygen) was required to deliver up energy that was needed for growth-independent "maintenance" functions.[27] Hence, the extrapolated substrate uptake rate at zero growth rate could be taken as a direct measure of this maintenance energy requirement and could be subtracted from the actual rate of substrate consumption to derive an evaluation of the "true" growth-associated substrate requirement. Thus:

$$q_{actual} = q_{growth} + q_{maintenance}$$

And dividing by μ (= D at steady state with a chemostat culture):

$$q_a/\mu = q_g/\mu + q_m/\mu$$

However, since q/μ is the reciprocal of the yield value, then:

$$1/Y = 1/Y_g + q_m \cdot 1/\mu \qquad (2)$$

where Y is the measured yield value and Y_g the "true" growth yield constant (that is, the yield value corrected for maintenance energy losses). Therefore, plotting 1/Y against $1/\mu$ (or $1/D$) should give a straight line with a slope equal to q_m that intersects the ordinate (when $1/D = 0$) at a value $1/Y_g$.

However, it is important to emphasize that both Y_g and q_m are essentially *mathematical* constants that derive from, and depend on, there being a linear relationship between q and μ. In fact, Y_g is the reciprocal of the slope of the line of regression of q on μ (Figure 1), and q_m is the intercept on the ordinate. Although this linearity suggests that cell synthesis proceeds at all growth rates with the same basic efficiency (represented by Y_g), and that Y_g is therefore a true *biological* constant, it does not prove that such is the case. Indeed, that the basic efficiency of cell synthesis well may vary with growth rate follows from the fact that, compositionally and metabolically, microbial cells vary markedly with growth rate (see, for example, Herbert,[6] Tempest and Herbert,[28] O'Brien et al.[29]). Clearly, these physiological changes must have energetic consequences. Equally clearly, if the energetic demands of cell synthesis do change progressively with growth rate, then the growth-unassociated energy losses (represented by q_m at zero growth rate) also must vary. Thus, Equation 2, though formally correct, is not directly and unequivocally interpretable in physiological terms.

That the maintenance rate of substrate consumption by carbon-sufficient cultures *must* vary with growth rate is evident from the growth-rate-linked changes in oxygen consumption rate as compared with those expressed by carbon-limited cultures (Figure 6). Here it is clear that at growth rates close to μ_{max} (which in these cultures was about 0.85/hr) the specific rates of oxygen consumption were closely similar. Indeed, if a dilution rate of 0.85/hr defines a situation in which the organisms are close to being no longer nutrient limited, then one would expect the oxygen consumption rate at this growth rate to be approximately constant, irrespective of the nominal limitation that was imposed on the culture. Of course, the situation is very different at low growth rates since carbon-limited cultures cannot oxidize the carbon source at a faster rate than it is being supplied to the culture, whereas in carbon-sufficient cultures the rate of carbon substrate oxidation will depend largely upon the metabolic potential of the organisms. The only constraint is the rate of penetration of substrate into the cells and possible feedback regulation of glycolysis and respiration. From Figure 6, it is clear that at low growth rates the oxidative potential of carbon-sufficient organisms is vastly greater than that which would be needed to meet the cells' minimum energetic demands (as indicated by the rate of oxygen uptake in carbon-limited cultures). However, if at growth rates close to μ_{max} the organisms were growing with a similar efficiency (which seems reasonable since their Y_O values were similar), then it follows that the amount of oxygen consumed in processes not associated with growth must be greatly decreased. That is to say, the maintenance rate of oxygen consumption must decrease progressively with growth rate. If this is indeed the case, then clearly the Y_O^{max} value derived from Equation 2 (that is, the Y_g value for oxygen) would be a gross overestimate.

The use of Y^{max} values in assessments of the energetic efficiency of microbial cell synthesis is a commonplace (see Stouthamer and Bettenhaussen,[4,30] Hempfling and Mainzer,[31] Downes and Jones,[32] Jones et al.[33]), and hence, it is important to emphasize the influence which a varying maintenance rate would exert on assessments of this parameter. Thus, making the simple assumption that *if* the maintenance rate varies with growth rate, it does so progressively and linearly, then Equation 2 may be modified by substituting

$$'m' = q_m^*(1 \pm c\mu)$$

for q_m. Hence:

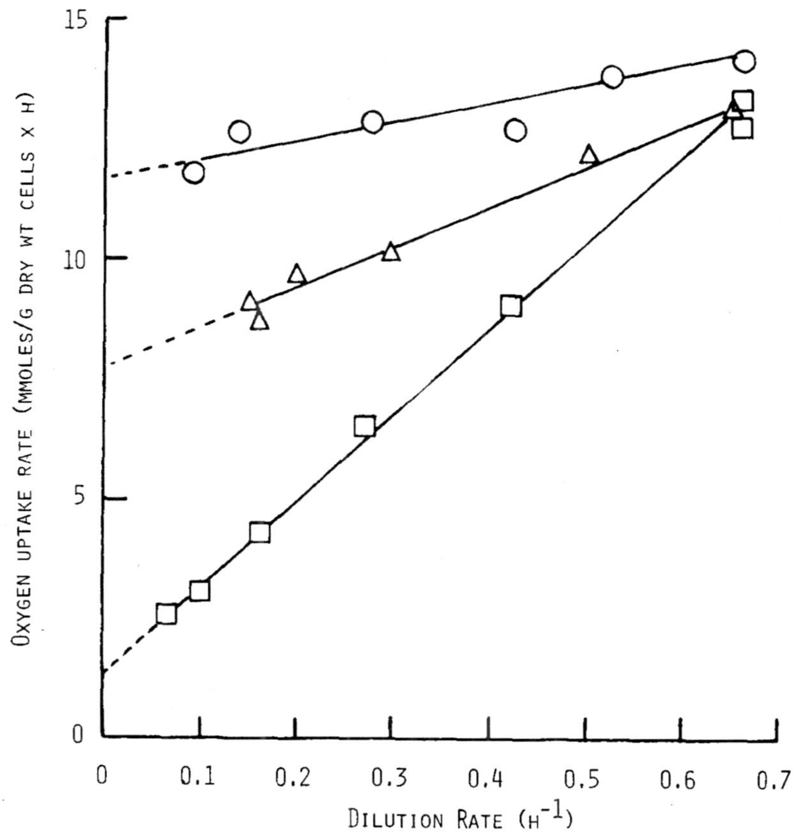

FIGURE 6. Relationship between the specific growth rate and specific rate of oxygen consumption in variously limited chemostat cultures of *Klebsiella aerogenes* growing in a glucose-containing medium. Cultures were, respectively, glucose-limited (□), sulphate-limited (△), and phosphate-limited (○). Data of Neijssel and Tempest.[14]

$$1/Y = 1/Y_g + m/\mu = 1/Y_g \pm q_m^*(1 + c\mu)/\mu$$

where "c" is a second constant that defines the variation in the maintenance rate with growth rate (that is, defines the slope of the line of regression of m on μ and q_m* is a constant that is equal to the maintenance rate at zero growth rate. Rearranging:

$$1/Y = 1/Y_g \pm c \cdot q_m^* + q_m^* \cdot 1/\mu \quad (3)$$

and from this it is clear that plotting $1/Y$ against $1/\mu$ still would give a straight line, but the intercept on the ordinate no longer would be $1/Y_g$, but $(1/Y_g + c \cdot q_m^*)$. It follows, therefore, that if "c" is a negative number, as suggested for carbon-sufficient cultures, the Y_g (or Y^{max}) will be overestimated. Conversely, if "c" is a positive number (that is, the maintenance rate *increases* with growth rate) then Y_g will be underestimated. Should the maintenance rate not vary with growth rate, then "c" would be zero, and Y_g would be correctly evaluated by Equation 2.

The utility of Equation 3 is clearly limited by the fact that, as yet, there is no way of determining the variation of m with μ. Nevertheless, it serves to reveal the uncertainties embodied in using Equation 2 for an evaluation of the "true" growth yield in its physiological sense.

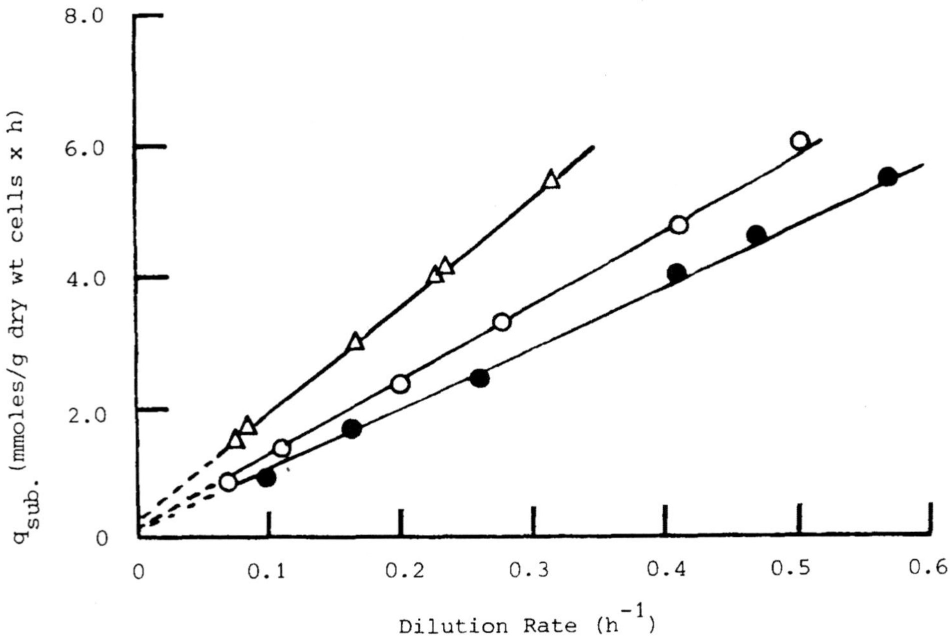

FIGURE 7. Specific rates of carbon-substrate consumption as a function of the growth rate found with chemostat cultures of *Klebsiella aerogenes* that were growing in an ammonia-containing medium either (●) glucose limited, or (○) gluconate limited. The rate of glucose consumption by glucose-limited cultures growing in a medium containing nitrate as the sole utilzable nitrogen source is also shown (△).

B. Comparison and Interpretation of Yield Data

It is clear from the discussion contained in the foregoing sections that interpretation of yield data in physiological and/or bioenergetic terms is far from straightforward. Moreover, a survey of the published literature reveals that the differences between the results of various workers using supposedly the same organism often are greater than those found between that organism when growing on different carbon substrates. This lack of consistency, which probably stems from small differences in the growth conditions employed, serves to undermine confidence in assessing the significance of small differences in reported yield values. Nevertheless, some trends are consistent and can be rationalized in physiological terms.

1. Yield Values With Respect to Carbon Substrate

It is commonly found that the rate of carbon-substrate uptake and assimilation by growing cultures is a linear function of the growth rate (Figure 7). With carbon-substrate-limited chemostat cultures, the substrate uptake rate declines to a small, but finite, value at zero growth rate. Under these latter conditions, the yield value obtained at growth rates close to μ_{max} is not very different from the theoretical Y^{max} value, and hence, the latter values may be used for comparative purposes.

The data contained in Figure 7 further reveal that the specific rate of substrate consumption, at any fixed growth rate, varies with the chemical nature of the carbon substrate and of the utilizable nitrogen source. Thus, the yield values expressed at any fixed growth rate (μ/q) vary as do the Y^{max} values (Table 2). Here it should be noted that glucose and gluconate are both C_6 compounds that only differ in the level of oxidation of one carbon atom, yet the yield values are significantly different. More marked is the difference in glucose consumption rate found between glucose-limited cultures of *K. aerogenes* that were provided with, respectively, ammonia and nitrate

TABLE 2

Comparison of Maximum Yield Values for Different Organisms Growing on Different Carbon Substrates

Substrate	N-source	$Y_{substrate}$	Y_C	Y_O	C_N*[a]	C:O Ratio**[b]	Ref.
Klebsiella aerogenes							
Mannitol	Ammonia	109	18.1	21.7	0.75	1.20	54, 55
Glucose	"	100	16.7	25.8	0.70	1.55	54, 55
Gluconate	"	89	14.8	27.5	0.62	1.86	54, 55
Glucose	Nitrate	62	10.3	20.1	0.43	1.96	54, 55
Glucose	"	64	11.3	22.0	0.47	1.95	4
Glycerol	Ammonia	47	15.7	20.0	0.65	1.27	Unpublished data
Glycerol	"	46	15.5	23.2	0.65	1.50	33
Lactate	"	40	13.2	14.5	0.55	1.10	Unpublished data
Lactate	"	37	12.4	15.5	0.52	1.25	33
Acetate	"	17	8.7	5.9	0.36	0.68	Unpublished data
Paracoccus denitrificans							
Glucose	Ammonia	111	18.5	45.0	0.77	2.43	35
Glycerol	"	56	18.7	36.5	0.78	1.95	35
Lactate	"	53	17.5	20.7	0.73	1.18	35
Acetate	"	25	12.2	13.3	0.51	1.09	35
Mannitol	"	81	13.6	19.7	0.57	1.45	36
Gluconate	"	78	13.0	21.6	0.54	1.66	36
Candida utilis							
Glucose	Ammonia	101	16.8	27.9	0.66	1.66	6
Ethanol	"	38	19.1	16.0	0.75	0.84	6

[a] C_N is the carbon conversion efficiency (i.e., gram cell carbon per gram substrate carbon metabolized).

[b] C:O Ratio is milliatoms substrate carbon metabolized per milliatom oxygen consumed.

as the sole utilizable nitrogen source. The latter culture consumed glucose, at each growth rate, at about twice the rate of the former culture, and the yield values were proportionately different (Table 2).

To account for the marked effect of nitrate on the yield value it is necessary to bear in mind the fact that the nitrogen content of bacterial cells is substantial (about 14% of the dry weight in glucose-limited *K. aerogenes*), and that nitrate must be reduced to ammonia before being assimilated into cell substance.

$$HNO_3 + 4(2H) \longrightarrow NH_3 + 3\,H_2O$$

Thus, if the complete catabolism of glucose generates 12 pairs of reducing equivalents, then one might anticipate that the molar growth yield observed with glucose-limited cultures growing on nitrate would be about 75% that found with cultures growing at a corresponding rate on ammonia (that is, $Y_{glucose}/1.333$). However, the actual difference in yield values was substantially larger, suggesting that other factors were influencing the efficiency with which glucose could be assimilated into cell substance when the cultures were growing on nitrate.

This raises again the question of the precise nature of a carbon substrate limitation. As mentioned previously, when a carbon substrate is supplied to a heterotrophic organism, it is used to serve three functions: (1) as a source of intermediates required for biosynthesis, (2) as a source of reducing equivalents, and (3) as a source of utilizable energy (ATP). These functions are closely interrelated since the synthesis of pentoses, for example, or citric acid cycle intermediates, will also generate energy and reducing equivalents. A carbon substrate limitation therefore represents a complex situation in which, depending on the nature of the carbon source and the redox state of other key nutrients, the culture basically may be either (1) carbon-limited (that is, the nutrient carbon supply is limiting growth, and energy generation occurs in excess of the biosynthetic demands), or (2) energy-limited (including a limitation imposed by the organisms' requirement for reducing equivalents). A carbon *plus* energy dual limitation may be theoretically possible, but one might assume, rare. On the basis of these considerations, it seems likely that, when growing on nitrate, glucose-limited cultures are essentially energy limited, and that the low yield values observed reflect not only the demand for extra reducing equivalents (required for nitrate reduction), but also the additional requirement for ATP necessary to assimilate the ammonia so formed via the glutamine synthetase-glutamate synthase reaction (see Brown[34]).

Similar considerations may also account for the difference observed between the yield values expressed by *K. aerogenes* when growing, respectively, glucose- and gluconate-limited in chemostat culture (Figure 7; Table 2). Under a true carbon limitation, one would expect the respiratory quotient to be relatively low since the q_{CO_2} value would be minimal and the excess reducing equivalents generated would be oxidized by oxygen. The spillage of excess energy generated by respiration would cause the Y_O value (gram organisms synthesized per gram-atom oxygen consumed) to be relatively low, but the $Y_{carbon\ substrate}$ value would be maximal. In contrast with an energy-limited culture, a greater proportion of the carbon substrate would be catabolized completely to CO_2. Thus, the q_{CO_2} would be relatively high and the $Y_{carbon\ substrate}$ relatively low. On the other hand, the Y_O value would be maximal since energy spillage would be minimal. In this connection, it seems possible that gluconate-limited cultures are essentially energy limited, but whether glucose-limited cultures are similarly energy limited when growing in media containing excess ammonia is not immediately obvious. However, when other parameters are considered (see following section) and when compared with cultures growing on mannitol (which is more reduced than glucose), there are good grounds for supposing that glucose-limited *K. aerogenes* cultures are effectively carbon limited, as are mannitol-limited cultures (see Table 2).

The yield values expressed by cultures growing on closely related compounds may be directly compared and meaningfully assessed. However, difficulties arise when the values obtained with organisms growing on C_6 compounds are compared with those found for growth on C_3 and C_2 compounds. Molar yield values (gram organisms synthesized per mole carbon substrate consumed) are confusing in that the molecular weights and carbon contents (g-atom/mol) are widely different. Thus, for comparative purposes, it is more useful to express yields in terms of substrate carbon (Y_C equals gram organisms synthesized per gram-atom substrate carbon metabolized). Alternatively, one may use as a basis for comparison the carbon conversion efficiency (C_N equals gram cell-carbon produced per gram substrate carbon metabolized). Since the carbon content of carbon-substrate-limited organisms does not vary significantly with the growth condition, these two yield values are closely proportional.

The data contained in Table 2 show that although the molar growth yield values of organisms growing on C_3 compounds are about one half those observed with the same organism growing on C_6 compounds, the carbon yield values, and carbon conversion efficiencies, are not so markedly different. Thus, with cultures of *K. aerogenes* growing on glycerol, the carbon conversion efficiency is about 65% as compared with 62% for organisms growing on gluconate and 70% for organisms growing on glucose. Again, however, there are consistent trends in that glycerol carbon is assimilated into cell substance with a greater efficiency than the more highly oxidized C_3 compound, lactate; and a similar situation was found with cultures of *Paracoccus denitrificans*[35] (Table 2). The yield values reported for the growth of *P. denitrificans* on different carbon substrates by Edwards et al.[35] are consistently higher than those found for corresponding cultures of *K. aerogenes*. On the other hand, van Verseveld and Stouthamer[36] presented data showing that the yield values expressed by *P. denitrificans* growing on mannitol and gluconate were substantially less than those of *K. aerogenes* growing on the same carbon substrates. Hence, it is not possible to draw firm conclusions regarding relative efficiencies with which cell synthesis proceeds in these two organisms.

By way of further comparison, Table 2 contains data obtained with two cultures of the yeast *Candida utilis*, one that was glucose limited and the other ethanol limited. Clearly, the yield value on glucose was closely similar to that found with the bacterium, *K. aerogenes*, though the carbon conversion efficiency was slightly less. Yet again, growth on ethanol, which is substantially more reduced than glucose, proceeded with a markedly increased efficiency with respect to carbon conversion, though the yield with respect to oxygen was, comparatively, very low. As might be anticipated, an ethanol limitation seemingly is a carbon limitation, and the excess reducing equivalents delivered up by substrate metabolism are oxidized through to water, thus depressing the Y_O value.

2. Yield Values With Respect to Oxygen and ATP

In order to come to an assessment of the energetics of microbial growth on different substrates and in different environments, it is necessary to determine the specific rate of energy (ATP) generation and to relate this to the specific rate of cell synthesis. That is

$$Y_{ATP} = \mu / q_{ATP}$$

The problem here is that, generally speaking, no direct measure can be made of the rate of ATP synthesis in growing bacteria. The best that can be done is to determine the rate of change in some other (measurable) parameter and to derive from this an evaluation of q_{ATP}. For example, with organisms growing anaerobically and producing only CO_2 and ethanol as end-products (e.g., some yeasts), the q_{ATP} should equal the $q_{ethanol}$ since each mole of glucose yields 2 mol of ethanol and generates (net) 2 mol of ATP by substrate-level phosphorylation reactions. With anaerobic bacterial cultures growing in nutritionally complex media with a single fermentable carbon-substrate energy source, yield values of about 10 (gram organisms synthesized per mole ATP concomitantly generated) have been obtained.[37] When corrected for energy dissipation in maintenance reactions, values of about 14 have been reported for the "true" yield constant (Y^{max}_{ATP})[4]

Aerobes, and facultative anaerobes growing aerobically, generate the bulk of their ATP by oxidative phosphorylation; and since the complete oxidation of 1 mol of glucose consumes 6 mol of oxygen and can generate, maximally, 38 mol of ATP (from ADP), it might be thought possible to derive an evaluation of the energetics of aerobic growth from a measurement of the oxygen yield value (Y_O). Thus, it might be argued that since, overall, 3.17 molecules of ATP are generated per atom of oxygen reduced to water, the growth yield value for ATP should be equal to the yield value for oxygen (Y_O) divided by 3.17. However, when this calculation is made with cultures of *K. aerogenes* growing on glucose, one comes to a Y^{max}_{ATP} value of about 8, which is suspiciously low, though possibly realistic. In this connection, recent studies on the respiratory chain components of various bacteria[18] suggest that *K. aerogenes* may contain only two proton translocating loops in its respiratory chain and, therefore, would be able to generate maximally only 26 mol of ATP per mole of glucose oxidized. Thus the ATP/O ratio would be maximally 2.17 and the Y^{max}_{ATP} for aerobic growth on glucose 11.9. In contrast to *K. aerogenes*, *P. denitrificans* reportedly possesses a mitochondrial-type respiratory chain[38] and, thus when growing on glucose, could generate maximally 3.17 mol of ATP per gram-atom oxygen reduced. Interestingly, the Y_O value for this organism, when growing on glucose, was found to be 45 (Table 2), suggesting a Y^{max}_{ATP} value of 14.2.

Relevant to this question of microbial growth energetics is the point raised previously. That is, are glucose-limited cultures basically energy limited or carbon limited? With cultures of *K. aerogenes*, the latter seemingly is the case since cultures growing on gluconate expressed a significantly higher Y_O value at all growth rates (see Figure 8). Thus, it can be calculated that if *K. aerogenes* possesses only two sites of energy conservation on its respiratory chain, then the complete oxidation of 1 mol of gluconate would generate maximally 23 mol of ATP and consume 5½ mol of oxygen. Hence, the derived Y^{max}_{ATP} value (based on a Y_O^{max} value of 27.5, Table 2) would be 13.1, as opposed to 11.9 for growth on glucose. Similarly, the complete oxidation of 1 mol of mannitol would generate maximally 28 mol of ATP and consume 6½ mol of oxygen. Thus, a Y_O^{max} value of 20.1 (Table 2) would indicate a Y^{max}_{ATP} value of 10.1. However, irrespective of the actual number of energy conservation sites, if one assumes that growth proceeds with no more *basic* efficiency when utilizing gluconate than when utilizing either glucose or mannitol, then it follows that neither glucose-limited cultures nor mannitol-limited cultures were primarily energy limited. In other words, cultures growing aerobically on either glucose or mannitol generate (or potentially generate) an energy overplus which must be disposed of in some way. If this does indeed occur, then Y_o values give no reliable measure of the concomitant rate of ATP synthesis.

Once more it is necessary to draw attention to the widespread use of Y_o measurements as indicators of the energetic efficiency of aerobic cell synthesis (for a recent review, see Stouthamer[39]) and to emphasize the assumptions they embody. One example serves to make the point. *Klebsiella aerogenes* is able to grow both aerobically and anaerobically in a simple salts medium containing glucose as the sole source of carbon and energy. Thus, it was reasoned, one could estimate the specific rate of ATP synthesis associated with cells growing at a specified rate anaerobically (from the rates of formation of fermentation products) and compare this with the specific rate of oxygen consumption expressed by cultures growing aerobically at an identical rate in the same medium.[4] Low values were found for the ratio q_{ATP} (anaerobic)/q_{O_2} (aerobic) from which it was concluded that the P/O ratio expressed aerobically was less than 1.5 and that this organism possessed maximally two sites of energy conservation. However, one must ask, were the authors comparing like with like? For it can be argued (as above) that aerobic glucose-limited cultures are not energy limited, whereas it seems highly probable that anaerobic glucose-limited cultures will be so. If the excess energy necessarily generated aerobically is simply wasted by means of energy-spilling reactions, then the actual q_{ATP} expressed aerobically well may be substantially in excess of that delivered up by corresponding cultures growing anaerobically. Hence, the P/O ratio will be underestimated. In this connection, some caution also is needed in interpreting changes in adenine nucleotide levels associated with oxygen uptake in those experiments where starved anaerobic suspensions of organisms are injected into aerated buffer solution (see Baak and Postma[40] and Hempfling[41]). For here it is clear that one can only measure the *net* increase in, say, ATP. That is, the difference between its rate of formation and rate of breakdown. However, the procedure is nevertheless valid if one assumes the derived P/O ratio to be a minimum estimated value, as generally is done.

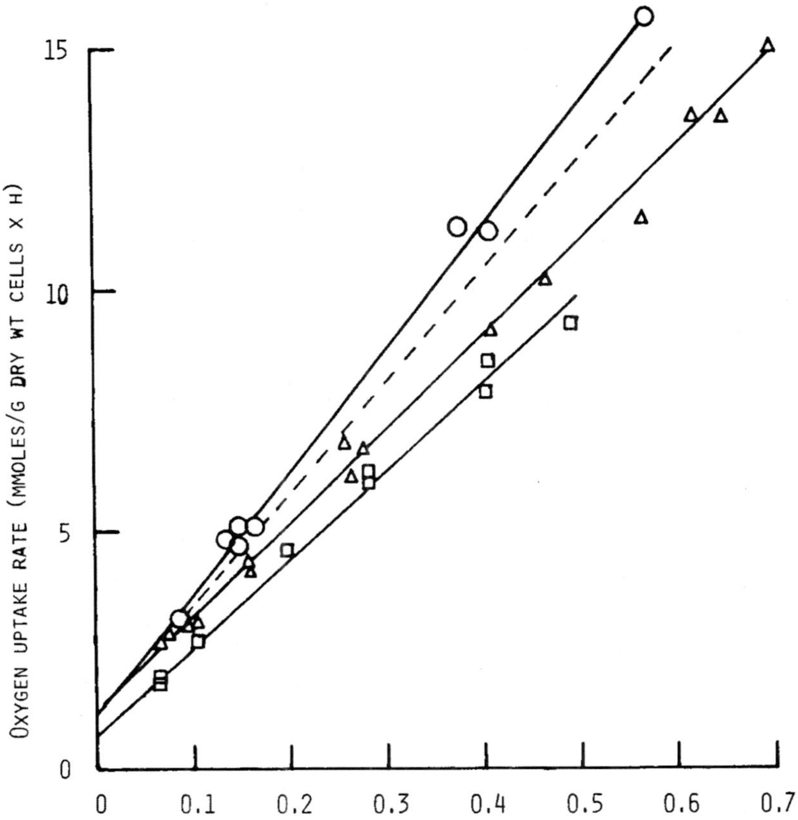

FIGURE 8. Relationship between the specific rate of oxygen uptake and dilution rate found with cultures of *Klebsiella aerogenes* that were (○) glycerol limited, (△) glucose limited, and (□) gluconate limited. The broken line represents the changes found with a mannitol-limited culture (datum points not shown for clarity). Data of Neijssel and Tempest.[54]

So far, we have concentrated on the interpretation of yield values found with carbon-substrate-limited cultures since, under these conditions, organisms generally grow with a markedly higher overall energetic efficiency than when the carbon substrate is present in excess of the growth requirement. However, data obtained with chemostat cultures in which growth is limited by components other than the carbon and energy source are relevant to the question of regulation of energy generation and its coupling to biosynthesis in actively growing cells, and therefore, will be briefly considered next.

3. *Yield Values of Carbon-Sufficient Cultures*

Fundamental to the interpretation of yield values expressed by carbon-substrate-sufficient chemostat cultures is the extent to which the uptake of carbon substrate can be modulated when growth is limited by the supply of some other essential nutrient (such as ammonia, phosphate, sulphate, potassium, or magnesium). With cultures of *K. aerogenes* growing at a moderately low rate (0.17/ hr), the rate of glucose uptake was not extensively modulated (Table 1), and the same held true when glucose was replaced by other carbon substrates (Table 3). Thus, when either glycerol, mannitol, or lactate were present in excess of the growth requirement, and growth limited by the supply of either sulphate, ammonia, or phosphate, carbon substrate was consumed at a substantially higher rate than when cultures were carbon-substrate limited. Similarly, oxygen

TABLE 3

Carbon Substrate and Oxygen Uptake Rates and the Corresponding Molar Yield Values Found with Variously Limited Chemostat Cultures of *Klebsiella aerogenes* NCTC 418 Growing at a Fixed Dilution Rate[a] on Different Carbon Substrates

		q(mmol/g dry wt cell · hr)		Y(g cell formed/mol C-substrate or oxygen)	
Substrate	Limitation	C-substrate	Oxygen	C-substrate	Oxygen
Glucose	Carbon	2.10	4.12	81.0	40.8
	Sulphate	4.97	7.39	34.2	23.0
	Ammonia	5.90	7.39	28.8	23.0
	Phosphate	6.30	9.77	27.0	17.4
Glycerol	Carbon	4.02	4.70	42.3	36.2
	Sulphate	6.16	11.33	27.6	15.0
	Ammonia	5.13	10.24	33.1	16.6
	Phosphate	6.16	10.49	27.6	16.2
Mannitol	Carbon	1.87	5.25	91.0	32.4
	Sulphate	3.46	11.49	49.1	14.8
	Ammonia	3.59	9.24	47.3	18.4
	Phosphate	2.34	7.59	72.8	22.4
Lactate	Carbon	4.61	6.85	36.9	24.8
	Sulphate	9.44	17.00	18.0	10.0
	Ammonia	11.11	17.00	15.3	10.0
	Phosphate	7.00	12.50	24.3	13.6

[a] Dilution rate D = 0.17/hr, 35°C, and pH 6.8.

uptake rate was increased under all conditions of carbon-substrate excess as compared with carbon-substrate-limited cultures growing at the same rate. Clearly, as mentioned previously, these differences cannot be expressed at μ_{max} (since here all cultures are only nominally limited by the chosen nutrient), and indeed, they are not (Figure 6). This change in substrate uptake rate, which presumably reflects a progressive change in substrate uptake potential with growth rate in the case of carbon-substrate-sufficient cultures, has important consequences for the interpretation of yield data, particularly in the interpretation of Y^{max} values.

The point was made earlier that Y^{max} is defined as the reciprocal of the line of regression of q on μ, and that the maintenance rate (q_m) is the extrapolated value of q when $\mu = 0$. Thus, with carbon-substrate-sufficient cultures, the high rate of substrate uptake and catabolism expressed at low growth rates causes the line of regression of q on μ to be flattened (Figure 6), leading to a marked increase in the Y^{max} value. Taken at face value, this could be interpreted as indicating that carbon-substrate-sufficient cultures express a vastly increased maintenance energy requirement, but nevertheless, use the remaining energy with a greatly increased efficiency in cell synthesis. The fault here resides in a failure to take into account the essential difference between a carbon-substrate-limited and a carbon-substrate-sufficient culture. In the former, the rate of substrate uptake is limited by its rate of supply to the growing culture, whereas in the latter, it is not. Thus, with carbon substrate-sufficient cultures, Y^{max} values specify no more than the change in substrate uptake *potential* with growth rate. They, therefore, provide no basis on which to evaluate the energetic requirements of cell synthesis, nor do they provide a clue as to the efficiency with which oxidative phosphorylation proceeds in organisms so cultured.

A clear illustration of this point is to be found in the work of Hempfling and Mainzer.[31] Here, a culture of *E. coli* was grown in a glucose-limited chemostat culture in,

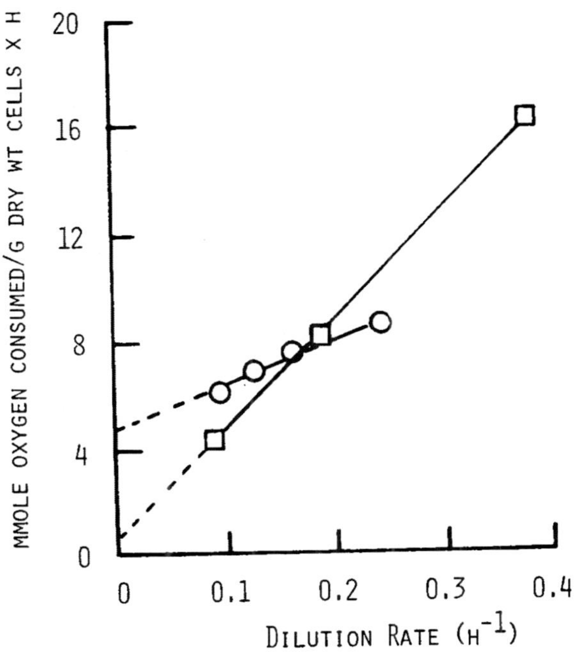

FIGURE 9. Influence of cyclic AMP on the relationship between oxygen consumption rate and growth rate found with a glucse-limited culture of *Escherichia coli* B. Cultures growing in the absence (□) and presence (O) of 2.7 m*M* cAMP. Plotted from the data of Hempfling and Mainzer.[31]

respectively, the absence and presence of cyclic AMP (cAMP). The steady-state bacterial concentrations and oxygen consumption rates were determined at a number of different dilution rates and the data compared. It was found that the presence of cAMP provoked a substantial increase in oxygen consumption rate at the lower growth rate values causing the plot of q_{O_2} vs. growth rate to be flattened (Figure 9) and, thus, the derived $Y_{O_2}^{max}$ value to be considerably increased. Surprisingly, and in spite of the fact that organisms grew in the presence of cAMP with a markedly lower expressed efficiency with respect to glucose and oxygen at the lower growth rates, it was concluded that cAMP relieved partial catabolite repression of energy conservation sites. This allowed the cells both to generate ATP and to consume it in biosynthesis with a greatly *increased* efficiency! This paradox undoubtedly arises from a failure to appreciate the profound influence that a change in the so-called maintenance rate can exert on the value of Y^{max}. They are not independent, in a physiological sense, though they are so, mathematically (Equation 2).

The actual interdependence of the constants q_m and Y^{max} is further illustrated in experiments where carbon substrate is added discontinuously to chemostat cultures. In this connection, it might be mentioned that it is common practice with most chemostat cultures to add medium drop-wise (see Herbert et al.[42] and Evans et al.[43]), which means that at low dilution rates the supply of growth-limiting nutrient becomes markedly discontinuous. That this can have a significant effect on the different growth parameters is evident from the data contained in Figure 10. In this experiment, a concentrated glucose solution was added to the culture at a rate of 1 drop/ 2 min and the dilution rate was varied by altering the rate of addition of bulk (glucose free) medium. The oxygen uptake rate was found to oscillate with a periodicity of 2 min (as expected),

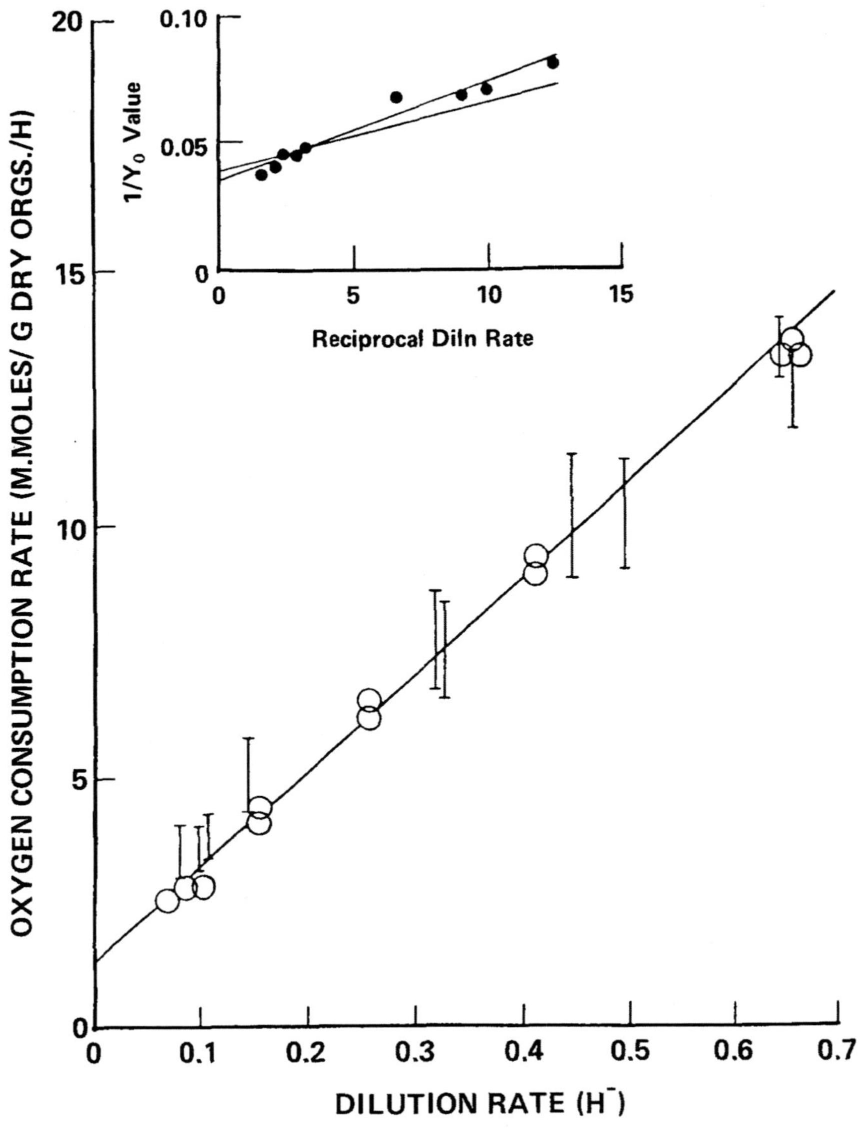

FIGURE 10. The relationship found between the specific rates of oxygen consumption and dilution rate with a glucose-limited chemostat culture of *Klebsiella aerogenes* in which the glucose was added either along with the bulk medium (O) or separately in regular small increments (1 drop every 2 min) as a concentrated solution (I). The upper and lower limits of the vertical bars show the maximum and minimum oxygen consumption rates, which oscillated with a periodicity of about 2 min. The insert shows the relationship between the reciprocal Y_O value and reciprocal dilution rate obtained with the pulsed culture (●) as compared with the nonpulsed culture (symbols omitted for clarity). In this latter graph, the Y_O values are mean values derived from the mean of the oxygen uptake rate at each dilution rate. These data indicate that the pulsed culture not only expressed a higher maintenance rate (steeper slope), but also had an increased Y_O^{max} value (changes intercept on the ordinate). Data of Neijssel and Tempest.[54]

but the maximum and minimum rates at each growth rate could be easily calculated. The data obtained (represented by the bars in Figure 10) showed that, as compared with minimally pulsed cultures, the oxygen uptake rate was significantly increased at low dilution rates, but not at the higher growth rate values. This suggested that pulsed cultures expressed a higher maintenance energy requirement, but nevertheless, grew with an increased basic efficiency (as represented by an increased Y_O^{max}). Again, such a conclusion fails to take into account the influence that a change in the maintenance rate (m) exerts on Y^{max}; m is a complex function which, at least in part, reflects the degree to which the organism can integrate transient fluxes in nutrient (and, hence, in pool intermediary metabolites) without loss of efficiency with respect to energy generation and to assimilation of substrate carbon. Clearly with cultures of *K. aerogenes* (Figure 10), such integration could be better effected at high growth rates than at the lower ones.

The data contained in Figure 10 raises once more the question of the regulation of carbon-substrate uptake and assimilation under nonrestricting (carbon-substrate sufficient) conditions. In particular, how do carbon-substrate-limited cultures respond when the growth limitation is suddenly and totally relieved? With glucose-limited cultures of *K. aerogenes* NCIB 418, it was found that excess glucose pulsed into the culture was consumed at a high rate, irrespective of the rate at which organisms were growing (and assimilating glucose) prior to the pulse (Figure 11). Interestingly, growth rate was not concomitantly increased (see also Harvey[44]), and much of the excess glucose was catabolized to products other than CO_2 which accumulated in the medium. Thus, the $Y_{glucose}$ and Y_O values expressed by these transient-state cultures were extremely low and much energy must have been dissipated as heat.

The primary uptake mechanism for glucose in *K. aerogenes* is the PEP-glucose phosphotransferase system,[45] and Figure 12 shows that slowly growing glucose-limited cultures of this organism possessed extremely high levels of the enzymes of this uptake system. At progressively higher growth rates, the PEP-glucose phosphotransferase activity progressively diminished such that, with cultures growing at dilution rates above 0.7/hr, it could not account totally for the expressed rate of glucose uptake, and auxiliary glucose uptake mechanisms must have been invoked. However, the significance of these changes in glucose uptake potential with growth rate becomes apparent when they are considered in relation to the behavior of continuously pulsed glucose-limited cultures (Figure 10). For it is clear that, at low growth rates, the difference between the glucose uptake potential (as reflected by the PEP-glucose phosphotransferase activity) and the uptake rate that can be expressed under glucose-limiting conditions is enormous, whereas at the higher growth rate values it is minimal. Since organisms seemingly cannot immediately accelerate their rate of cell synthesis following a pulse of glucose, overflow metabolism is likely to be more marked at low growth rates than at high ones as is clearly evident (Figure 11). It is reasonable to assume that continuously pulsed glucose-limited cultures will behave similarly, though to a lesser degree since the transient rises in glucose concentration would be of a much smaller magnitude. This would explain the loss of carbon conversion efficiency at low growth rates (expressed as an increase in the maintenance energy rate).

Whether the discontinuous addition of medium at low dilution rates, commonly occurring with chemostat cultures, contributes significantly to the "maintenance" rate is not known, but it seems highly likely that it does with this particular organism. On the other hand, glucose-limited cultures of *E. coli* have been found to modulate their glucose uptake potential with growth rate such as to maintain a constant differential,[46] and it remains to be shown whether this mode of modulation of the uptake system or that found with *K. aerogenes* is the most common among bacterial species. Whichever proves to be the case, it is clear that Y^{max} values, as defined by Equation 2, are not simple biological constants and cannot be meaningfully compared without reference to other parameters.

4. Chemolithotrophs and Phototrophs

Up to this point, attention has been focused on quantitative aspects of the aerobic growth of heterotrophic organisms. Underlying the various considerations of the en-

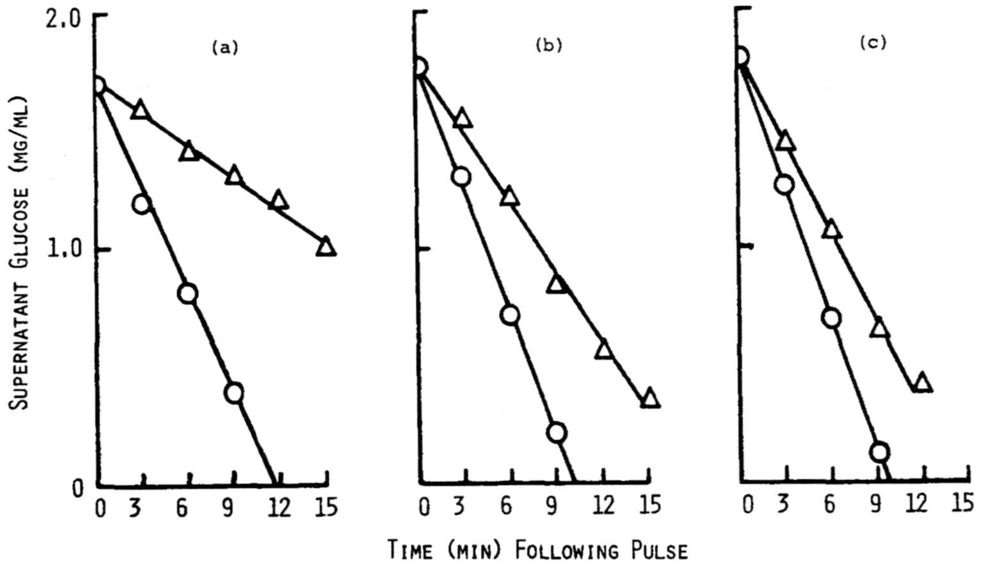

FIGURE 11. Changes in the extracellular glucose concentration (O) and total oxidizable carbon expressed as glucose equivalents (Δ) following the addition of glucose (10 m*M* end concentration) to a glucose-limited culture of *Klebsiella aerogenes* NCIB 418 growing at a dilution rate of (a) 0.11/hr, (b) 0.54/hr, and (c) 0.70/hr. Data of O'Brien et al.[29]

ergetics of growth of such organisms has been the general thesis, first tested by Rosenberger and Elsden,[47] that energy-yielding catabolic reactions are not directly controlled by the energy-consuming processes of anabolism. This lack of strict coupling no doubt accounts for the marked differences observed in the measured yield values when cultures that have been grown under different conditions are compared. Nevertheless, the high degree of reproducibility generally achieved with cultures growing under a fixed set of prescribed conditions suggests that some measure of coupling does exist between anabolic and catabolic processes.

With chemolithotrophs and phototrophic organisms, the situation can be much more easily analyzed since with these the energy-generating reactions are separate from those effecting synthesis of intermediary metabolites. The pathways of carbon and energy flow in these organisms can be broadly represented as follows:

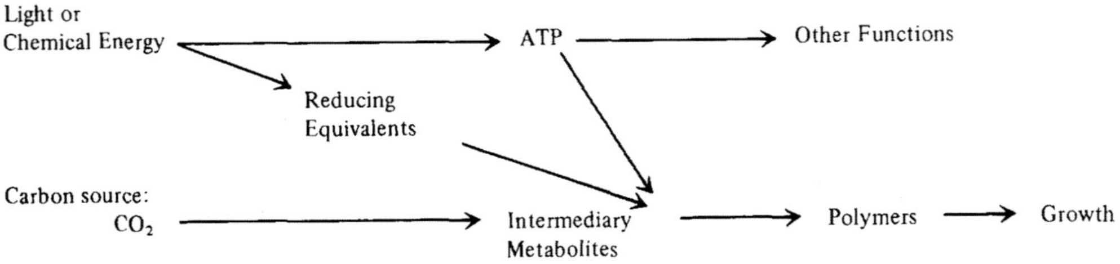

Clearly, if growth is tightly coupled to the rate of ATP synthesis, then a fixed relationship should exist between q_{ATP} and the rate of CO_2 assimilation. This has not been extensively tested, but sufficient data are present in the literature to support the conclusion that coupling is just as loose in these organisms as in the heterotrophs. Thus,

Lewenstein and Bachofen[48] found with chemostat cultures of *Chlorella fusca* that the rates of apparent ATP synthesis and CO_2 fixation did not seem to be strictly correlated when environmental conditions were varied. Similar observations have been made with cultures of the cyanobacterium *Oscillatoria agardhii*.[56] Moreover, overflow metabolism has been found to occur widely with cultures of phototrophs (*Anabaena cylindrica, Rhodopseudomonas capsulata*) and chemolithotrophs (*Desulfovibrio* sp., *Thiobacillus thiooxidans*), particularly when such are incubated under conditions where growth is impeded by the availability of essential nutrients like utilizable nitrogen or phosphorus (see Tempest and Neijssel[49]). This, again, attests to a lack of strict coupling between ATP synthesis and growth in this group of organisms.

IV. GENERAL CONCLUSIONS

Microbial growth is the product of an exceedingly large number of interconnected enzyme-catalyzed reactions, and the fact that cell synthesis proceeds with a more or less constant efficiency in any particular environment indicates that a substantial measure of control must be exercised over the fluxes of intermediary metabolites and precursor substances involved in polymer synthesis. However, these processes of regulation are further complicated in aerobic chemoheterotrophic organisms since the energy needed for biosynthesis necessarily must be derived from the breakdown of carbon substrate that is being simultaneously assimilated into cell substance. Hence, one might expect mechanisms to exist in these organisms that allow intermediary metabolites to be precisely partitioned between the catabolic (energy generating) and the anabolic (energy consuming) reactions of biosynthesis. Thus one could envisage control systems to exist within the cell that would act at specific branch points between, respectively, catabolic and anabolic pathways of metabolism and which, further, would be "tuned" to the overall energy status of the cell. That is, they would be "tuned" to the "energy charge" (see Atkinson[50,51]) or, more precisely, to the ratio of the rates of ATP production and turnover since the growing cell is in a dynamic state. That some such controls do indeed exist within the microbial cell is abundantly obvious from the extensively reported involvement of adenine nucleotides as control elements in intermediary metabolism (see Chapman and Atkinson[52]). The mode of action of these regulatory processes (i.e., allosteric effectors) leads not unreasonably to the concept of there being "coupling" between ATP synthesis and growth which manifests itself as a precise Y_{ATP} value. However, that such a concept is untenable is clear from observations extending back many years (see Rosenberger and Elsden[47]) in that energy (ATP) generation can occur at a high rate under conditions where cell synthesis is severely constrained, if not totally inhibited. Clearly there is no *obligatory* coupling between ATP synthesis and growth, and herein lies the source of much confusion and contradiction regarding the interpretation of yield data in bioenergetic terms. Thus, whereas under some conditions (e.g., carbon-substrate limited, anaerobic) it is reasonable to assume that growth is basically energy limited, and therefore, organisms will grow with an optimum energetic efficiency, under many other circumstances this well may not be the case. In particular, the capacity of the respiratory chain to generate ATP at a high rate renders it improbable that growth rate, aerobically, will be commonly limited by the rate of ATP synthesis. It is obvious, therefore, that further progress in evaluating the energetics of microbial growth, particularly of aerobes, hinges critically on the acquisition of a better understanding of those energy-spilling processes extant within the cell.

In this chapter, an attempt has been made to identify and analyze those processes that are fundamental to considerations of yield and to view these in the context of those theories that have been put forward to account for variations in this parameter. Such an exercise allows the broad conclusion to be drawn that yield values per se are

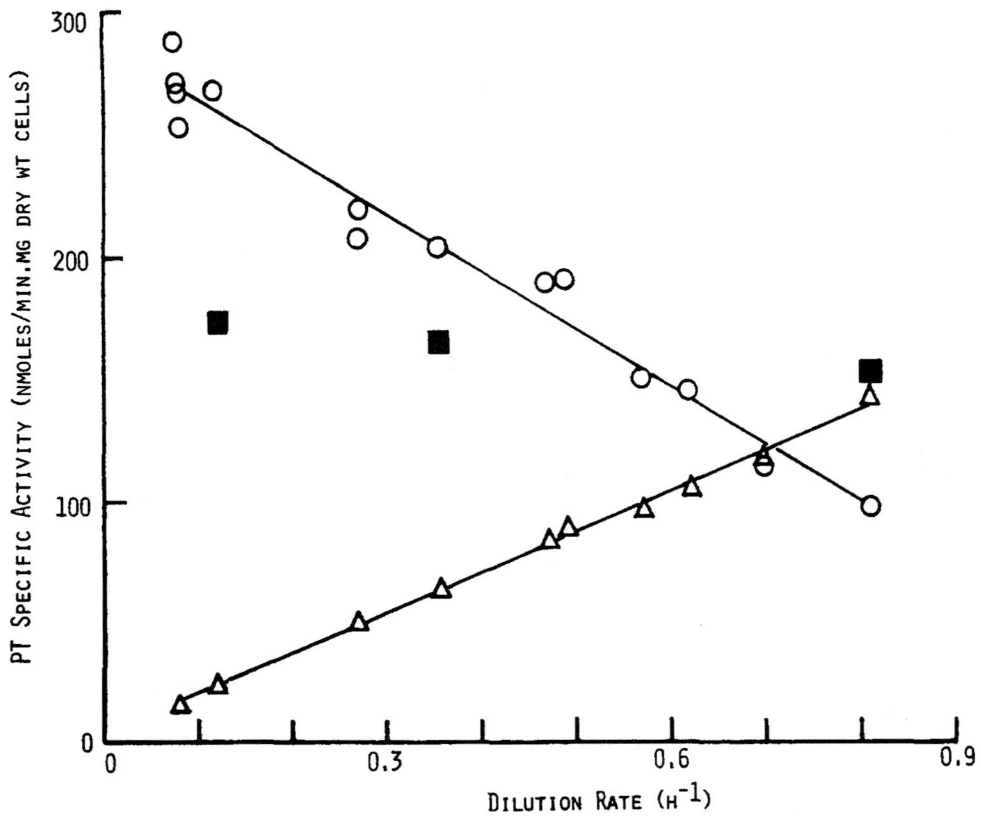

FIGURE 12. Glucose-PEP phosphotransferase activities and glucose uptake rates of *Klebsiella aerogenes* NCTC 418 growing in a glucose-limited chemostat culture at different dilution rates. (○) Phosphotransferase activity of decryptified cell; (△) actual rate of glucose utilization in the growing culture; and (■) rate of glucose consumption following the addition of a pulse of glucose (10 m*M* end concentration). Data of O'Brien et al.[29]

not readily interpretable in precise bioenergetic and/or physiological terms, and unless treated with extreme circumspection, they may lead to the formulation of concepts that are, at best, dubious. It is not inappropriate, therefore, to reiterate briefly those facts (or factors) that any interpretation of yield values should embrace. These are as follows:

1. Microbial cells must consume carbon substrate in order to grow, but the reverse is not true. Cells do not have to grow in order to consume carbon substrate.
2. Similarly, growth demands an expenditure of energy (ATP hydrolysis), but cells do not have to grow to expend energy. ATP can be synthesized and turned over at a high rate when growth is impeded. There is no obligatory coupling between ATP synthesis and growth.
3. Carbon-substrate-limited cultures are not necessarily energy limited, particularly when growing aerobically.
4. Organisms may, under certain conditions, delete sites of energy conservation. More to the point, however, they clearly can invoke energy-spilling reactions that undermine assessments of q_{ATP} and, thereby, bedevil assessments of associated parameters such as P/O ratios and Y_{ATP} values.
5. The maintenance energy rate and maximum growth yield value are essentially mathematical constants and not biological constants. In a strict physiological

sense, both may vary with growth rate. This clearly is the case with carbon-substrate-sufficient cultures. Both parameters also may be influenced by the mode of addition of substrate to continuously growing cultures.

6. Carbon-substrate-sufficient cultures are not directly comparable with carbon-substrate-limited cultures, and comparisons of Y_O values (or Y_O^{max} values) form no firm basis upon which to assess the relative efficiencies of aerobic energy conservation, nor the energetic requirements of cell synthesis.

This apparent capacity of microorganisms to dissociate catabolism from anabolism raises a final question regarding the importance of yield as a selective force in the evolution of microbial species. Have, in fact, those organisms that can effect conversion of substrate carbon to biomass with an increased efficiency generally been selected in the course of evolution? Eco-physiological considerations render this improbable (see Tempest and Neijssel[49]). Many natural environments are severely nutrient limited, and under these conditions, it will be those organisms that can scavenge traces of nutrient effectively, rather than those that can utilize it optimally, that will survive and flourish. This "scavenging" activity well may place heavy energetic demands upon the organism (as clearly is seen with cultures that are, say, ammonia or phosphate limited), and it is obvious that they would be at a considerable disadvantage if the rate of energy (ATP) generation was, concomitantly, severely constrained. As compared with many essential nutrients, energy is not in short supply in the biosphere. It is not surprising, therefore, that organisms in general tend to be inefficient convertors of the energy (see Wilkie[53]).

REFERENCES

1. **Herbert, D.,** Some principles of continuous culture, in *Recent Progress in Microbiology,* Symp. 7th Int. Congr. Microbiol., Stockholm, Tunevall, G., Ed., Almqvist and Wiksell, Stockholm, 958,381.
2. **Rogers, P.J. and Stewart, P.R.,** Energetic efficiency and maintenance energy characteristics of *Saccharomyces cerevisiae* (wild type and petite) and *Candida parapsilosis* grown aerobically and microaerobically in continuous culture, *Arch. Microbiol.,* 99, 25, 1974.
3. **Neijssel, O.M.,** The Significance of Overflow Metabolism in the Physiology and Growth of *Klebsiella aerogenes,* thesis, University of Amsterdam, Amsterdam, 1976.
4. **Stouthamer, A.H. and Bettenhaussen, C.W.,** Determination of the efficiency of oxidative phosphorylation in continuous cultures of *Aerobacter aerogenes, Arch. Microbiol.,* 102, 187, 1975.
5. **Krebs, H.A.,** The Pasteur effect and the relations between respiration and fermentation, in *Essays in Biochemistry,* Vol. 8, Campbell, P.N. and Dicken, F., Eds., Academic Press, New York, 1972, 1.
6. **Herbert, D.,** Stoichiometric aspects of microbial growth, in *Continuous Culture 6, Applications and New Fields,* Dean, A.C.R., Ellwood, D.C., Evans, C.G.T., and Melling, J., Eds., Ellis Horwood Ltd., Chichester, U.K., 1976, 1.
7. **Dawes, E.A. and Senior, P.J.,** The role and regulation of energy reserve polymers in micro-organisms, *Adv. Microb. Physiol.,* 10, 135, 1973.
8. **Herbert, D. and Kornberg, H.L.,** Glucose transport as rate-limiting step in the growth of *Escherichia coli* on glucose, *Biochem. J.,* 156, 477, 1976.
9. **Neijssel, O.M., Hueting, S., and Tempest, D.W.,** Glucose transport capacity is not the rate-limiting step in the growth of some wild-type strains of *Escherichia coli* and *Klebsiella aerogenes* in chemostat culture, *FEMS Microbiol. Lett.,* 2, 1, 1977.
10. **Neijssel, O.M. and Tempest, D.W.,** The regulation of carbohydrate metabolism in *Klebsiella aerogenes* NCTC 418 organisms, growing in chemostat culture, *Arch. Microbiol.,* 106, 251, 1975.
11. **Mitchell, P.,** Chemiosmotic coupling in oxidative and photosynthetic phosphrylation, *Biol. Rev. Cambridge Philos. Soc.,* 41, 445, 1966.
12. **Haddock, B.A. and Jones, C.W.,** Bacterial respiration, *Bacteriol. Rev.,* 41, 47, 1977.

13. **John, P. and Whatley, F.R.,** Oxidative phosphorylation coupled to oxygen uptake and nitrate reduction in *Micrococcus denitrificans, Biochim. Biophys. Acta,* 216, 342, 1970.
14. **Neijssel, O.M. and Tempest, D.W.,** Bioenergetic aspects of aerobic growth of *Klebsiella aerogenes* NCTC 418 in carbon-limited and carbon-sufficient chemostat culture, *Arch. Microbiol.,* 107, 215, 1976.
15. **Neijssel, O.M.,** The effect of 2,4-dinitrophenol on the growth of *Klebsiella aerogenes* NCTC 418 in aerobic chemostat cultures, *FEMS Microbiol. Lett.,* 1, 47, 1977.
16. **Dicks, J.W. and Tempest, D.W.,** Potassium-ammonium antagonism in polysaccharide synthesis by *Aerobacter aerogenes* NCTC 418, *Biochim. Biophys. Acta,* 136, 176, 1967.
17. **Harrison, D.E.F.,** A study of the effect of growth conditions on the cytochromes of chemostat-grown *Klebsiella aerogenes* and kinetic changes of a 500 nm absorption band, *Biochim. Biophys. Acta,* 275, 83, 1973.
18. **Jones, C.W.,** Aerobic respiratory systems in bacteria, in *Microbial Energetics,* 27th Symp. Soc. Gen. Microbiol., Haddock, B.A. and Hamilton, W.A., Eds., Cambridge University Press, Cambridge, 1977, 23.
19. **Poole, R.K. and Haddock, B.A.,** Effects of sulphate-limited growth in continuous culture on the electron transport chain and energy conservation in *Escherichia coli* K12, *Biochem. J.,* 152, 537, 1975.
20. **Rainnie, D.J. and Bragg, P.D.,** The effect of iron deficiency on respiration and energy-coupling in *Escherichia coli, J. Gen. Microbiol.,* 77, 339, 1973.
21. **Light, P.A. and Garland, P.A.,** A comparison of mitochondria from *Torulopsis utilis* grown in continuous culture with glycerol, iron, ammonium, magnesium or phosphate as the growth-limiting nutrient, *Biochem. J.,* 124, 123, 1971.
22. **Aiking, H., Sterkenburg, A., and Tempest, D.W.,** Influence of specific growth limitation and dilution rate on the phosphorylation efficiency and cytochrome content of mitochondria of *Candida utilis* NCYC 321, *Arch. Microbiol.,* 113, 65, 1977.
23. **Brown, C.M. and Stanley, S.O.,** Environment-mediated changes in the cellular content of the "pool" constituents and their associated changes in cell physiology, *J. Appl. Chem. Biotechnol.,* 22, 363, 1972.
24. **Tempest, D.W.,** The biochemical significance of microbial growth yields: a reassessment, *Trends Biochem. Sci.,* 3, 180, 1978.
25. **Neijssel, O.M., Sutherland-Miller, T.O., and Tempest, D.W.,** Pyruvate reductase and D-lactate dehydrogenase: a possible mechanism for avoiding energy conservation at site 1 of the respiratory chain in *Klebsiella aerogenes, Proc. Soc. Gen. Microbiol.,* 5, 49, 1978.
26. **Hueting, S. and Tempest, D.W.,** Influence of acetate on the growth of *Candida utilis* in continuous culture, *Arch. Microbiol.,* 115, 73, 1977.
27. **Pirt, S.J.,** The maintenance energy of bacteria in growing cultures, *Proc. Roy. Soc. London Ser. B,* 163, 224, 1965.
28. **Tempest, D.W. and Herbert, D.,** Effect of dilution rate and growth-limiting substrate on the metabolic activity of *Torula utilis* cultures, *J. Gen. Microbiol.,* 41, 143, 1965.
29. **O'Brien, R.W., Neijssel, O.M., and Tempest, D.W.,** Glucose phosphoenolpyruvate phosphotransferase activity and glucose uptake rate of *Klebsiella aerogenes* growing in chemostat culture, *J. Gen. Microbiol.,* 116, in press, 1980.
30. **Stouthamer, A.H. and Bettenhaussen, C.W.,** Utilization of energy for growth and maintenance in continuous and batch cultures of microorganisms. A reevaluation of the method for the determination of ATP production by measuring molar growth yields, *Biochim. Biophys. Acta,* 301, 53, 1973.
31. **Hempfling, W.P. and Mainzer, S.E.,** Effects of varying the carbon source limiting growth on yield and maintenance characteristics of *Escherichia coli* in continuous culture, *J. Bacteriol.,* 123, 1976, 1975.
32. **Downs, A.J. and Jones, C.W.,** Energy conservation in *Bacillus megaterium, Arch. Microbiol.,* 105, 159, 1975.
33. **Jones, C.W., Brice, J.M., and Edwards, C.,** The effect of respiratory chain composition on the growth efficiencies of aerobic bacteria, *Arch. Microbiol.,* 115, 85, 1977.
34. **Brown, C.M.,** Nitrogen metabolism in bacteria and fungi, in *Continuous Culture 6, Applications and New Fields,* Dean, A.C.R., Ellwood, D.C., Evans, C.G.T., and Melling, J., Eds., Ellis Horwood Ltd., Chichester, U.K., 1976, 170.
35. **Edwards, C., Spode, J.A., and Jones, C.W.,** The growth energetics of *Paracoccus denitrificans, FEMS Microbiol. Lett.,* 1, 67, 1977.
36. **van Verseveld, H.W. and Stouthamer, A.H.,** Oxidative phosphorylation in *Micrococcus denitrificans.* Calculation of the P/O ratio in growing cells, *Arch. Microbiol.,* 107, 241, 1976.
37. **Bauchop, T. and Elsden, S.R.,** The growth of microorganisms in relation to their energy supply, *J. Gen. Microbiol.,* 23, 457, 1960.

38. **John, P. and Whatley, F.R.**, *Paracoccus denitrificans* and the evolutionary origin of the mitochondrion, *Nature (London)*, 254, 495, 1975.
39. **Stouthamer, A.H.**, Energetic aspects of the growth of micro-organisms, in *Microbial Energetics*, 27th Symp. Soc. Gen. Microbiol., Haddock, B.A. and Hamilton, W.A., Eds., Cambridge University Press, Cambridge, 1977, 285.
40. **Baak, J.M. and Postma, P.W.**, Oxidative phosphorylation in intact *Azotobacter vinelandii*, *FEBS Lett.*, 19, 189, 1971.
41. **Hempfling, W.P.**, Studies of the efficiency of oxidative phosphorylation in intact *Escherichia coli*, *Biochim. Biophys. Acta*, 205, 169, 1970.
42. **Herbert, D., Phipps, P.J., and Tempest, D.W.**, The chemostat: design and instrumentation, *Lab. Pract.*, 14, 1150, 1965.
43. **Evans, C.G.T., Herbert, D., and Tempest, D.W.**, The continuous cultivation of microorganisms. II. Construction of a chemostat, in *Methods in Microbiology*, Vol. 2, Norris, J.R. and Ribbons, D.W., Eds., Academic Press, New York, 1970, 277.
44. **Harvey, R.J.**, Metabolic regulation in glucose-limited chemostat cultures of *Escherichia coli*, *J. Bacteriol.*, 104, 698, 1970.
45. **Kundig, W., Ghosh, S., and Roseman, S.**, Phosphate bound to histidine in a protein as an intermediate in a novel phosphotransferase system, *Proc. Natl. Acad. Sci. U.S.A.*, 52, 1067, 1964.
46. **Hunter, I.S. and Kornberg, H.L.**, Glucose transport in *Escherichia coli* growing in glucose-limited chemostat culture, *Biochem. J.*, in press.
47. **Rosenberger, R.F. and Elsden, S.R.**, The yields of *Streptococcus faecalis* grown in continuous culture, *J. Gen. Microbiol.*, 22, 727, 1960.
48. **Lewenstein, A. and Bachofen, R.**, CO_2-fixation and ATP synthesis in continuous cultures of *Chlorella fusca*, *Arch. Microbiol.*, 116, 169, 1978.
49. **Tempest, D.W. and Neijssel, O.M.**, Eco-physiological aspects of microbial growth in aerobic nutrient-limited environments, in *Advances in Microbial Ecology*, Vol. 2, Alexander, M., Ed., Plenum Press, New York, 1978, 105.
50. **Atkinson, D.E.**, The energy charge of the adenylate pool as a regulatory parameter. Interaction with feedback modifiers, *Biochemistry*, 7, 4030, 1968.
51. **Atkinson, D.E.**, Regulation of enzyme function, *Annu. Rev. Microbiol.*, 23, 47, 1969.
52. **Chapman, A.G. and Atkinson, D.E.**, Adenine nucleotide concentrations and turnover rates. Their correlation with biological activity in bacteria and yeast, *Adv. Microb. Physiol.*, 15, 253, 1977.
53. **Wilkie, D.R.**, Thermodynamics and the interpretation of biological heat measurements, *Prog. Biophys. Mol. Biol.*, 10, 259, 1960.
54. **Neijssel, O.M. and Tempest, D.W.**, The role of energy-spilling reactions in the growth of *Klebsiella aerogenes* NCTC 418 in aerobic chemostat culture, *Arch. Microbiol.*, 110, 305, 1976.
55. **Neijssel, O.M.**, unpublished results.
56. **van Liere, E.**, personal communication.

Chapter 2

ELECTRON-TRANSFER-DRIVEN SOLUTE TRANSLOCATION ACROSS BACTERIAL MEMBRANES

W.N. Konings and P.A.M. Michels

TABLE OF CONTENTS

I. INTRODUCTION

Electron transfer in cytochrome-linked electron transfer systems has been recognized for several decades to play a major role in the energy metabolism of respiring and phototrophically growing bacteria. More recently, it was shown that under anaerobic conditions electron transfer systems with alternative terminal electron acceptors perform a similar function in many bacteria.[1-4] Even in cells grown under so-called "glycolytic" conditions, anaerobic electron transfer systems contribute in certain organisms to the energy supply of a cell.[5] For a long time, it was widely accepted that the only form in which the redox energy became available for a cell was ATP. Many studies have been concerned with the question of how the energy released during electron transfer is converted into the synthesis of ATP by membrane bound Ca^{2+} - Mg^{2+} - stimulated ATPase. Several mechanisms have been proposed to explain this energy coupling, and numerous studies have been performed to disprove these models. One model which up to the present time still has managed to stand up to the criticisms raised is the chemiosmotic model proposed by Mitchell.[6-8] This model is clearly distinct from many other models proposed because it postulates an energy coupling of the electron transfer systems not only to ATP-synthesis, but also to other energy requiring functions in bacterial membranes. One of these energy consuming processes is the translocation of solutes across the cytoplasmic membranes against a concentration gradient.

The currently available information about the mechanism of energy transduction to the accumulation of solutes is in corroboration with the chemiosmotic concept. It is the aim of this chapter to review the information about bacterial transport systems with special attention to "active" transport processes coupled to electron transfer systems. We have focussed our attention especially towards the information supplied by studies with bacterial membrane vesicles. Only in those situations where information obtained from studies with intact cells was helpful or essential for clarifying or substantiating certain aspects will reference be made to these studies. We therefore do not pretend to cover in this chapter all the relevant, or even the most pertinant, information about solute transport in bacteria. For those aspects which have not received full attention, we direct the reader to one of the reviews which have appeared in the last few years.[9-14]

II. THE BACTERIAL CELL ENVELOPE

For cellular metabolism, a specific cytoplasmic composition is required which is distinctly different from that of the cell's surroundings. Such separation between the cytoplasm and the environment is mainly achieved by the diffusion barrier of the inner layer of the cell envelope, the cytoplasmic membrane. It consists of a liquid-crystalline bilayer of phospholipids in which proteins are embedded. It is in this cytoplasmic membrane that the energy transducing electron transfer systems and ATPase complex are located.

An important feature of the cytoplasmic membrane is its impermeability for many solutes. This property, which is required for maintaining the specific cytoplasmic composition as well as for chemiosmotic energy transduction (see below) is achieved by the hydrophobic properties of its components. This also results in membranes with a low electrical capacity, so that even small charge translocations across the membrane generate a substantial electrical potential.[15] Controlled influx of metabolizable solutes and ions and efflux of end products of the metabolism across the cytoplasmic membranes can occur by specific translocation systems.

Exterior of the cytoplasmic membrane lies the cell wall, which is mainly composed of peptidoglycan. Peptidoglycan consists of a network of polysaccharides cross-linked by short peptides. It is this rigid macromolecule which determines the volume and the shape of the cell. The thickness varies considerably between Gram-negative (2 to 3 nm) and Gram-positive organisms (15 to 18 nm).[16] Most solutes used by bacteria can freely diffuse through the pores of the network.

Outside the cell wall is a third layer in Gram-negative organisms, the outer membrane. This membrane consists of phospholipids and lipopolysaccharides in which proteins are embedded. The lipopolysaccharides are exclusively located in the outer leaflet of the bilayer. They are responsible for the major antigenic properties of the bacterial cell surface. The outer membrane forms, just like the cytoplasmic membrane, a barrier for solutes. The outer membrane differs from the cytoplasmic membrane by the presence of proteins which form nonspecific hydrophilic pores through which solutes can penetrate with a molecular weight of up to 600.[17] In addition to these nonspecific pores, various outer membrane proteins which function as receptor proteins for phages and colicines appear to be involved in a selective permeation of solutes.[18]

The peptidoglycan layer and the outer membrane are connected by a specific molecule, the so-called Braun's lipoprotein. The area located between the cytoplasmic membrane and the outer membrane is termed the periplasmic space. It contains several proteins, some of which are involved in the perception of chemotactic stimuli and in the transport of certain solutes across the cell envelope.

The three layers of the cell envelope thus differ both in structure and function. Of the three layers, the cytoplasmic membrane forms the most important diffusion barrier for solutes of small molecular weight. Insight into the mechanism of solute transport through the cell envelope therefore requires knowledge about the mechanism of solute translocation across the cytoplasmic membrane and the coupling of this process to the cell's energy generating machinery.

III. SOLUTE TRANSPORT ACROSS CYTOPLASMIC MEMBRANES

A. The Chemiosmotic Concept

The chemiosmotic model of energy coupling offers an explanation for the coupling between energy generating and energy consuming systems in the cytoplasmic membrane. According to this model,[8] energy transducing systems like electron transfer systems, ATPase-complexes, and transport systems are anisotropically incorporated in the membrane. Electron carriers can use the redox energy for the translocation of protons across the cytoplasmic membrane, normally from the inner to the outer aqueous compartment. Translocation of protons is equivalent to the movement of OH^- in the opposite direction, so that electron transfer results in the distribution of protons and hydroxyl ions on opposite sides of the membrane. Because the membrane is essentially impermeable to ions, and in particular to protons and hydroxyl ions, electron transfer will establish an electrochemical proton gradient ($\Delta\tilde{\mu}_{H^+}$) which consists of a proton gradient (ΔpH) and a charge gradient, the electrical potential ($\Delta\psi$). The sum of these forces constitutes the proton motive force:

$$\Delta\tilde{\mu}_{H^+} = \Delta\psi - Z\ \Delta pH$$

$\Delta\tilde{\mu}_{H^+}$ is the proton motive force, $\Delta\psi$ the electrical potential; and ΔpH the pH difference between both sides of the membrane, $Z = 2.3\ RT/F$ in which R is the gas constant, T the absolute temperature, and F the Faraday constant. Z has a numerical value of about 60 mV at 25°C.

The chemiosmotic hypothesis postulates that this proton motive force or one of its components is the driving force for energy consuming processes in the membrane (Figure 1). Membrane-bound systems like ATPase complexes, solute carriers, and the flagellar rotor can reverse the flow of protons and thus convert the energy of $\Delta\tilde{\mu}_{H^+}$ into energy-rich compounds such as ATP, or drive osmotic work such as the formation of solute gradients or mechanical work such as flagellar movement. The energy-transducing systems, e.g., ATPase and solute carriers and electron transfer chains, can act reversibly. This enables the cell to generate, for instance, a $\Delta\tilde{\mu}_{H^+}$ at the expense of ATP in the absence of electron transfer.

B. Definitions of Translocation Processes Across Cytoplasmic Membranes

Transport across the cytoplasmic membrane of bacteria is mediated by various translocation systems which differ in the mechanism of translocation and/or energization.

Historically, the processes which lead to translocation of solutes across the cytoplasmic membrane in an unmodified form have been classified into three distinct groups:

1. Passive diffusion: the solute crosses the membrane without specific interactions with membrane components down its concentration gradient as a result of molecular motion.
2. Facilitated diffusion: the solute combines reversibly with a specific carrier molecule in the membrane. The carrier or carrier-solute complex oscillates between inner and outer surfaces of the membrane and releases or binds the solute on either side. This process results in solute translocation down its concentration gradient.
3. Active transport: the solute combines with a specific carrier molecule in the membrane and the accumulation or extrusion of the solute occurs against its own gradient at the expense of metabolic energy.

Since the postulation of the chemiosmotic hypothesis for solute transport, it became increasingly clear that classification of transport processes into one of these groups is not always possible. The classification mentioned above was based on the conception that the driving force for translocation in both passive diffusion and facilitated diffusion is the chemical solute gradient only. Thus, transport would occur down the solute concentration gradient until equilibration across the membrane has been reached. However, when a charged solute is translocated by facilitated diffusion, the driving forces are the chemical gradient of the solute and the electrical gradient. The sum of these forces will determine the overall direction of translocation. Equilibrium will be reached when the total driving force is zero. When this facilitated diffusion process leads to accumulation of the solute against its concentration gradient, we are dealing in fact with an "active transport" process.

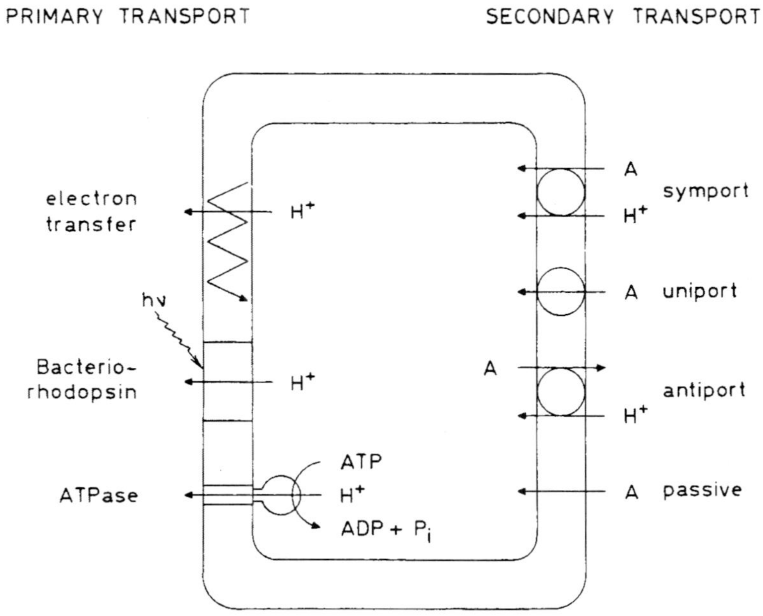

FIGURE 1. Schematic presentation of primary and secondary transport systems in bacteria.

In order to avoid this confusion, we have chosen for the following classification of the bacterial transport systems (Figure 1):

1. Primary transport: transport by enzyme systems which convert chemical or light energy into electrochemical energy. These transport systems comprise e.g., the electrogenic proton pumps: the electron transport systems, the Ca^{2+}-Mg^{2+}-stimulated ATPase and the light-driven proton pump bacteriorhodopsin.
2. Secondary transport: transport systems which are driven by electrochemical gradients. These transport systems are defined as *passive* when transport occurs without specific interactions with membrane components and as *facilitated* when transport is mediated by specific carrier molecules in the membrane. Mitchell[8] visualized three different systems for facilitated secondary transport. These are (1) Uniport: only one solute is translocated by a carrier protein, (2) Symport: different kinds of solutes are translocated by the same carrier protein; the solutes can be a metabolizable substrate, a proton or another ion, and (3) Antiport: the carrier translocates different solutes in opposite directions.
3. Group translocation: the solute is substrate for a specific enzyme system in the membrane. The enzymatic reaction results in a chemical modification of the solute and a release of the product at the other side of the membrane.
4. ATP-driven transport: solute transport is mediated by a specific membrane protein and the energy for translocation is supplied by ATP or phosphate-bound energy directly. In essence, this type of transport is a special form of primary transport. Mixed forms of transport mechanisms may also occur in which both ATP and a $\Delta\tilde{\mu}_{H^+}$ across the membrane are required for solute accumulation against its (electro)chemical gradient. This is, in our terminology, a combination of primary and facilitated secondary transport.

C. Secondary Transport

Secondary transport of solutes across the cytoplasmic membrane can occur passively, without specific interactions with membrane components, or facilitated, mediated by specific carrier molecules. The simplest translocation process is the passive transport of a neutral solute. This transport will be driven by the chemical gradient of the solute only: $\Delta\tilde{\mu}_A = Z \log A_{in}/A_{out}$; equilibrium will be reached when $\Delta\mu_A = 0$. In equilibrium (when no net transport of solute occurs), the internal solute concentration will equal the external solute concentration.

Transport of a charged solute will not only depend on the chemical concentration gradient of the solute, but also on its electrical potential gradient across the membrane. For a monovalent cation, the driving force, therefore, will be $\Delta\tilde{\mu}_A = Z \log A^+_{in}/A^+_{out} + \Delta\psi$ and a steady state will be reached when $Z \log A^+_{in}/A^+_{out} = -\Delta\psi$. In this steady state, the solute concentration internally and externally need not to be equal. In the absence of other electrical charge translocating systems (primary or secondary transport systems), net transport will stop already when the internal solute concentration is lower than the external concentration. However, when a $\Delta\psi$ (interior negative) is generated by other transport systems, accumulation of solute can occur.

Some weak acids can cross the membrane passively in an undissociated form, which in fact is a cotransport of the anion with a proton(s). For a monoprotic acid (HA), the proton/anion stoicheiometry equals 1. In the interior, the acid will dissociate according to the internal pH. In this transport, no net charge is translocated, and the driving force will depend only on the chemical gradients of the components involved, i.e., HA, A^-, and H^+. The total driving force, therefore, will be $Z (\log A^-_{in}/A^-_{out} + \log HA_{in}/HA_{out} + \log H^+_{in}/H^+_{out})$. This can be rewritten as $Z (\log A^-_{in}/A^-_{out} + \log HA_{in}/HA_{out}) - Z \Delta pH$. Equilibrium will be reached when: $Z \Delta pH = Z (\log A^-_{in}/A^-_{out} + \log HA_{in}/HA_{out})$. In many bacterial cells, the internal pH is higher than the external pH, and accumulation of the acid can occur.

In Table 1, the driving forces for passive secondary transport processes are given.

In a similar way, the driving forces for solute transport by active secondary transport systems can be derived (Table 1). The driving forces are given for uniport, symport, and antiport systems with various proton/solute stoicheiometries. The proton/solute stoicheiometry might depend on the net charge of the carrier. Mitchell[8] depicted the carrier molecules as being electroneutral. In later studies, mainly from Kaback's laboratory,[19-21] it was postulated that the carriers are negatively charged and that the carrier-solute (proton) complex is electroneutral during translocation across the membrane. Furthermore, it was postulated that the charge of the carrier was not fixed, but was determined by a pH-dependent dissociation of functional groups.[22,23] These postulates were based on the observations that in membrane vesicles of *E. coli* the ratio of the chemical potential gradient for a solute like proline over the electrochemical potential gradient for protons changes with the external pH. This ratio equals about one at an external pH of 5.5 and increases to two at an external pH of 7.5.[22,23]

Based on these considerations, Rottenberg[24] proposed a modified chemiosmotic model for transport (see Figure 2).

An essential feature of the chemiosmotic concept of transport is the interconversion of solute gradients. Recently, Skulachev[25] extended this idea by putting forward the hypothesis that some of the solute gradients, established at the expense of the electrochemical proton gradient, can be considered as energy buffers for a cell. The energy storage capacity of $\Delta\psi$ and ΔpH is relatively small, and additional storage capacity may be found in solute gradients, especially in transmembrane gradients of K^+ and Na^+. Potassium is usually accumulated into cells via uniport systems in response to the $\Delta\psi$, while sodium appears to be extruded by Na^+/H^+ antiport systems. Thus, electrophoretic K^+ influx causes conversion of a $\Delta\psi$ into a potassium gradient and, by addi-

TABLE 1

Stoicheiometry and Driving Forces for Secondary Solute Transport

Transport mechanism	Solute	Charge of carrier	Stoichiometry	Driving force
Passive	A^{o}			$Z \log (A_{oin}/A^{o}_{out})$
	H^{+}			$-Z\,\Delta pH + \Delta\psi$
	A^{+}			$Z \log (A^{+}_{in}/A^{+}_{out}) + \Delta\psi$
	AH^{+}		−1	$Z \log (AH^{+}_{in}/AH^{+}_{out}) + Z\,\Delta pH^{a}$
	A^{2+}			$Z \log (A^{2+}_{in}/A^{2+}_{out}) + 2\,\Delta\psi$
	A^{-}			$Z \log (A^{-}_{in}/A^{-}_{out}) - \Delta\psi$
	A^{-}		1	$Z \log (A^{-}_{in}/A^{-}_{out}) - Z\,\Delta pH^{b}$
	A^{2-}			$Z \log (A^{2-}_{in}/A^{2-}_{out}) - 2\,\Delta\psi$
Facilitated				
Uniport	A^{+}	C^{o}	–	$Z \log (A^{+}_{in}/A^{+}_{out}) + \Delta\psi$
Proton-solute symport	A^{o}	C^{-1}	1	$Z \log (A^{o}_{in}/A^{o}_{out}) + \Delta\psi - Z\,\Delta pH$
		C^{-2}	2	$Z \log (A^{o}_{in}/A^{o}_{out}) + 2\,(\Delta\psi - Z\,\Delta pH)$
	A^{+}	C^{-1}	1	$Z \log (A^{+}_{in}/A^{+}_{out}) + 2\,\Delta\psi - 2\,Z\,\Delta pH$
	A^{-1}	C^{o}	1	$Z \log (A^{-}_{in}/A^{-}_{out}) - Z\,\Delta pH$
		C^{-1}	2	$Z \log (A^{-}_{in}/A^{-}_{out}) + \Delta\psi - 2\,Z\,\Delta pH$
	A^{-2}	C^{o}	2	$Z \log (A^{-2}_{in}/A^{-2}_{out}) - 2\,Z\,\Delta pH$
		C^{-1}	3	$Z \log (A^{-2}_{in}/A^{-2}_{out}) + \Delta\psi - 3\,Z\,\Delta pH$
Cation — (M^{+}) solute			H^{+}:M^{+}:A	
symport	A^{o}	C^{-1}	0:1:1	$Z \log (A^{o}_{in}/A^{o}_{out}) + Z \log (M^{+}_{in}/M^{+}_{out}) + \Delta\psi$
		C^{-2}	1:1:1	$Z \log (A^{o}_{in}/A^{o}_{out}) + Z \log (M^{+}_{in}/M^{+}_{out}) + 2\,\Delta\psi - Z\,\Delta pH$
	A^{-1}	C^{o}	0:1:1	$Z \log (A^{o}_{in}/A^{o}_{out}) + Z \log (M^{+}_{in}/M^{+}_{out})$
		C^{-1}	1:1:1	$Z \log (A^{o}_{in}/A^{o}_{out}) + Z \log (M^{+}_{in}/M^{+}_{out}) + \Delta\psi - Z\,\Delta pH$
Proton-solute antiport			H^{+}/A	
	A^{+}	C^{o}	1	$Z \log (A^{+}_{in}/A^{+}_{out}) - Z\,\Delta pH$
		C^{-1}	2	$Z \log (A^{+}_{in}/A^{+}_{out}) - 2\,Z\,\Delta pH$
	A^{2+}	C^{o}	2	$Z \log (A^{2+}_{in}/A^{2+}_{out}) - 2\,Z\,\Delta pH$
		C^{-1}	3	$Z \log (A^{2+}_{in}/A^{2+}_{out}) - 3\,Z\,\Delta pH$

[a] When A is small with respect to AH^{+}.
[b] When AH is small with respect to A^{-}.

tional proton translocation (by primary transport), into a ΔpH. This ΔpH can, subsequently, be converted into a Na^{+} gradient by the Na^{+}/H^{+} antiport.

D. Methods for Determination of Δψ and ΔpH

Several methods have been developed to measure the membrane potential and the pH gradient across the membrane. These methods will be surveyed here briefly. For, more detailed information, the reader is directed to other review articles.[26,27]

1. Electrical Membrane Potential (Δψ)

a. Distribution of Membrane-Permeable Ions

Membrane-permeable ions will distribute across the membrane according to the membrane potential. The distribution of these ions over the compartments at both sides of the membrane can be measured by chemical, radiochemical, or spectroscopical assays, or by means of ion selective electrodes.

Measurements are done either discontinuously after physical separation of the membranes from the external medium, or continuously with ion-selective electrodes[28] or by the application of flow dialysis.[29]

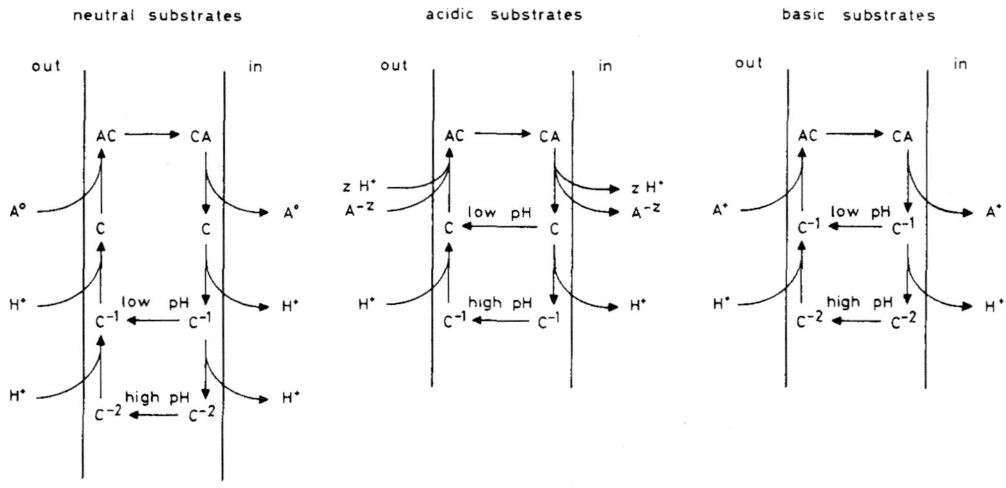

FIGURE 2. Models for cotransport carrier in bacterial cells. C, carrier; A, solute; Z, negative charge of solute. (According to Rottenberg, H., *FEBS Lett.*, 87, 171, 1978. With permission.)

The ions used most commonly are lipophilic compounds like triphenylmethylphosphonium ($TPMP^+$), tetraphenylphosphonium (TPP^+), dibenzyldimethylammonia (DDA^+), phenyldicarbo-undecaborane (PCB^-),[30,31] the permeable ions thiocyanate, chloride, and iodide, and potassium or rubidium in the presence of valinomycin. Also, fluorescent ions have been used such as 8-anilinonaphtalenesulphonic acid (ANS)[32,33] and cyanine dyes.[34] However, the reliability of these latter probes for quantitative measurements have been questioned.[35-37]

b. Absorbance Changes of Membrane-Bound Components

Absorption spectra of membrane-bound components, e.g., carotenoids, chlorophylls, or membrane-incorporated components like merocyanins appear to be sensitive to changes in the electrical field across the membrane.[38,39] The absorbance changes of carotenoids have been used frequently for membrane potential measurements in membrane particles from phototrophic bacteria.[40,41] The necessary calibrations have been made using known, artificially induced membrane potentials. However, it has been questioned whether the absorbance changes really reflect the membrane potential.[42]

2. pH Gradients (ΔpH)

a. Direct pH Measurements

The ΔpH can be calculated from direct measurements of the pH changes in the external medium of cells and membrane vesicles by means of pH electrodes or pH indicators when the internal buffer capacity is known.[43-45]

b. Distribution of Membrane-Permeable Weak Acids or Bases

Weak acids or bases which are membrane permeable in neutral form will distribute across the membrane according to the pH gradient. The most commonly used compounds for pH gradients, internally alkaline, are the acids acetate and 5,5-dimethyl-2,4-oxazolidedione (DMO); for pH gradients internally acid the weak bases ammonia and methylamine are used.[26] Usually, these compounds are used in radioactively labeled form. Also, fluorescent amines such as 9-aminoacridine and atebrine have been used. The fluorescence quenching of these compounds upon energization of membrane particles has been ascribed to accumulation of these probes inside the membrane particles.[46] However, evidence has been presented that this interpretation is not valid.[47,48] The correct relationship between fluorescence quenching and pH gradient has still to be established.

c. ^{31}Phosphor Nuclear Magnetic Resonance Measurements

The chemical shift of the ^{31}P resonance of a phosphate has been shown to be a reliable indicator for the internal pH in cells and cell organelles.[49,50]

E. Model Systems for Transport Studies

Initially, the knowledge of transport systems in bacteria was obtained from studies with whole cells. These studies demonstrated the existence of specific transport systems and supplied information about the specificity and kinetic constants. In particular, the isolation of transport-deficient mutants established the functional role of specific transport systems.[51] The information about molecular aspects of the transport processes which can be obtained from studies with intact cells is, however, limited. Not only the cytoplasm, but also the cell envelope layers, the outer membrane, the cell wall, and the periplasm can affect the transport properties of an intact cell. This restricts a clear-cut interpretation of the experimental data about the properties at the cytoplasmic membrane level.

These considerations urged the development of well-defined biochemical model systems. Ideally, such a system should consist of a phospholipid bilayer in which the transport systems, and possibly the energy generating systems, are located. Furthermore, in order to allow measurements of transport activities, an additional requirement is that the phospholipid bilayer separates two closed compartments. The isolation of cytoplasmic membrane vesicles, first described by Kaback,[52] came very close to these requirements. The vesicles consist of cytoplasmic membranes of bacterial cells which form closed spherical structures and which retain the physiologically active integrated membrane functions. Transport activities in these membrane vesicles can be studied in a similar way as in whole cells by measuring the concentration changes of solutes in the external and/or internal compartment.

Membrane vesicles are isolated in essence by a two-step procedure: (1) the organism is converted into an osmotically sensitive form and, subsequently, (2) this form is lysed under controlled conditions in the presence of nucleases and a chelating agent. The osmotically sensitive form, termed spheroplast for Gram-negative organisms and protoplast for Gram-positive organisms, is usually obtained by the lysozyme-EDTA procedure. For detailed information about the isolation procedures, the reader is directed to Kaback[53] and Konings.[13,54] For some organisms, extensive modifications of the lysozyme-EDTA method have been developed.[55-59] For most Gram-positive bacteria, a one-step procedure can be employed in which conversion of the cells into an osmotically sensitive form is combined with the lysis step.[60] The structures observed in electron micrographs of ultrathin sections of membrane vesicles are almost exclusively closed membraneous sacs surrounded by a single trilaminar layer 6.5 to 7.0 nm thick.[53,60] The diameter of the vesicles varies between 0.1 and 0.5 μm, and the inner volume of the vesicles lies between 2 and 4 μl/ mg membrane protein.[61,62]

The vesicles are essentially devoid of cytoplasmic constituents and periplasmic enzymes. They contain less than 5% of the cell's DNA and RNA content, but 10 to 15% of the protein and at least 60% of the phospholipids initially present in the intact cells.[53] The vesicles also contain very low contaminations of endogenous energy sources such as NADH, succinate, or D-lactate.[60,61] The membrane vesicles consist mainly of protein (60 to 70% of dry weight), phospholipid (30 to 40%) and carbohydrate (1%).[53] In vesicles from Gram-negative bacteria, some lipopolysaccharide can be present.[63,64]

Membrane vesicles have retained a number of membrane-associated enzymes and perform several integrated membrane functions, e.g.,

1. Phospholipid synthesis[65-68]
2. Nucleotide metabolism[69-70]
3. Peptidoglycan synthesis[71]
4. Lipopolysaccharide synthesis[71]

and of special interest for this chapter,

5. Ca^{2+}-Mgs^{2+}-activated ATPase[72-74]
6. Electron transfer[13,63]
7. Group translocation[13,63]
8. Secondary transport[13,63]

The systems (6) and (7) will be discussed in more detail in the following sections.

An important feature of membrane vesicles as a model system for transport studies in bacteria is the orientation of the membrane with respect to the orientation of the cytoplasmic membrane in intact cells. A large amount of evidence has accumulated indicating that this orientation is right-side out.[13,75] The major lines of evidence are obtained from: (1) transport studies showing that essentially all vesicles perform secondary transport with the same characteristics as intact cells,[13,75] (2) freeze-etch electron microscopy showing that the particle distribution in the outer and inner fracture faces are the same in membrane vesicles as in intact cells,[60,75,76] and (3) studies with antibodies against membrane proteins.[77,78] Recently, convincing evidence was obtained from crossed immunoelectrophoresis[79] with antibodies prepared against vesicle membranes.

Besides the isolation of right-side out membrane vesicles, also inside-out membrane preparations have been isolated. From phototrophic organisms, invaginations of the cytoplasmic membranes can be isolated by French press treatment of intact cells followed by differential centrifugation. These so-called chromatophores have been used in the last 2 decades for studies on light-dependent electron transfer and photophosphorylation (for review see Jones[80]). Their inside out orientation is demonstrated by freeze-etch electronmicroscopy, the location of Ca^{2+}-Mg^{2+} ATPase at the outer surface, and the direction of the light-induced electrochemical proton gradient.[81,82] Cytoplasmic membrane particles with inverted orientations have also been isolated from other organisms. Herzberg and Hinkle[83] described an isolation procedure by French press treatment of intact *E. coli* cells. These membrane vesicles perform oxidative phosphorylation, a function which cannot be demonstrated in right-side-out membrane vesicles due to the location of Ca^{2+}-Mg^{2+} ATPase at the inner surface of the membrane. Furthermore, the demonstration that these vesicles perform transport of calcium ions, an ion which is normally extruded from intact cells, is evidence for an inverted orientation of these vesicles.[84]

A model system with many promising applications for the study of transport processes is the artificial vesicle prepared from purified lipids and primary or secondary transport systems. The physicochemical properties of lipid vesicles have been studied extensively.[85]

Several techniques are available for the incorporation of primary transport systems in lipid vesicles.[86-88] These reconstitution experiments are facilitated by the fact that the activity of many of these primary transport systems can be assayed biochemically (ATP-hydrolysis,[89] oxidoreduction,[90] etc.). Most promising for the preparation of homogeneous vesicles are those techniques by which the transport systems are incorporated in preformed lipid vesicles.[91] The reconstitution of vesicles with purified primary transport systems from eukaryotic origin (Na^+/K^+-, Ca^{2+}-, and H^+-ATPases, H^+-oxidoreductases, and H^+-transhydrogenase) has been described.[92] Also, reconstitutions with a prokaryotic H^+-ATPase[89] and a light-driven proton pump[87] have been performed. The structure of vesicles reconstituted with bacteriorhodopsin has been extensively characterized.[93-95] Reconstitution of secondary facilitated transport systems has been hampered by the lack of biochemical assay procedures for these systems. Despite this difficulty, the reconstitution of alanine carriers, purified from a thermophilic bacterium and from *B. subtilis*, has been reported.[96,97] Also, reconstitution experiments with secondary facilitated transport systems from eukaryotic cells have been described.[92]

The coreconstitution of two transport systems in one liposome opens possibilities for studies on transport-system interactions. Reconstitution of oxidative and photophosporylation has been demonstrated,[90,91,98] and the results of these studies contributed significantly to the appreciation of the chemiosmotic theory.[99] Analogous experiments in which a secondary transport system is reconstituted with a primary transport system have also been reported.[100] These experiments will allow the study of energy coupling between primary and secondary transport systems in the absence of intervening ion translocation systems.

IV. PRIMARY TRANSPORT SYSTEMS

A. Aerobic Electron Transfer Systems

The respiratory chain in bacteria is located in the cytoplasmic membrane or its invaginations. The electron carriers are tightly incorporated in the membrane, and functional respiratory chains are found in membrane vesicles isolated from many bacteria. These respiratory chains contain different electron carriers which usually include dehydrogenases, flavins, quinones, nonhaem-iron proteins, several types of cytochromes, and terminal oxidases. The nature of the dehydrogenases varies in different organisms and depends largely on the growth conditions in many organisms.[101-104] Membrane vesicles from aerobically grown *E. coli* contain high activities of D-lactate dehydrogenase, succinate dehydrogenase, and NADH dehydrogenase. In membrane vesicles from *B. subtilis*, high activities of NADH dehydrogenase and succinate dehydrogenase are present. Growth of *B. subtilis* on glycerol results in the induction of L-α-glycerol phosphate dehydrogenase and growth on L-lactate on the induction of L-lactate dehydrogenase.[61,101] Other dehydrogenases have been found in vesicles, e.g., L-malate dehydrogenase in *Azotobacter vinelandii*[105] and D-glucose dehydrogenase in *Pseudomonas aeruginosa*.[55]

Most dehydrogenases are coupled very effectively to the respiratory chain, as is evident from the observation that the corresponding substrates are aerobically oxidized by the membrane vesicles at a high rate. In membrane vesicles from *E. coli*, oxidation of D-lactate, L-lactate, succinate, or α-glycerol phosphate results in a stoicheiometric conversion to pyruvate, fumarate, or dihydroxyacetone phosphate, respectively.[106,107] Upon addition of the substrates, an extensive reduction of respiratory chain intermediates (including flavins and cytochromes) is observed.

In membrane vesicles from *E. coli*, addition of the substrates D-lactate, succinate, or NADH results in reduction of flavoprotein, and cytochromes *b*, *a*, and a_2.[61] Together with cytochrome *o*, these cytochromes constitute all classes of cytochromes known to be present in *E. coli*.[108] Similar observations have been made with membrane vesicles from *B. subtilis*.[62] Further evidence for the involvement of the respiratory chain in oxidation of the substrates is obtained from inhibition experiments with respiratory chain inhibitors. The sites of inhibition by amytal, 2-heptyl-4-hydroxyquinoline-*N*-oxide (HQNO), and cyanide have been well established in *E. coli*[109] and *B. subtilis.*[110] These inhibitors severely block oxidation of the electron donors. Membrane vesicles of a number of bacteria also oxidize nonphysiological electron donors such as the reduced form of phenazine methosulphate (PMS),[101] sulphonated phenazine methosulphate (MPS),[111] pycocyanine,[75] and tetramethyl-1,4-phenyldiamine dihydrochloride (TMPD).[112] These electron donors mediate electrons to a site of the respiratory chain closer to oxygen than the dehydrogenases.

Recently, it was demonstrated that in membrane vesicles from B. *subtilis* W 23 ascorbate-PMS reduces exclusively the terminal oxidase cytochrome a_{601} and that oxidation of ascorbate via membrane incorporated PMS is inhibited b cyanide, but not by HQNO. It was concluded that the site of interaction of PMS with the respiratory chain of *B. subtilis* was at the level of the terminal oxidase.[113] For *E. coli*, evidence was presented for a site of interaction of PMS at the level of cytochrome b_1.[114]

Of particular interest is the location of the electron carriers in the cytoplasmic membrane since such information can supply information about the mechanism of proton translocation across the membrane. Moreover, knowledge about the location of electron carriers, in particular of dehydrogenases, would be useful for the determination of the membrane orientation in cytoplasmic membrane vesicles.

With crossed immunoelectrophoresis it was demonstrated that at least two NADH dehydrogenases are located at the inner surface of the membrane. Furthermore, these studies confirmed the conclusions derived previously that D-lactate dehydrogenase is located at the inner surface of the vesicle membrane.[79] The results do not offer an explanation for the puzzling observation that membrane vesicles from *E. coli* oxidize NADH at the same rate or faster than D-lactate, while D-lactate oxidation drives solute transport more effectively than NADH. When added to the outside of the vesicles, both compounds can apparently reach the corresponding dehydrogenases and are oxidized rapidly enough to support secondary transport in membrane vesicles and in energy-deprived intact cells.[328] For D-lactate, a facilitated secondary transport system has been demonstrated in *E. coli.*[115] Such a translocation system has not been described for NADH, and it seems questionable whether NADH can cross the membrane of *E. coli* rapidly enough by passive secondary transport to explain the high oxidation rates by a NADH dehydrogenase located at the inner surface of the membrane. It seems to be more likely that NADH dehydrogenase spans the membrane and can oxidize NADH at the outer surface. This could explain the lower efficiency of NADH oxidation in energizing transport when such oxidation results in the translocation of protons from the outside to the inside by the NADH-dehydrogenase-linked proton translocating site, thus reducing the total number of protons extruded during NADH-oxidation.

Such an explanation would be consistent with the finding of Futai[116] that NADH oxidized internally is as effective in energizing transport as D-lactate. The observation of Stroobant and Kaback[117] that externally added ubiquinone increases the efficiency of NADH oxidation in energizing transport to the level of D-lactate oxidation would fit such an explanation.

At this moment, the information available about the sidedness of cytochromes in bacterial membranes is surprisingly lacking. Studies with ferricyanide as terminal electron acceptor in membrane vesicles from *B. subtilis* W 23 indicated a location of cytochrome c_{553} on the outer surface of the membrane.[118] Prince et al.[119] and Dutton et al.[120] presented evidence for a location of cytochrome c_2 at the periplasmic site of the membrane in *Rhodopseudomonas sphaeroides.*

1. Generation of Electrochemical Proton Gradient by Respiration

The electrochemical proton gradient or one of its components $\Delta\psi$ or ΔpH has been recorded in a number of aerobically grown bacteria (see for reviews References 11 and 12). The $\Delta\tilde{\mu}_{H^+}$ is usually between −100 and −200 mV (inside negative and alkaline). It consists of a $\Delta\psi$ of maximally −140 mV and a pH gradient up to −100 mV. The components of the $\Delta\tilde{\mu}_{H^+}$ depend strongly on the environmental conditions, e.g., the external pH[50,121] and the ion composition.[121] In most bacteria, the internal pH is maintained around pH 7.5. In organisms that live in extreme environments, the composition of $\Delta\tilde{\mu}_{H^+}$ can be different. For instance, in alkalophilic bacteria, the $\Delta\psi$ can reach high values while the ΔpH is reversed (inside acid).[122] In acidophilic bacteria, the ΔpH can be high (inside alkaline) while the $\Delta\omega$ is reversed (inside positive).[123-128] The $\Delta\tilde{\mu}_{H^+}$ has also been measured in sphaeroplasts of *E. coli.*[126] At an external pH of 6.5, the $\Delta\psi$ was −130 mV (inside negative) and the ΔpH was −100 mV (inside alkaline), yielding a $\Delta\tilde{\mu}_{H^+}$ of −230 mV.

The generation of $\Delta\tilde{\mu}_{H^+}$ or one of its components has been studied most extensively in membrane vesicles from *E. coli.*[10,127] Using isotopically labeled lipophilic cations [i.e., dimethyldibenzyl ammonium (DDA^+) in the presence of tetraphenylborate, or triphenylmethylphosphonium ($TPMP^+$)], or a fluorescent lipophilic cation (safranine-*O*) or rubidium in the presence of valinomycin, it was demonstrated that *E. coli* membrane vesicles generate a $\Delta\psi$ (interior negative) of approximately −75 mV in the presence of Asc-PMS or D-lactate.[128,130] Determination of the ΔpH in membrane vesicles has been hampered by the lack of a reliable procedure for measuring the uptake of membrane-permeable weak acids or bases. This problem was solved by the introduction of the flow-dialysis technique.[29] Ramos and Kaback[127] measured in membrane vesicles during Asc-PMS oxidation the ΔpH, $\Delta\psi$, $\Delta\tilde{\mu}_{H^+}$, and the internal pH (see Figure 3). From pH 5.0 to 5.5, ΔpH remains almost constant at −115 to −120 mV, decreases drastically above pH 5.5, and is negligible at pH 7.5 and above. Despite marked variations in ΔpH as a function of external pH, the internal pH and $\Delta\psi$ remain essentially constant at pH 7.8 and −75 mV, respectively. As a result of the variation in ΔpH, $\Delta\tilde{\mu}_{H^+}$ exhibits a maximal value of about −195 mV at pH 5.5 and a minimal value of about −75 mV at pH 7.5 and above. The variation in ΔpH with external pH is in part the result of variations in the rates of oxidation of reduced PMS. Since the relatively low rates of reduced PMS oxidation observed above pH 5.5 produce increasingly lower ΔpH values, but constant $\Delta\psi$ values, it seems reasonable to suggest that relatively low rates of electron flow are sufficient to generate $\Delta\psi$, while relatively high rates are necessary to support a significant ΔpH .

In order to generae a significant ΔpH across the vesicle membrane without producing an increase in $\Delta\psi$, anions must move with the protons ejected from the vesicles, or alternatively, cations must move in the opposite direction (i.e., into the vesicles). Such ion movements have not yet been studied in detail, but recent studies in chromatophores from *Rps. sphaeroides* demonstrate that these ion movements occur and affect strongly the generation of ΔpH and $\Delta\psi$.[131]

In previous studies, it was demonstrated that there is little relationship between the ability of the vesicles to oxidize an electron donor and the ability of that electron donor to drive solute transport.[61] However, a qualitative relationship exists between the ability of various electron donors to drive transport and their ability to generate a $\Delta\tilde{\mu}_{H^+}$.[22] The $\Delta\psi$, ΔpH, and $\Delta\tilde{\mu}_{H^+}$ are maximal in the presence of Ascorbate-PMS or D-lactate, while succinate and, especially, NADH produce much weaker effects (Table 2).

In Figure 4 the effects are shown of increasing concentrations of the ionophores valinomycin and nigericin and the proton conductor CCCP on ΔpH, $\Delta\psi$, and $\Delta\tilde{\mu}_{H^+}$ in membrane vesicles from *E. coli.*[127] With increasing concentrations of valinomycin, the

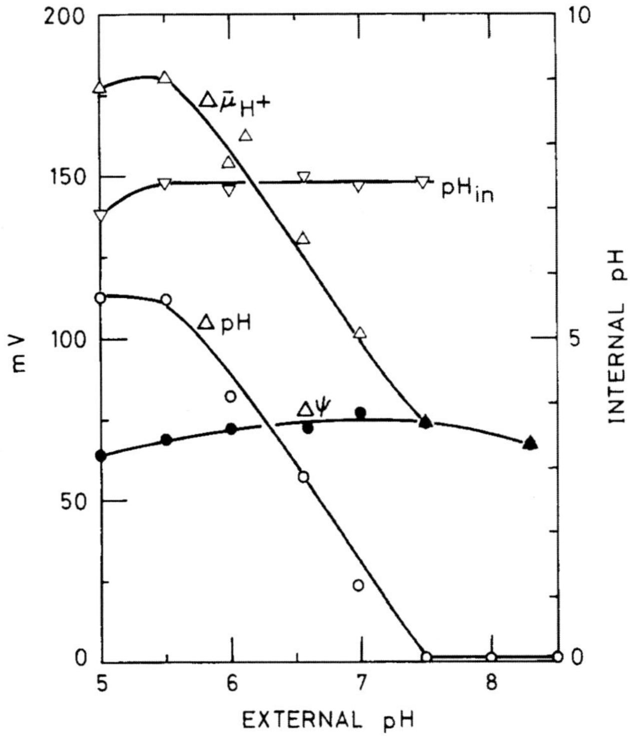

FIGURE 3. Effect of external pH on internal pH, ΔpH, Δψ, and $\Delta\tilde{\mu}_{H^+}$ in membrane vesicles of *E. coli* ML 308-225. The electron donor was ascorbate (20 m*M*) + PMS (0.1 m*M*). The internal pH (pH_{in}) was calculated from flow dialysis experiments with radioactively labeled acetate at each pH given. The ΔpH values were calculated from the difference between pH_{in} and external pH (corrected for the change induced by ascorbate-PMS oxidation). Δψ values were calculated from filtration assays carried out with [^{3}H] $TPMP^+$. $\Delta\tilde{\mu}_{H^+}$ is the sum of ΔpH and Δψ. (From Ramos, S., Schuldiner, S., and Kaback, H.R., *Proc. Natl. Acad. Sci. U.S.A.*, 73, 1892, 1976. With permission.)

Δψ decreases from about −70 mV (in the absence of valinomycin) to −20 mV (at 5μ*M* valinomycin). The ΔpH, however, increases to about 130% of the control value at 1 μ*M* valinomycin and remains constant at higher valinomycin concentrations. The effect of valinomycin on the $\Delta\tilde{\mu}_{H^+}$ is relatively small, producing only about 20% loss at 5 μ*M* valinomycin. The effects of nigericin are opposite to those of valinomycin (Figure 4B). The ΔpH decreases to 0 as nigericin concentration is increased from 0 to 0.1 μ*M*, while Δψ increases from about −60 mV to −90 mV over the same concentration range. The $\Delta\tilde{\mu}_{H^+}$ in the presence of 0.1 μ*M* nigericin decreases with 45%. At pH 7.5 where the ΔpH is 0, nigericin has no effect on Δψ.

The proton conductor CCCP inhibits both Δψ and ΔpH and diminishes at a concentration of 1 μ*M* the $\Delta\tilde{\mu}_{H^+}$ at pH 5.5 by approximately 60% (Figure 4C).

The ionophores and the proton conductor have no significant effect on the oxidation rates of Asc-PMS or D-lactate, indicating the lack of respiratory control in bacterial membrane vesicles.

2. Solute Transport Coupled to Respiration

The role of the respiratory chain in energizing secondary transport processes has

TABLE 2

Effects of Different Electron Donors on ΔpH, Δψ, and $\Delta\tilde{\mu}_{H^+}$ in Membrane Vesicles of *Escherichia coli* ML 308-225

Electron donor	ΔpH[a]	Δψ[b] (mV)	$\Delta\tilde{\mu}_{H^+}$
Ascorbate (20 m*M*) + PMS (0.1 m*M*)	−115	−74	−189
D-lactate (20 m*M*)	−102	−70	−172
Succinate (20 m*M*)	0	−64	−64
NADH (5 m*M*)	0	0	0

[a] ΔpH was calculated from flow-dialysis experiments carried out with sodium [^{3}H] − acetate.
[b] Δψ was determined from filtration assays carried out with [^{3}H] − TPMP$^+$ after 5 and 10 min incubations.

Taken from Kaback, H.R., *J. Cell. Physiol.*, 89, 575, 1976. With permission.

been studied in detail in isolated bacterial membrane vesicles.[10] Membrane vesicles from *E. coli* oxidize the electron donors D-lactate, succinate, and NADH at a high rate. Kaback and Milner[106] observed that, especially, the oxidation of D-lactate stimulated markedly the transport of amino acids in membrane vesicles from *E. coli* (Figure 5). Other electron donors, such as succinate, L-lactate, D,L-hydroxybutyrate, and NADH, also could energize transport, but these electron donors were less effective energy sources for transport than D-lactate.[61] Furthermore, in membrane vesicles from cells that had been induced to synthesize L-α-glycerol phosphate dehydrogenase, formate dehydrogenase, or D-alanine dehydrogenase, L-α-glycerol phosphate, formate, or D-alanine, respectively, also stimulated amino acid transport.[102,104] In later publications, it was demonstrated that oxidation of these electron donors could energize transport of a wide variety of solutes, but the highest initial rates of transport were always observed with D-lactate as energy source.[61,63,132-134] Similar effects of electron donors on transport of solutes were observed in membrane vesicles from many other Gram-negative as well as Gram-positive bacteria.

These studies have been reviewed recently[13] and the reader is directed to this article for further information. In general, it can be concluded that oxidation of an electron donor *via* the respiratory chain supplies the energy for transport of solutes. In a number of membrane vesicles, transport of solutes was also energized by the nonphysiological electron donor system, ascorbate plus phenazine methosulphate (PMS).[62,101,135] Ascorbate alone caused only a small stimulation of transport while PMS had no effect at all.

Accumulation of solutes in right-side-out membrane vesicles is only observed in the presence of electron donors. No other intermediate metabolites or cofactors like ATP, phosphoenolpyruvate, glucose, hexose monophosphate, and many others energized transport to any extent whatsoever.[62,106]

The role of the electron transfer chain in the energization process has been thoroughly established in membrane vesicles from a number of organisms. In short, the evidence presented[13,63] is

1. The energy sources reduce cytochromes present in the membranes.
2. Respiratory chain inhibitors effectively block transport energized by the energy sources.

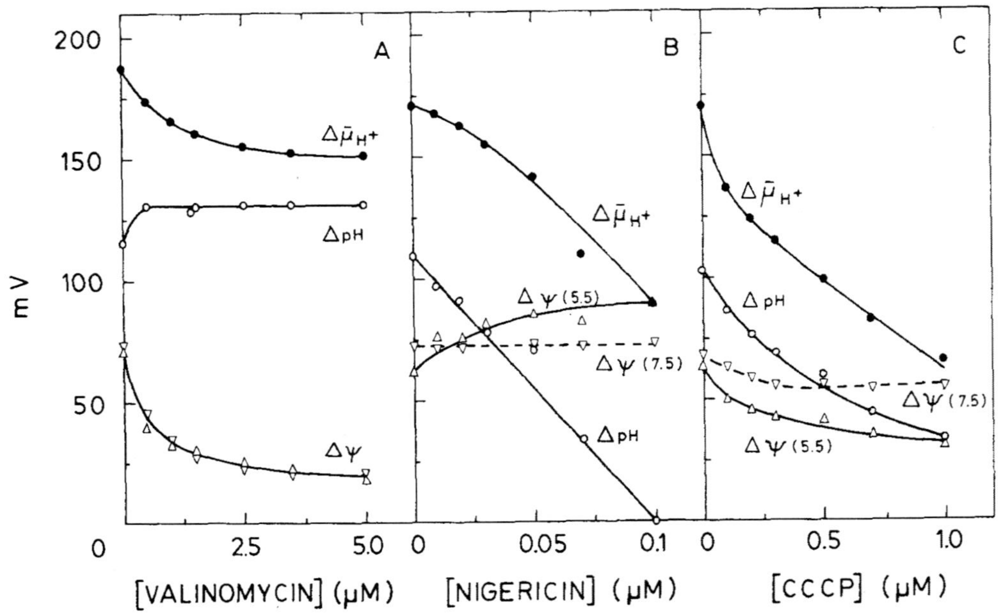

FIGURE 4. Effects of valinomycin (A), nigericin (B), and CCCP (C) on ΔpH (O), Δψ, (Δ and ∇), and $\Delta\tilde{\mu}_{H^+}$ (●) in membrane vesicles from *E. coli* ML 308-225. Steady-state levels of ΔpH and $\Delta\tilde{\mu}_{H^+}$ were determined at an external pH of 5.5. Steady-state levels of Δw were determined at pH 5.5 (Δ) and pH 7.5 (∇). The electron donor was ascorbate (20 m*M*) + PMS (0.1 m*M*). ΔpH, Δψ, and $\Delta\mu_{H^+}$ were determined as described in the legend to Figure 3. (According to Ramos, S., Schuldiner, S., and Kaback, H.R., *Proc. Natl. Acad. Sci. U.S.A.*, 73, 1892, 1976. With permission.)

3. Transport cannot be energized in mutants defective in essential respiratory chain intermediates like cytochromes or quinones.

Energization of transport does not always require a complete respiratory chain. Recent studies of Bisschop et al.[113] demonstrated that reduced PMS donates electrons at the terminal end of the respiratory chain of *B. subtilis*, most likely at the level of cytochrome a_{601}. Reduced PMS oxidation *via* this terminal end of the respiratory chain results in the generation of an electrochemical proton gradient. Also, oxidation of an electron donor *via* the initial part of the respiratory chain can drive transport, as is demonstrated by the oxidation of NADH with ferricyanide as terminal electron acceptor. Ferricyanide was shown to accept electrons at the level of cytochrome *c* in membrane vesicles from *B. subtilis.*[118] It has been mentioned before that in many cases no correlation exists between the rate of oxidation of the electron donor and the capacity to drive solute transport. An explanation has to be sought at the level of the generation of an electrochemical proton gradient. In this respect, it is important to realize that transport of the solute is not the only energy consuming process occurring in the membrane vesicles. Also, other transport processes usually occur. Several electron donors have to be transported prior to oxidation. The energy requirements of these transport processes can differ. In addition, translocations of ions, present in the internal or external medium, occur. For instance, phosphate[136] and K^+ are accumulated while other ions such as chloride and Na^+ are extruded from the vesicles.[12,131,137]

3. *Role of Quinones*

Information about the role of menaquinones in the respiratory chain of *B. subtilis* has been obtained from reconstitution studies with the menaquinone analogue mena-

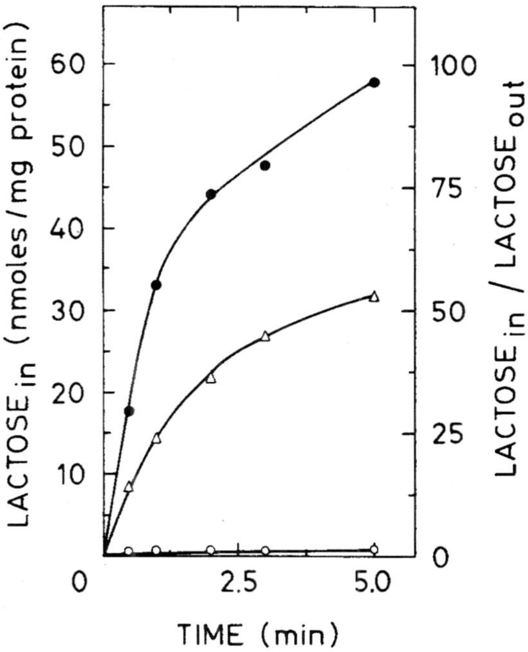

FIGURE 5. Lactose transport by *E. coli* ML 308-225 membrane vesicles. The electron donors used are (●) ascorbate (20 m*M*) + PMS (0.1 m*M*), (Δ) D-lactate (20 m*M*), and (●) No additions. (Taken from Kaback, H.R., Rudnick, G., Schuldiner, S., Short, S.A., and Stroobant, P., in *The Structural Basis of Membrane Function*, Academic Press, New York, 1976, 107. With permission.)

dione in membrane vesicles from the menaquinone-deficient strain *B. subtilis aro D*.[138] These membrane vesicles oxidize NADH at a low rate and, consequently, NADH does not drive amino acid transport. Supplementation of the membrane vesicles with menadione results in incorporation of menadione in the membranes. The amount of menadione incorporated increases with the external menadione concentration up to a maximum of 7 nmol menadione bound per milligram membrane protein. The same amount of menaquinone-7 is found in membranes from mutant cells grown under conditions of optimal menaquinone synthesis.[139] The oxidation rate of NADH increases linearly with the menadione content, and this oxidation is blocked by respiratory chain inhibitors in the same way as in membrane vesicles from wild type strains. The initial rate of amino acid transport increases with the NADH-oxidation rate up to a maximal value (Figure 6). It is most likely that the rate of amino acid transport is determined by $\Delta\tilde{\mu}_{H^+}$ and that the increase of the rate of amino acid transport reflects the increase of $\Delta\tilde{\mu}_{H^+}$. Evidence in support of this explanation has been presented by Robertson et al.[140] These investigators showed that the rate of lactose transport in *E. coli* membrane vesicles increases linearly with ΔpH (varied by titration with nigericin) and $\Delta\psi$ (varied by titration with valinomycin).

The rate of amino acid transport in vesicles from *B. subtilis aro D* was optimal at submaximal rates of NADH oxidation, indicating that $\Delta\tilde{\mu}_{H^+}$ was maximal at submaximal rates of NADH oxidation.[138] Because proton efflux will most likely increase with the oxidation rate of NADH, this would mean that energy dissipating processes (aspecific proton fluxes, ion fluxes) prevent an increase of $\Delta\tilde{\mu}_{H^+}$ above a certain maximal level. Stroobant and Kaback[10,117] studied the effect of ubiquinone-1 added externally

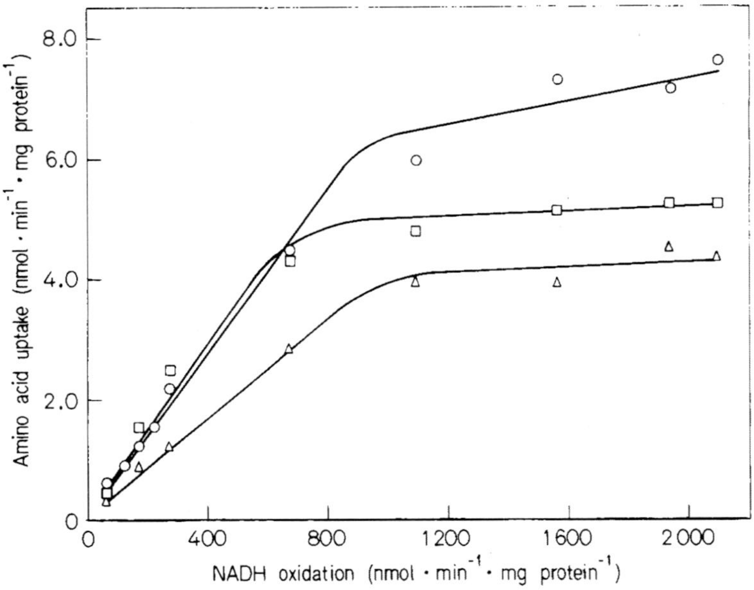

FIGURE 6. Initial rates of NADH-driven amino acid transport as a function of the NADH-oxidation rate in membrane vesicles from *Bacillus subtilis aroD*. (□) L-Glutamate, (△) L-serine, and (○) L-alanine. Initial rates of amino acid transport were measured after 30 sec incubation with 10 m*M* NADH as electron donor in membrane vesicles reconstituted to various levels of NADH oxidase activities with menadione. (Taken from Bisschop, A. and Konings, W.N., *Eur. J. Biochem.*, 67, 357, 1976. With permission.)

to the membrane vesicles on NADH oxidation and the accumulation of lactose and amino acids by membrane vesicles from *E. coli* ML 308-225. Ubiquinone-1 increased the initial rate and steady state levels of accumulation to levels comparable to those observed during D-lactate oxidation, but hardly affected the NADH-oxidation rate. However, the increase of lactose and amino acid transport could be correlated with the increase of $\Delta\tilde{\mu}_{H^+}$. In the absence of ubiquinone-1, hardly any $\Delta\psi$, ΔpH or $\Delta\tilde{\mu}_{H^+}$ was generated, while in the presence of 0.08 m*M* ubiquinone-1 the steady state levels of $\Delta\psi$, ΔpH and $\Delta\tilde{\mu}_{H^+}$ were −59 mV, −62 mV and −131 mV, respectively.

It has been discussed before (Section IV A.) that an explanation for these observations might be that NADH oxidation takes place at the outer surface of the vesicle membrane. In untreated vesicles, NADH dehydrogenase may translocate protons from the outer surface to the inner surface, thus reducing the number of protons ejected during NADH oxidation, while in the presence of ubiquinone-1 these protons are released at the outer surface.

B. Anaerobic Electron Transfer Systems

Many facultatively and obligately anaerobic bacteria can use a wide variety of compounds as terminal electron acceptors. Among them are the nitrogen compounds nitrate and nitrite, the sulphur compounds sulphate and sulphite,[141-143] thiosulphate and tetrathionate,[144] and the organic compounds fumarate and carbonate. The electrons are transferred to these acceptors via a series of electron carriers which are comparable to those present in respiratory chains. In some obligate anaerobes, there are very simple systems that do not contain cytochromes as electron carriers.[145]

In contrast to the many studies that have been done on electron transfer in the respiratory chain, the information available on anaerobic electron transfer systems is often limited. Even less information is available about the role of these electron transfer systems in energy metabolism. This rather restricted information has certainly contributed to an undervaluation of the role of anaerobic electron transfer systems in the metabolic machinery of anaerobically grown bacteria.

The information about anaerobic electron transfer systems has been reviewed recently.[1-4,146-148] Studies on anaerobic secondary transport coupled to electron transfer systems have been performed mainly in *E. coli*.[1] These will be reviewed in the following sections.

1. Nitrate Respiration in E. coli

Anaerobic electron transfer with nitrate as terminal electron acceptor is termed nitrate respiration.[149] In *E. coli*, the system is induced by anaerobic growth in the presence of nitrate. Although the system allows anaerobic growth on a variety of nonfermentable substrates (e.g., D-lactate, succinate, L-α-glycerol phosphate, and formate),[150] it is likely that formate is the best physiological electron donor.[151] Formate is oxidized by formate dehydrogenase, a molybdoprotein of about 600,000 daltons which contains selenium, nonheme iron, acid-labile sulphide, and heme.[152,153] By analogy to other aldehyde dehydrogenases, the molybdenum is the intermediate electron acceptor in the enzyme complex.[154,155] The electrons are subsequently transferred to ubiquinone possibly via selenium, nonheme iron, and a cytochrome of the *b*-type.

The cytochromes involved in nitrate respiration are of the *b*-type, i.e., a formate dehydrogenase-linked (cyt b_{558}) and a nitrate-reductase-linked cytochrome. The latter cyt $b_{556}^{NO_3^-}$ is specifically induced by nitrate[151] and is genetically[156] and kinetically[157] distinct from other b-type cytochromes of *E. coli*. Cyt $b_{556}^{NO_3^-}$ is closely associated with nitrate reductase and is probably involved in the attachment of the enzyme to the membrane.[158-161] Solubilized, purified nitrate reductase is a molybdenum-containing iron-sulphur protein composed of three nonidentical subunits which are present in a 1:1:1 ratio.[152,159,160,162,163] One of the subunits has been identified as cytochrome b_{556}. Molybdenum is involved directly in electron transfer,[154] but the role of the iron-sulphur centers has not yet been elucidated. The location of the nitrate reductase components in the cytoplasmic membrane has been studied extensively.[150,158,164-167] In membrane preparations with a defined sidedness, spheroplasts (same orientation as the cytoplasmic membrane of intact cells) and membrane vesicles prepared by sonication (inside-out orientation), the presence of externally exposed tyrosine residues was studied with a specific nonpenetrant label (lactoperoxidase/H_2O_2 mediated incorporation of ^{125}I). Furthermore, the location at the outer surface of polypeptides of cyt $b_{556}^{NO_3^-}$ (the γ-subunit) and nitrate reductase the α- and β-subunits) was investigated with antibodies against these subunits. These studies demonstrated that the γ-subunit was located at the outer surface of the membrane, while the α-subunit of the nitrate reductase was located at the inner surface. The relative sidedness of the β-subunit could not be determined with certainty.[165,168] These results clearly demonstrate that the nitrate reductase complex spans the membrane structurally. From the oxidation rates of reduced benzyl- and methyl-viologen in whole and broken cells with nitrate as terminal electron acceptor, it was concluded that the viologen dyes are oxidized at the inner surface of the membrane. Furthermore, studies in membrane preparations from cytochrome-deficient cells indicated that the oxidation of the viologen dyes is cytochrome independent and that the electrons are donated directly to the α- and β-subunits of the nitrate reductase.[158,164] Nitrate reductase, therefore, has to be accessible from the inner surface of the membrane. On the other hand, nitrate reductase is also accessible from the outer surface of the membrane because the γ-subunit (cyt b_{556} NO_3^-) was shown to be located at the outer surface (see above), and electrons from nitrate reductase can be accepted by the membrane impermeable electron acceptor ferricyanide from the outer surface.[169] These observations indicate that the nitrate reductase complex spans the membrane also functionally.

Evidence for a location of formate dehydrogenase at the inner surface was also presented.[172] Formate is a weak acid which is membrane permeable in undissociated form as is shown by the dissipating effect of high concentrations of formate on the ΔpH across the vesicle membrane. Oxidation of formate at the inner surface of the vesicle membrane is suggested by the observation that formate oxidation is inhibited by nigericin and stimulated by calinomycin.

The available information suggests that the nitrate respiration system is arranged in the membrane as shown in Figure 7.

a. Generation of Electrochemical Proton Gradient by Nitrate Respiration

Energy generation by nitrate respiration has been demonstrated in a wide variety of bacteria.[1-4,146] Indirect evidence has been obtained from the increase of the molar growth yields under anaerobic conditions by the addition of nitrate. It was concluded that 1 mol of ATP is formed per mol of nitrate reduced.[170] Furthermore, in cell-free extracts from *E. coli* $P/2e^-$ ratios (molecules of ATP formed per electron pair transferred to the terminal electron acceptor) measured with the electron donors NADH, glutamate, and citrate, and with nitrate as terminal electron acceptor were 0.55, 0.65, and 1.1, respectively.[171] Direct evidence for the generation of a $\Delta\tilde{\mu}_H{}^+$ by nitrate respiration has been obtained from measurements of $H^+/2e$ ratios. In spheroplasts from *E. coli* H^+/NO^-_3 ratios ($H^-/2e$ ratios in which nitrate served as terminal electron acceptor) of 4 were recorded during L-malate oxidation and of 2 during the oxidation of glycerol, succinate, or D-lactate.[150] Estimation of H^+/NO^-_3 ratios with formate as electron donor was difficult because formate is accumulated in response to a ΔpH.[172] Nevertheless, estimations indicated that the H^+/NO^-_3 ratio was 4.[150,168] The generation of a ΔpH during nitrate respiration with NADH, formate, D-lactate and D,L-α-glycerol-*P* as electron donors was indicated by the quenching of atebrin fluorescence in inverted membrane vesicles from *E. coli*.[173] More detailed information has been obtained from studies of the uptake of the lipophilic cation $TPMP^+$ and of the weak acid acetate by membrane vesicles from *E. coli* ML 308-225 induced for nitrate respiration.[172]

At a medium pH of 6.6, the $\Delta\tilde{\mu}_H{}^+$ generated by nitrate respiration was about −165 mV consisting of a $\Delta\psi$ of −90 mV and a ΔpH of −75 mV. This $\Delta\tilde{\mu}_H{}^+$ is very similar to the $\Delta\tilde{\mu}_H{}^+$ generated aerobically in membrane vesicles by respiration.[127] High external concentrations of formate dissipate the ΔpH. In the presence of 10 m*M* formate, the $\Delta\tilde{\mu}_H{}^+$ consists almost solely of a $\Delta\psi$ of about −90 mV.

b. Solute Transport Coupled to Nitrate Respiration

Transport of lactose has been studied under anaerobic conditions in whole cells of *E. coli* ML 308-225, a strain which is constitutive for the M-protein, the lactose permease. Cells grown on glucose in the presence of nitrate, i.e., conditions which induce formate dehydrogenase and nitrate reductase, exhibit a markedly increased lactose transport in the presence of formate and nitrate.[58] On the other hand, cells grown

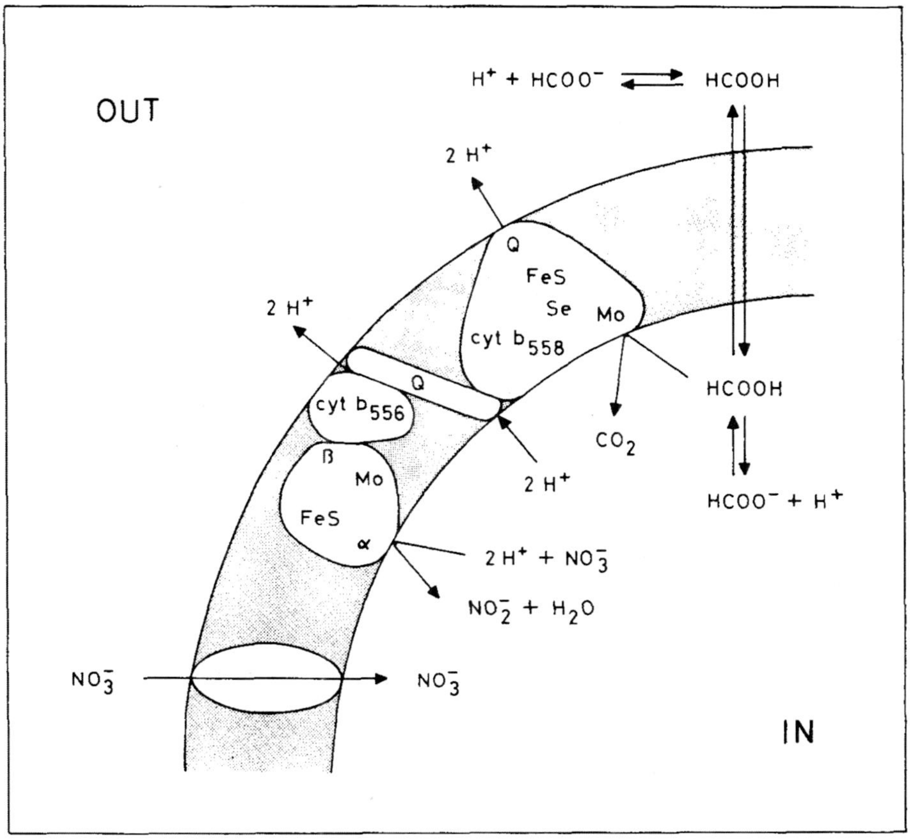

FIGURE 7. Topography of the nitrate respiration pathway of *E. coli.* Mo, molybdenum; Se, selenium; FeS, iron-sulphur center; Cyt, cytochromes; Q, ubiquinone; α and β are subunits of nitrate reductase. The figure is based on the schemes proposed by Jones[167] and Boonstra.[327]

anaerobically on glucose in the absence of an electron acceptor fail to show an increase in lactose transport upon the addition of formate and nitrate. These data indicate a coupling of lactose transport to the electron transfer system formate dehydrogenase-nitrate reductase.

More evidence for such a coupling has been obtained from studies in membrane vesicles from *E. coli* grown anaerobically on glucose in the presence of nitrate.[58,174]

These membrane vesicles contain high formate dehydrogenase and nitrate reductase activities and reduce nitrate rapidly in the presence of formate. Anaerobic transport of lactose and amino acids is coupled to this formate dehydrogenase-nitrate reductase electron transfer system as is demonstrated by the marked stimulation of uptake in the presence of both the electron donor formate and the electron acceptor nitrate (Figure 8). Moreover, a high rate of amino acid uptake is observed with chlorate as electron acceptor. Ferricyanide, which most likely accepts electrons from the electron transfer system at a level beyond cytochrome b_{556}, can also replace nitrate.[169]

Further evidence for the involvement of electron transfer in solute transport has been obtained from studies with electron transfer inhibitors. The formate-plus-nitrate-dependent transport of amino acids and lactose is almost completely inhibited by 2-*n*-heptyl-4-hydroxyquinoline-*N*-oxide (HQNO), an inhibitor at the level of cytochrome *b*, and by cyanide, an inhibitor of nitrate reductase itself.[58,174]

The membrane vesicles also contain a functional respiratory chain, and transport of

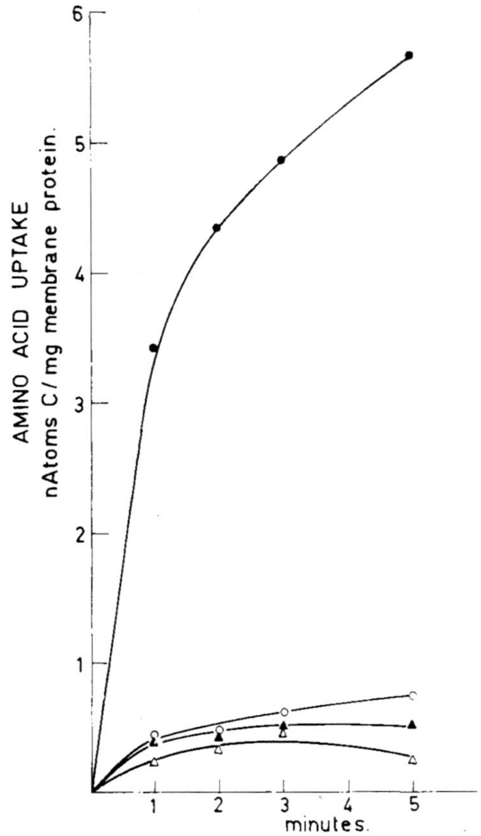

FIGURE 8. Uptake of amino acids under anaerobic conditions in membrane vesicles from *E. coli* ML 308-225 grown anaerobically on glucose in the presence of nitrate. (▲) No additions, (O) Na formate (10 m*M*), (Δ) KNO_3 (10 m*M*), and (●) Na formate (10 m*M*) + KNO_3 (10 m*M*). (Taken from Konings, W.N. and Boonstra, J., *Curr. Top. Membr. Transp.*, 9, 177, 1977. With permission.)

amino acids and lactose can be energized by electron transfer to oxygen as terminal electron acceptor. Effective electron donors are NADH and the nonphysiological electron donor system ascorbate plus phenazine methosulphate (Asc-PMS). Formate can also effectively energize transport under aerobic conditions. Under these conditions, the addition of nitrate has no significant effect on the rate of uptake. The electron donors NADH and Asc-PMS, however, fail to stimulate transport under anaerobic conditions in the presence of nitrate. This indicates that in these membrane vesicles only formate dehydrogenase is coupled effectively to nitrate reductase.[58,174]

Under other growth conditions, however, other electron donors also donate electrons to this electron transfer system. In membrane vesicles from *E. coli* grown anaerobically on glycerol in the presence of nitrate, L-α-glycerol phosphate plus nitrate stimulates amino acid transport, but the extent of stimulation is lower than with formate plus nitrate.[174]

A similar coupling between transport and nitrate respiration has been demonstrated in strictly anaerobic organisms. Membrane vesicles from the strict anaerobe *Veillonella alcalescens*, grown on lactate in the presence of nitrate, catalyze active transport of L-glutamate and other amino acids under anaerobic conditions in the presence of the electron donor L-lactate and the electron acceptor nitrate. L-lactate alone, or nitrate alone, has hardly any effect on L-glutamate uptake. L-lactate could be replaced by NADH, L-α-glycerol phosphate, formate, and L-malate, indicating that in these membrane vesicles several dehydrogenases are coupled effectively to nitrate respiration. None of these electron donors could energize transport under aerobic conditions, as expected since *Veillonella alcalescens* does not contain a functional respiratory chain.[175]

2. Fumarate Reduction in E. coli

Anaerobic electron transfer systems in which fumarate functions as the final electron acceptor have not been studied as much as the nitrate respiration systems (for reviews see References 3 and 148). Electron-transfer-linked fumarate reduction has been demonstrated in several facultative organisms. Fumarate reduction has been found also in facultative and strict anaerobes grown in the absence of fumarate,[3,5,148,176] indicating that the fumarate reductase system is commonly present in anaerobic bacteria. In most organisms, the activity of fumarate reduction is increased by growth under anaerobic conditions in the presence of fumarate. Other electron acceptors, e.g., oxygen and nitrate, repress the formation of this anaerobic electron transfer system.

The terminal oxidase of fumarate reduction is fumarate reductase, which catalyzes the reduction of fumarate to succinate. The other components of this electron transfer system vary from organism to organism. In some organisms, like *Streptococcus faecalis*,[145] very simple systems are present in which dehydrogenases, flavins, quinones, and nonheme-iron proteins participate. In other organisms cytochromes, usually of the *b*-type, are also electron transfer intermediates. Depending on the growth conditions, several substrates such as L-α-glycerol phosphate, NADH, L-malate, formate, lactate, and molecular hydrogen can donate electrons to this electron transfer system. The components of fumarate reduction have been found in the particulate fraction of cell extracts, and the system can be isolated as a functional complex from this fraction.

E. coli grown anaerobically on glucose possesses a fumarate reductase system in which electrons are transferred from formate or NADH via menaquinone and cytochrome b to fumarate reductase.[5] Growth of *E. coli* under anaerobic conditions with glycerol as carbon source and fumarate as electron acceptor results in the induction of anaerobic L-α-glycerol phosphate dehydrogenase coupled to fumarate reductase. These two enzymes constitute a functional complex which is membrane-bound[58,174,177] and which catalyzes the dehydrogenation of L-α-glycerol phosphate at the expense of fumarate without any added cofactors.[177]

Fumarate reductase is a membrane-bound enzyme[178] whose synthesis is repressed under aerobic conditions and derepressed anaerobically, and to some extent aerobically, in the presence of glucose.[179,180] In contrast to succinate dehydrogenase, this enzyme oxidizes succinate at about the same rate as it reduces fumarate, but the K_m for fumarate is much lower than for succinate.[179] Besides its function as a terminal oxidase during anaerobic growth, fumarate reductase can also provide succinate for biosynthesis when the tricarboxylic acid cycle enzymes are repressed.[181,182] Mutants have been isolated which lack functional fumarate reductase and are unable to use fumarate as an anaerobic electron acceptor.[180]

The complex of anaerobic L-α-glycerol phosphate dehydrogenase and fumarate reductase is present in particulate fractions of cell extracts and in isolated membrane vesicles.[58,174] It catalyzes the anaerobic oxidation of L-α-glycerol phosphate in the presence of fumarate as terminal electron acceptor. No stimulation of the coupled activity is observed upon the addition of FAD or FMN, and the complex is probably saturated with flavins.[177] Singh and Bragg[183] demonstrated in a cytochrome-deficient (hem A^-) mutant of *E. coli* that electron transfer from NADH or α-glycerol phosphate to fumarate can also occur in the absence of cytochromes.

Growth of *E. coli* anaerobically in the presence of fumarate results in an increase in fumarate reductase activity. It is of interest that these cells contain, in addition, the respiratory chain and the nitrate respiration system. Studies with isolated membrane vesicles demonstrated that these electron transfer systems can be linked to the same dehydrogenase such as L-α-glycerol phosphate dehydrogenase,[174] indicating that some electron carriers may be shared by all three electron transfer systems.

The topography of the fumarate reductase system of *E. coli* is not yet elucidated. By analogy with other dehydrogenases, it is likely that L-α-glycerol-*P* dehydrogenase is located at the inner surface of the membrane.[184] Furthermore, the observation that fumarate added to the medium has to be translocated before it is reduced[185,186] suggests that fumarate reductase, too, is located at the inner surface. The presence of cytochromes is essential for the generation of a $\Delta\tilde{\mu}_{H^+}$ by fumarate reduction.[5,183,187] These results, and the observation that the membrane-impermeable ferricyanide functions as an electron acceptor, suggest a location of the fumarate reductase system in the membrane as presented in Figure 9.

a. Generation of Electrochemical Proton Gradient by Fumarate Reduction

The generation of metabolic energy by fumarate reduction is evident from the observation that *E. coli* can grow anaerobically on L-malate and molecular hydrogen.[188] In contrast to related organisms, e.g., *Proteus rettgeri,*[3] *E. coli* is not able to grow anaerobically on fumarate as sole carbon and energy source, but the addition of fumarate results in an increase of the growth yield during anaerobic growth on glucose.

The role of fumarate reduction in the generation of metabolic energy was also demonstrated for *E. coli* grown anaerobically on glucose without exogenous electron acceptor.[5] The growth yields and maximal growth rates of mutants deficient in components of the fumarate reductase system, i.e., menaquinone (AN 843), cytochromes (AN 704), or fumarate reductase (AN 472) are considerably smaller than those of the wild type (AN 248), a ubiquinone-deficient strain (AN 750), or a strain uncoupled for oxidative phosphorylation (*uncB*, AN 283). The uptake of glutamine and proline has been studied in these *E. coli* mutants.

Evidence has been presented that in *E. coli* ATP is the direct energy source for glutamine uptake, while $\Delta\tilde{\mu}_{H^+}$ is the direct energy source for proline uptake.[189,190] The transport activity of glutamine, therefore, can supply information about the formation of ATP while the uptake of proline can be used as indication for the generation of a $\Delta\tilde{\mu}_{H^+}$. Anaerobically, in the presence of glucose, all strains accumulate glutamine and thus form ATP. This ATP can be formed by substrate-level phosphorylation and, in the wild type, also by phosphorylation coupled to fumarate reduction. Under these conditions, all strains also accumulate proline and thus form a $\Delta\tilde{\mu}_{H^+}$. This $\Delta\tilde{\mu}_{H^+}$ can be generated by fumarate reduction and/or by ATP hydrolysis in the wild-type cells, by fumarate reduction only in the *uncB* cells (AN 283), and by ATP-hydrolysis only in the cytochrome-deficient cells (AN 704). Moreover, the growth parameters and the transport data indicate that $\Delta\tilde{\mu}_{H^+}$ is preferably generated by fumarate reduction and not by ATP-hydrolysis. The coupling of fumarate reduction to phosphorylation is more directly demonstrated by studies with particulate subcellular fractions of *E. coli.* Electron transfer from L-α-glycerol phosphate to fumarate resulted in the formation of high-energy phosphate, and $P/2e^-$ ratios of 0.1 were measured.[184] Also, more direct evidence for the generation of an electrochemical proton gradient by fumarate reduction has been presented. In inverted membrane vesicles, the generation of a ΔpH upon

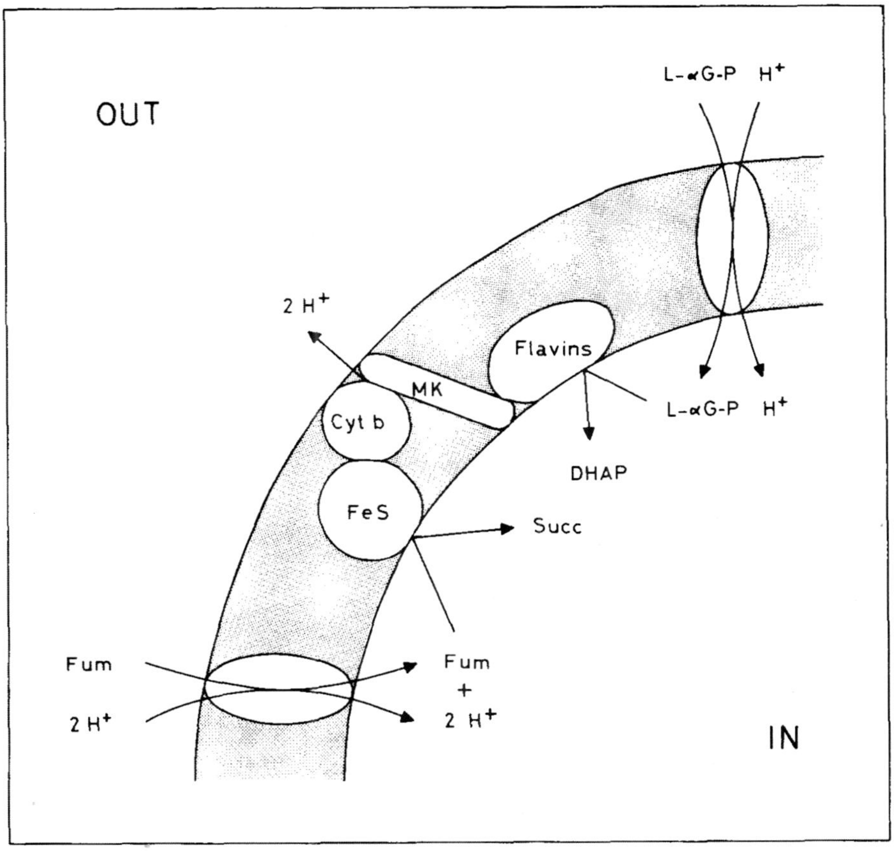

FIGURE 9. Topography of the fumarate reductase pathway of *E. coli*. Mk, menaquinone; FeS, iron-sulphur center; Cyt, cytochromes; L-α-G-P, L-α-glycerol phosphate; DHAP, dihydroxy acetone phosphate; Fum, fumarate; and Succ, succinate.

fumarate reduction was shown by atebrin fluorescence studies.[173,183] In whole cells, a fumarate-dependent oxidation of endogenous substrates resulted in proton translocation with $H/2e^-$ ratios of about 1.[185,186,191] Fumarate reduction leads also to the generation of $\Delta\psi$, as was shown by the accumulation of the lipophilic cation $TPMP^+$ by right-side-out membrane vesicles from *E. coli*.[169] In membrane vesicles from *E. coli* AN 283 (*unc B*), steady-state membrane potentials of −90 mV were recorded during fumarate reduction.[5]

b. Solute Transport Coupled to Fumarate Reduction

A coupling of solute transport to the anaerobic electron transfer system with fumarate as terminal electron acceptor has been suggested by uptake experiments in whole cells. Butlin[192] and Rosenberg et al.[193] demonstrated that mutants of *E. coli* deficient in Ca^{2+}- and Mg^{2+}-stimulated ATPase (*unc A*) can catalyze secondary facilitated transport of serine and phosphate under anaerobic conditions with fumarate as electron acceptor. In whole cells of *E. coli* ML 308-225 grown anaerobically on glycerol in the presence of fumarate, a marked stimulation of lactose uptake is observed upon the addition of L-α-glycerol phosphate plus fumarate. Under these conditions, L-α-glycerol phosphate dehydrogenase and fumarate reductase are induced. Such a stimulatory effect of L-α-glycerol phosphate plus fumarate is not observed in cells grown anaerobically on glucose in the presence of nitrate, or in cells grown anaerobically on glycerol.

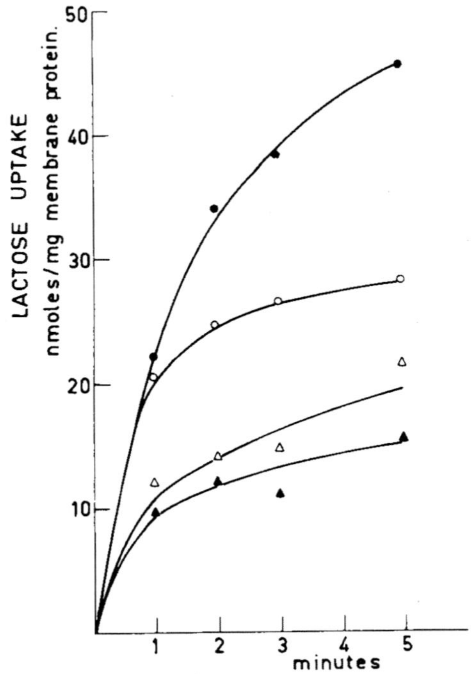

FIGURE 10. Uptake of lactose under anaerobic conditions in membrane vesicles from *E. coli* ML 308-225 grown anaerobically in the presence of glycerol and fumarate. (▲) No additions, (△) Na-fumarate (10 m*M*), (○) Na-L-α-glycerol phosphate (10 m*M*), (●) Na-L-α-glycerol phosphate (10 m*M*) + Na-fumarate (10 m*M*). (Taken from Konings, W.N and Boonstra, J., *Curr. Top. Membr. Transp.*, 9, 177, 1977. With permission.)

More evidence for a coupling between amino acid transport and anaerobic electron transfer to fumarate has been obtained with membrane vesicles from cells grown on glycerol in the presence of fumarate. These membrane vesicles, isolated with the same procedure as used for vesicles from glucose-nitrate grown cells,[58] have a high endogenous rate of lactose uptake, and the addition of the electron-donor L-α-glycerol phosphate alone, or of fumarate alone, causes significant stimulation of lactose uptake (Figure 10). In the presence of both L-α-glycerol phosphate and fumarate, however, a stimulation of amino acid and lactose uptake is observed which is significantly higher than the sum of the stimulations exerted by the electron donor or acceptor alone.[58,174]

In agreement with these observations, the membrane vesicles contain high activities of anaerobic L-α-glycerol phosphate dehydrogenase and fumarate reductase, and fumarate reduction occurs at a high rate in the presence of L-α-glycerol phosphate.[174] Further evidence for the involvement of electron transfer to fumarate is presented by the observation that HQNO inhibits transport energized by L-α-glycerol phosphate plus fumarate by more than 70%.[58]

Anaerobically in the presence of fumarate, transport in these membrane vesicles is also stimulated to some extent by D-lactate. This indicates that L-α-glycerol phosphate dehydrogenase and D-lactate dehydrogenase are coupled to fumarate reductase. It is of interest that these membrane vesicles reduce nitrate at a high rate in the presence of formate, and that formate plus nitrate catalyzes transport of lactose even better than L-α-glycerol phosphate plus fumarate. Formate plus fumarate, however, did not stimulate transport to a significant extent. The electron donor L-α-glycerol phosphate in these vesicles is also coupled to nitrate reductase, and the stimulation observed with this electron donor in the presence of nitrate is even higher than with fumarate.[174]

These observations indicate that in membrane vesicles from cells grown anaerobically on glycerol in the presence of fumarate two anaerobic electron transfer systems are present, both of which can provide energy for solute transport. The data obtained from the uptake experiments suggest that these electron transfer systems have some common electron transfer intermediates. Moreover, these membrane vesicles contain a functional respiratory chain, and solute accumulation is observed aerobically in the presence of Asc-PMS, NADH, and D-lactate.[58,174] In membrane vesicles isolated from *E. coli* grown anaerobically on glucose without exogenous electron acceptor, formate dehydrogenase is coupled to fumarate reductase. Membrane vesicles isolated from these cells contain formate-fumarate oxidoreductase activity and accumulate amino acids in the presence of formate and fumarate.[5]

C. Cyclic Electron Transfer Systems

The phototrophic bacteria, *Chromatiaceae, Rhodospirillaceae,* and *Chlorobiaceae* are characterized by a membrane-bound cyclic electron transfer system which converts light energy into a chemiosmotic form of energy. The cyclic electron transfer systems of the phototrophic bacteria are very similar except for differences in the nature of pigments. Electron donors can provide electrons for the formation of NADH. For *Chromatiaceae* and *Chlorobiaceae,* the electron donors are sulphur compounds (S^{2-}, $S_2O_3^{2-}$) while *Rhodospirillaceae* require organic compounds, e.g., malate or succinate.

The cyclic electron transfer chains are composed of a photoreaction center and a series of electron carriers. A schematic representation of a cyclic electron transfer chain is shown in Figure 11. Light energy, absorbed by the light-harvesting pigments, carotenoids, and bacteriochlorophyll, is transferred to the photoreaction center complex. The radiant energy is used to transfer an electron from a donor molecule (P_{870}) to an acceptor molecule within this complex. Electrons are subsequently transferred to the oxidized donor via a series of electron carriers. These electron carriers are ubiquinone and the cytochromes *b* and *c*. Cytochrome *c* is shown to be the direct electron donor for photo-oxidized P_{870}. In some species, two different membrane-bound cytochromes *c* have been demonstrated: a high potential cytochrome *c* (E_m = + 300 mV) and a low potential cytochrome c (E_m = 0 mV) (see for review References 80 and 194).

Light-induced electron transfer and its coupling to ATP-synthesis and solute transport processes have been studied most extensively in *Rhodopseudomonas sphaeroides* and *R. capsulata.* The mechanism of cyclic electron transfer in these organisms is described here in more detail.

The reaction center of *R. sphaeroides* and *R. capsulata* is a complex of about 70,000 daltons. It consists of three protein subunits, four molecules of bacteriochlorophyll, two molecules of bacteriophaeophytin, one equivalent nonheme iron, one or two molecules of ubiquinone, and one molecule of a specific carotenoid.[195] Light energy which is funneled to the reaction center, is used to oxidize P_{870}, a dimer of two bacteriochlorophyll molecules.[196] This oxidation occurs with a quantum yield of almost 1.0.[196] The midpoint oxidation-reduction potential (E_m) of P_{870}/P^+_{870} is about +450mV.[197] The electron is transferred to the primary electron acceptor which is a quinone-iron complex (QFe). This transfer occurs with a half time of 100 to 200 psec. The E_m of QFe/Q^-·Fe has been estimated to be −180 mV in the operational electron transfer chain.[198] Electron transfer in the reaction center is mediated by bacteriophaeophytin (BPh, E_m = −550 mV) with a half time of less than 10 psecs.[199-201] From the reaction center,

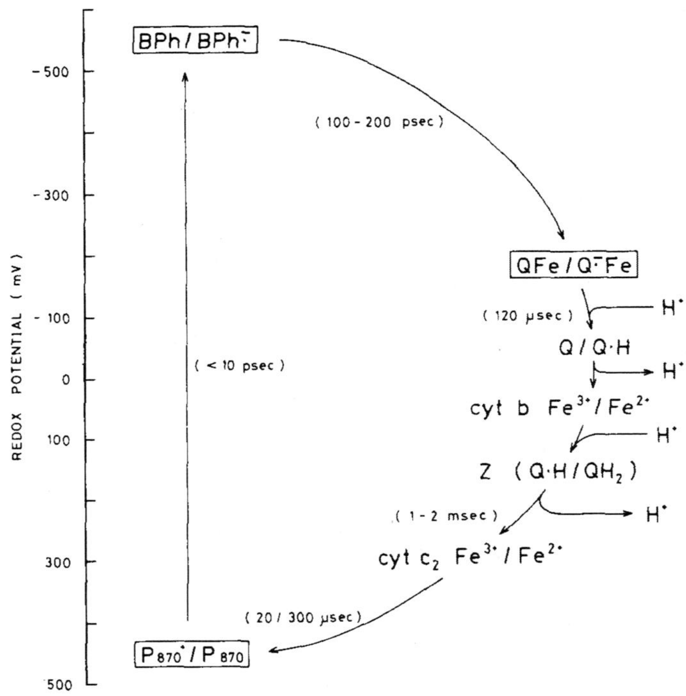

FIGURE 11. Cyclic electron transfer system of *R. sphaeroides.* P_{870}, photoreaction center bacteriochlorophyll; BPh, bacteriophaeophysin; and QFe, nonheme iron-quinone complex. Half times of reactions are indicated in parenthesis. BPh, QFe, and P_{870} are components of the photoreaction center.

electrons are transferred to a ubiquinone-10 pool of 10 to 20 ubquinones per reaction center. The E_m of this reaction has not yet been determined. The transfer of the electron in this reaction occurs in 120 psecs. The reaction is accompanied with the binding of one proton from the cytoplasm, yielding a semiquinone (Q · H).[202,203] Meanwhile, the photo-oxidized P_{870} is reduced by cytochrome c_2 (E_m = +295 mV in *R. sphaeroides* and +340 mV in *Rps. capsulata*). The reduction of P_{870} is biphasic with half times of 20 and 300 μsec. This indicates the involvement of two cytochromes c_2 with different orientations with respect to the reaction center.[120] Oxidized cyt c_2 is reduced by the semiquinone via Q-cytochrome *b*-c_2 oxidoreductase. The exact nature of these reactions is not clear. This antimycin-A-sensitive process involves, besides cyt *b* (E_m = +50 mV at pH 7.0), an unknown complex Z. Cytochrome *b* is reduced by a ubisemiquinone formed by the oxidation of the primary acceptor of the photo-oxidation. The reduction of cytochrome c_2 by Z (E_m = +155 mV at pH 7.0) occurs with a half time of 1 to 2 sec. This reaction is the rate-limiting step of the electron transfer. It has been proposed that Z is ubiquinol acting as the couple ubisemiquinone/ubiquinol (Q · H/Q · H_2), in which case cytochrome *b* would be reoxidized by ubisemiquinone.[204-208] Such a scheme resembles the Q-cycle hypothesis of Mitchell[209,210] in which ubiquinone acts both at the reducing and oxidizing site of cytochrome *b*.

The mechanism of proton extrusion during cyclic electron transfer has been studied most extensively in membrane particles, the so-called chromatophores, isolated from *R. sphaeroides* and *R. capsulata*. The photochemical reaction center spans the entire membrane.[211] Within this complex, P_{870} is located exterior with respect to the primary acceptor, and a physical separation of charges across a distance of 3 to 4 nm will occur upon illumination.[212,213] Cytochrome c_2 is located at the periplasmic side of the membrane, acting as reductant of photo-oxidized P_{870}.[119] Kinetic considerations led to the conclusion that quinones accept electrons from the reaction centr at the outer surface of the chromatophore membrane, i.e., the inner side of the membrane in intact cells. Upon this reduction, the quinones bind protons from the cytoplasm.[202,203,214,215] Two protons are bound per single turnover of the cyclic electron transfer chain. One proton is bound before, the other after, the antimycin A inhibition site.[293]

Other information about the vectorial character of electron transfer has been obtained from the kinetics of the carotenoid absorbance changes. The absorbance spectra of the pigments undergo changes under influence of the electric field alterations due to the spatial charge movements via the electron transfer chain. In the formation of these absorbance changes, three phases can be distinguished which are kinetically and potentiometrically related to events in the electron transfer.[204,216] Two models have been proposed for the cyclic electron transfer chain which offer an explanation for the observations described above (Figure 12).[203]

1. Generation of Electrochemical Proton Gradient by Cyclic Electron Transfer

The generation of $\Delta\tilde{\mu}_H{}^+$ has been studied in intact cells and in right-side-out and inside-out oriented membrane preparations from *Rhodospirillaceae*. Only qualitative data are available about the $\Delta\tilde{\mu}_H{}^+$ in whole cells. Light-induced changes of carotenoids and bacteriochlorophyll spectra,[217] of ANS-fluorescence,[218] and of atebrin fluorescence[218] have been interpreted as indications for proton efflux from the cells.

More quantitative information is available from studies in isolated membrane preparations.[45] Two different membrane preparations have been isolated. Right-side-out oriented membrane vesicles were obtained by osmotic lysis. Upon illumination, these vesicles generate at an external pH 7.0 a $\Delta\tilde{\mu}_H{}^+$ up to −110 mV consisting of a $\Delta\psi$ of −70 mV (inside negative) and a ΔpH of −40 mV (inside alkaline). In membrane vesicles isolated by passage of cells through a Yeda press followed by differential centrifugation, illumination resulted in the generation of a $\Delta\tilde{\mu}_H{}^+$ which, depending on the external pH, varied between −20 and −115 mV.[219] The $\Delta\psi$ was, at all external pH values, around −50 mV (inside negative); ΔpH varied between −60 mV at pH 5 and −20 mV at pH 9. However, these vesicles are apparently not devoid of cytoplasmic components because a substantial $\Delta\tilde{\mu}_H{}^+$ is generated in the dark, and a Donnan potential of −25 mV can be measured in the presence of FCCP.

Inside-out-oriented membrane preparations, so-called chromatophores, have also been isolated from *Rhodospirillaceae*. These chromatophores are mainly derived from invaginations of the cytoplasmic membrane formed during growth at low oxygen pressures and low light intensities.[81,82] Light-induced steady-state membrane potentials of 90 to 100 mV (inside positive) were calculated from the uptake of thiocyanate and from the fluorescence changes of oxocarbocyanine and ANS.[28,36,45,220-222] Dissipation of the ΔpH by nigericin or NH_4Cl resulted in an increase of the $\Delta\psi$ up to 150 to 165 mV (inside positive).

Changes in the carotenoid absorbtion bands have been demonstrated to reflect changes in the membrane potential. By comparison of these changes with those induced by known potassium diffusion potentials, light-induced steady-state potentials of up to 240 mV have been estimated.[223,224] However, the reliability of this method has been questioned.[26,42] The light-induced ΔpH in chromatophores has been determined from the uptake of the weak base methylamine (Figure 13).[45,220,222] Depending on the experimental conditions, values ranging from nearly zero to 108 mV have been reported. The ΔpH is large when membrane-permeable ions such as Cl^- are present (see Figure 13) or when the $\Delta\psi$ is dissipated by means of ionophores. The ΔpH is small

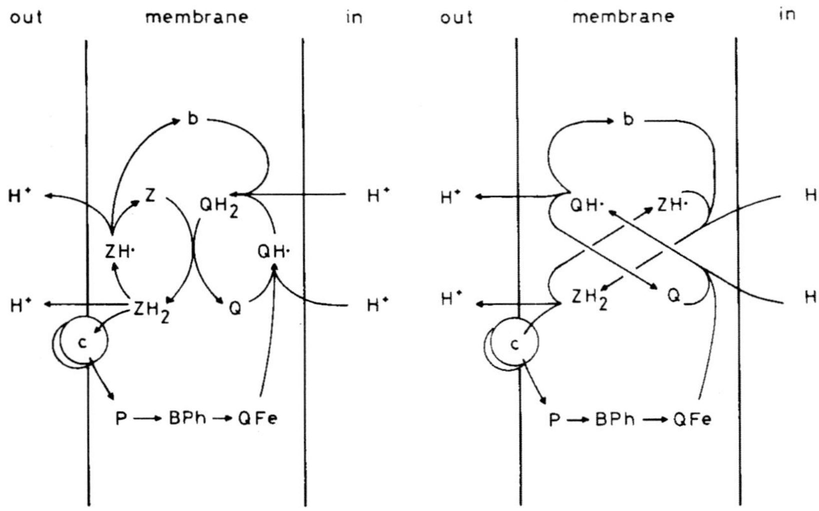

FIGURE 12. Two possible models for the mechanism of cyclic electron transfer and proton translocation in cell membranes of *Rhodopseudomonas sphaeroides*, according to Petty et al.[203] P, photoreaction center bacteriochlorophyll; BPh, bacteriophaeophysin; QFe, iron-quinone complex; Z, unknown component, most likely quinone; C, cytochrome c_2; b, cytochrome *b*.

when ΔpH-dependent secondary transport is high, or in the absence of membrane-permeable ions (see below).

The generation of a ΔpH in chromatophores has also been calculated from the fluorescence changes of 9-aminoacridine, according to Schuldiner et al.[46] These calculations are based on the assumption that 9-aminoacridine is accumulated in the chromatophores in response to a ΔpH. The ΔpH gradients calculated with this procedure are, however, significantly higher (ΔpH values of 150 to 200 mV have been reported[221,224,225]) than those calculated from the accumulation of methylamine. However, it has been demonstrated that the fluorescence changes of 9-aminoacridine are not solely the result of accumulation of the probe inside the chromatophores, but also of energy-dependent interactions of the probe with the membrane.[47,48]

2. *Solute Transport Coupled to Cyclic Electron Transfer*

The information on transport processes energized by cyclic electron transfer is limited. Intact cells of *R. sphaeroides* accumulate amino acids,[59] C_4-dicarboxylic acids, and pyruvate[226], and in the light intact cells of *R. capsulata* perform uptake of K^+,[227] Mg^{2+}, and Mn^{2+}.[228] Cells of the obligately halophilic phototrophic bacterium *Ectothiorhodospira halophila*, also accumulate amino acids upon illumination.[229] In cells of *Chromatium vinosum*, light-energy-dependent uptake of several amino acids has been observed.[230]

More detailed information of light-induced transport processes has been obtained from studies on isolated membrane preparations. Membrane vesicles of *Rps. sphaeroides* accumulated amino acids upon illumination (Figure 14).[59,82] In chromatophores isolated from *Rhodospirillaceae*, light-dependent accumulation of calcium[82] and sodium ions[131] was observed (Figure 14). The accumulation of Na^+ was stimulated by valinomycin and completely inhibited by nigericin, indicating the existence of a Na^+/H^+ antiport. Indirect evidence for other secondary transport systems in membranes of *Rhodopseudomonas sphaeroides, Rhodopseudomonas capsulata*, and *Rhodospirillum rubrum* comes from measurements of $\Delta\tilde{\mu}_H{}^+$ in chromatophores suspended in various

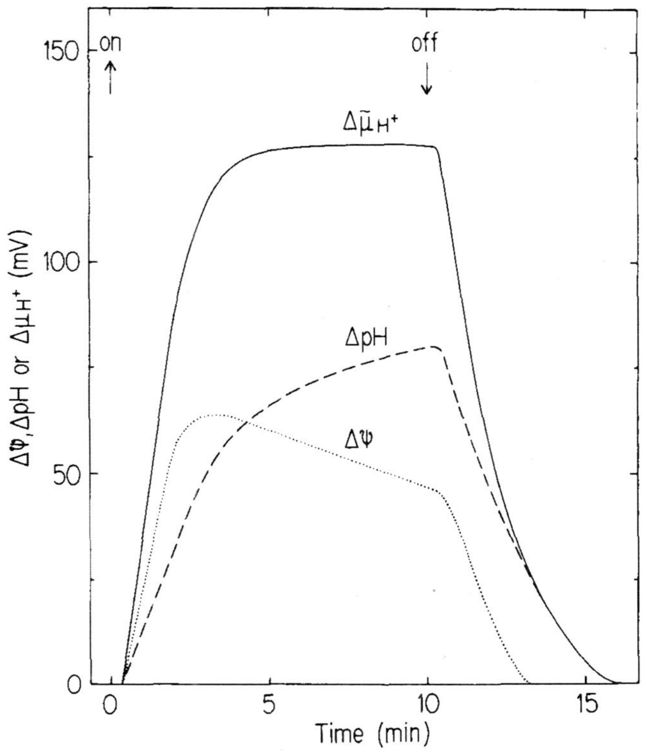

FIGURE 13. Time dependence of light-induced generation of electrochemical proton gradient in chromatophores of *R. sphaeroides*. The curves were deduced from the uptake data of methylamine (ΔpH) and thiocyanate (Δψ). (Taken from Michels, P.A.M. and Konings, W.N., *Eur. J. Biochem.*, 85, 147, 1978. With permission.)

media.[36,45,131,223,231] Secondary transport which involves the translocation of charge or protons will effect Δψ and ΔpH, respectively. By an additional electron-transfer-mediated proton translocation, a decrease of Δψ will be compensated by an increase of ΔpH and vice versa. This is illustrated in Figure 13 for chromatophores of *Rhodopseudomonas sphaeroides.* The Δψ which is generated rapidly due to the low electrical capacitance of the membrane is converted into a ΔpH after the discharging of the membrane potential as a result of chloride translocation.[45] The effects of different ions on the Δψ and ΔpH supply information about the fluxes of these ions through the membrane. Membranes of *Rhodospirillaceae* show an increasing anion permeability according to

$$SO_4^{2-} < Cl^- < NO_3^- < I^- < SCN^- = ClO_4^-$$

The cation permeability in *Rhodospirillum rubrum* increased, according to Pick and Avron,[36] in the following sequence:

$$\text{Choline}^+ < Li^+ = Na^+ << NH_4^+ = K^+ < Cs^+ < Rb^+$$

These membranes appeared to be more permeable for K^+ than for NO_3^-. In chromatophores of *Rhodopseudomonas sphaeroides*, the flux of potassium was smaller than the flux of Cl^-.[131]

Biological membranes are, in general, very permeable for small molecules of weak

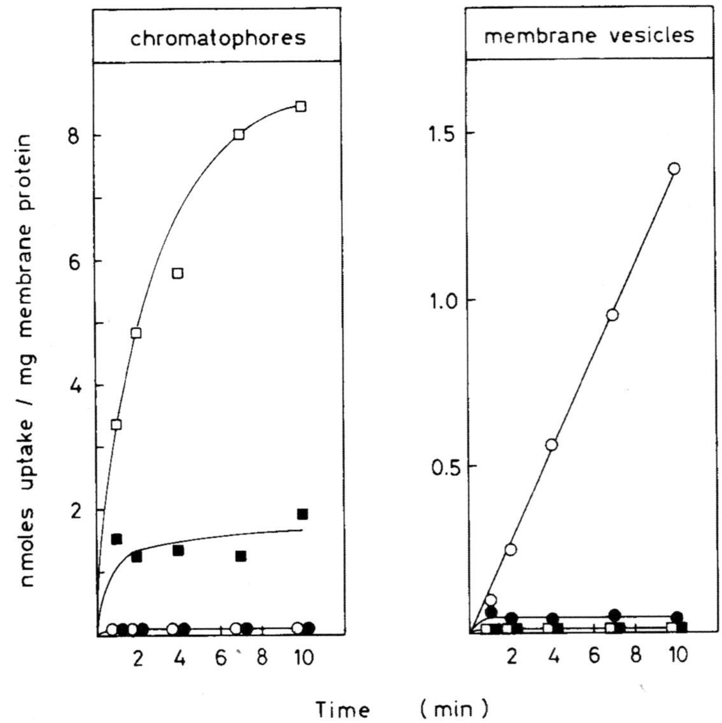

FIGURE 14. Light-induced uptake of L-alanine and Ca^{2+} by chromatophores and membrane vesicles of *R. sphaeroides.* (○) Alanine uptake in the light, (●) alanine uptake in the dark, (□) Ca^{2+} uptake in the light, (■) Ca^{2+} uptake in the dark. (Taken from Michels, P.A.M. and Konings, W.N., *Biochim. Biophys. Acta,* 507, 353, 1978. With permission.)

acids and bases. These solutes can cross the membranes in uncharged form. Internally, these acids or bases will dissociate according to the proton concentration and thereby change the internal pH. From their effects on $\Delta\psi$ at the expense of ΔpH, it was concluded that many bacterial membranes are permeable for the acids formate, acetate, propionate, butyrate, and benzoate and the bases ammonia, methylamine, imidazol, and *tris* (hydroxymethyl) aminomethane. Also, membranes of *R. sphaeroides* have been shown to be permeable for these compounds.[131]

D. Bacteriorhodopsin

With the description of the properties and function of bacteriorhodopsin, another primary transport system has been discovered. Bacteriorhodopsin is a membrane protein from the halophilic bacterium *Halobacterium halobium.*[232] In the membrane, it is present in the form of a two-dimensional hexagonal array which is called the purple membrane.[232,233] The cytoplasmic membrane of *H. halobium* is consequently differentiated in two regions: the purple membrane and the red membrane. The red membrane contains the enzymes usually present in cytoplasmic membranes (ATPase complex, electron transfer chain, etc.). Osmotic lysis of halobacterial cells in hypotonic media causes preferential rupturing of the cytoplasmic membrane in the regions connecting the purple and the red membranes. Differential and sucrose-density-gradient centrifugation can subsequently be used to separate the purple membrane from the red membrane.[234]

Bacteriorhodopsin is a complex composed of a protein with a molecular weight of

26,000 to which, covalently via a Schiff-base linkage, a vitamin A group is attached.[232] The complex has an absorption maximum at 568 nm.[232] The primary,[235] secondary,[236,237] and tertiary[237] structure of bacteriorhodopsin have been largely re-
secondary,[236,237] and tertiary[237] structure of bacreriorhodopsin have been largely re-
solved. Studies on the effect of illumination on the properties of bacteriorhodopsin revealed that absorption of a proton is followed by a cyclic process of absorbance, pH, and conformational changes[238,239] in this protein. These observations and the discovery of a photophosphorylation system in *H. halobium*, with an action spectrum identical to the absorbance spectrum of bacteriorhodopsin[240] led Oesterhelt and Stoeckenius to the postulation of a proton pump function for this chromoprotein.[241] Independent evidence for this proton pump function was obtained from reconstitution experiments. In a suspension of vesicular artificial membranes in which bacteriorhodopsin molecules are incorporated, light-induced pH changes can be observed, and these pH changes are affected by ionophores as would be predicted for a primary-proton pumping system.[98] An explanation for these pH changes based on the assumption that a large number of protons is bound by bacteriorhodopsin can be excluded because in the light a considerable $\Delta\tilde{\mu}_H{}^+$ can be maintained by bacteriorhodopsin even in the presence of significant amounts of valinomycin and nigericin.[242] Artificial membrane vesicles can be prepared in which bacteriorhodopsin has the same[94] or the opposite[98] orientation as in the cytoplasmic membrane of intact cells.

Because of its special and well-defined properties, bacteriorhodopsin is an attractive system for studies on the interaction of primary transport systems in reconstituted membranes. Such studies have been performed using bacteriorhodopsin plus an ATPase complex[98] or cytochrome c oxidase.[243] The technical difficulties associated with such an approach have hindered thus far a full utilization of its theoretical applications. However, the reconstitution of an ATP-synthesizing system from purple membranes, purified lipids, and an ATPase complex has contributed to an important extent to the appreciation of the chemiosmotic theory for energy transduction.[99]

1. Generation of Electrochemical Proton Gradient by Bacteriorhodopsin

In intact halobacterial cells, light-dependent $\Delta\tilde{\mu}_H{}^+$ formation has been demonstrated.[244,245] Quantitation of the contribution of bacteriorhodopsin to ΔpH and $\Delta\psi$ was difficult, partly because a rather large $\Delta\psi_H{}^+$ was maintained in the dark in these cells. The largest change in $\Delta\tilde{\mu}_H{}^+$ upon illumination (in the presence of DCCD) and the largest total $\Delta\tilde{\mu}_H{}^+$ were −170 and −270 mV, respectively.[245]

In membrane vesicles of *H. halobium*, a maximal steady state $\Delta\tilde{\mu}_H{}^+$ in the light of −226 mV has been reported[44] composed of a $\Delta\psi$ of −120 mV and a ΔpH of −106 mV. These membrane vesicles are prepared by sonication of intact cells.[246] Freeze-etch electron microscopy and determination of the NADH-menadione reductase activity revealed that 80 to 90% of these vesicles have a right-side-out orientation.[247]

In artificial bacteriorhodopsin-containing planar membrane systems, the generation of light-induced potentials has been reported.[248,249] The magnitude of these potentials varies with the clamp voltage across the planar membrane and can be as high as 500 mV.[249] In order to draw conclusions about the maximal $\Delta\tilde{\mu}_H{}^+$ generated across these membranes, more information about their physical properties is required.

In artificial membrane vesicles containing bacteriorhodopsin, the generation of light-dependent $\Delta\tilde{\mu}_H{}^+$ has been demonstrated.[242,250] Also in this system, accurate quantitation of the maximal $\Delta\tilde{\mu}_H{}^+$ has to wait until the physical properties are defined. On the assumption of homogeneity of the vesicles, Kagawa and co-workers reported[250] a maximal $\Delta\tilde{\mu}_H{}^+$ of +230 mV. This $\Delta\tilde{\mu}_H{}^+$ was calculated from the fluorescence changes of 9-aminoacridine (as probe for ΔpH) and 9-amino-6-chloro-2-methoxy-acridine

(ACMA, as probe for $\Delta\psi$). Measurements of the ΔpH and $\Delta\psi$ from the accumulation of radio-isotopically labelled compounds yields values for $\Delta\mu_H^+$ ranging from +90 to +130 mV.[242] These latter values roughly agree with the results of NMR-measurements of the resonances of vesicle-entrapped glucose-6-phosphate.[251] These experiments indicated maximum light-induced ΔpH values of about 150 mV. However, the NMR studies also revealed some inhomogeneity of the vesicles.

2. Solute Transport in Halobacterium halobium

In membrane vesicles from *Halobacterium halobium*, a facilitated secondary transport system catalyzing the exchange of "n" sodium ions against "m" protons has been demonstrated[252,253] (see Figure 15). The exact stoichiometry of this exchange reaction has not yet been determined. Eisenbach et al. have proposed that n equals m,[253] but Lanyi and co-workers presented evidence for m being larger than n.[252]

The Na^+-Ca^{2+} exchange system indicated in Figure 15 has been described by Belliveau and Lanyi.[254] The stoichiometry of this exchange reaction is most likely $k/1 > 2$ (see Figure 15). For the translocation reaction, potassium ions appear to be essential. From the work of Wagner et al., indications for the existence of a potassium-translocation system have been obtained.[255]

Membrane vesicles also translocate amino acids by facilitated secondary transport. Based on crossed-inhibition studies, nine different amino acid translocation systems have been detected,[256] i.e., arg-his-lys; gln-asn; val-leu, ile, met; thr-gly, ala, ser; trp, tyr-phe; asp; glu; and pro. Cysteine is not transported by a facilitated transport system. It inhibits methionine transport in a noncompetitive way.[257] All facilitated amino acid transport in *H. halobium* appears to occur in cotransport with sodium. The transport systems for gly and for leu have been studied in detail. The results for the two systems can be fitted into a model in which the two amino acids are cotransported with one sodium ion.[256] Of the two translocation reactions, therefore, only the leucine translocation is electrogenic.

For activity of some amino acid transport systems, low concentrations of potassium ions are required. However, no cotransport with K^+ or H^+ has been observed so far.

Besides light, oxidation/reduction reactions can also drive the accumulation of amino acids in cells and membrane vesicles from *H. halobium*.[258] In vesicles, the non-physiological electron donor dimethylene diamine can be used. In intact cells, NADH, L-α-glycerol phosphate and succinate can function as electron donors, whereas oxygen and nitrate can function as electron acceptors.[259]

Finally, it should be mentioned that the histidine transport in intact cells of *H. halobium* was reported to depend on ATP and $\Delta\tilde{\mu}_{Na}^+$.[260] However, this ATP-dependence of histidine translocation could not be demonstrated in membrane vesicles.[259]

V. MECHANISM OF SECONDARY TRANSPORT

A. Relationship Between $\Delta\tilde{\mu}_H^+$ and Secondary Transport

During the last 10 years, the essential role of $\Delta\tilde{\mu}_H^+$ in secondary transport has been well established. In intact cells, it has been demonstrated that accumulation of solutes is associated with the influx of protons[261,262] and is inhibited by proton and ion conductors.[263,264] Furthermore, artificially generated membrane potentials have been shown to drive solute transport in energy-depleted cells of *S. faecalis*,[265] *S. lactis*,[266,267] *Staphylococcus aureus*,[268] and *Clostridium pasteurianium*.[269] The uptake was shown to be sensitive to uncouplers, but insensitive to DCCD. Membrane potentials (inside negative) are generated by valinomycin-induced efflux of potassium or by imposing gradients of other ionic species across the cytoplasmic membrane. Robertson et al.[140]

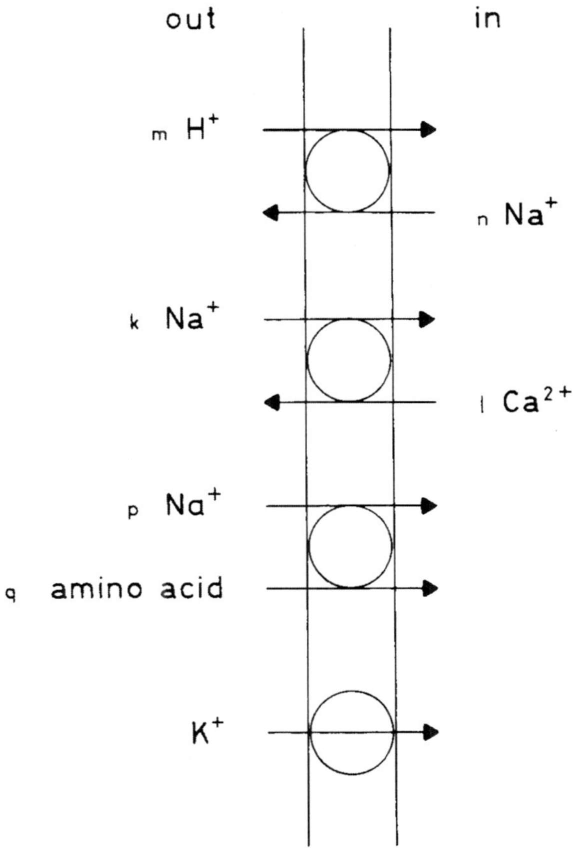

FIGURE 15. Facilitated secondary transport systems for Na^+, K^+, Ca^{2+}, and amino acids in *H. halobium*.

examined the effects of valinomycin-induced potassium diffusion potentials on carrier-mediated efflux of lactose. Membrane potentials, inside negative, slowed down the rate of lactose efflux, while membrane potentials, inside positive, enhanced this rate.

Also, artificially imposed pH gradients were shown to drive solute transport. Sudden exposure of energy-depleted cells of *S. lactis*,[267] *E. coli* DL 54,[270] and *S. aureus*[271] preincubated at pH 8 to a medium of pH 6 resulted in uptake of β-galactosides.

Similar results have been obtained from studies with membrane vesicles. Membrane vesicles of *E. coli* accumulate lactose, ammonia, $TPMP^+$ and DDA^+ in response to valinomycin-induced potassium diffusion potentials (inside negative).[128,130]

Artificially imposed gradients have been used to demonstrate that carriers act reversibly. The direction of transport was shown to be determined by the direction of the electrochemical proton gradient. Burnell et al.[272,273] studied the uptake of sulphate and phosphate into right-side-out (with respect to the orientation of the cytoplasmic membrane of intact cells) and inside-out membrane vesicles of *Paracoccus denitrificans*. Only in right-side-out vesicles was electron transfer driven transport of anions observed. In inverted vesicles, however, transient uptake could be induced by an artificial pH gradient (inside alkaline) generated by a pulse of KCl in the presence of nigericin (catalyzes electroneutral exchange of K^+ for H^+) or by the influx of ammonia. Similar results were obtained for lactose[274] and Ca^{2+} [275] transport in membrane vesicles of *E. coli*. Ca^{2+} is normally expelled by whole cells. Inside-out vesicles accumulated

Ca^{2+} upon respiration or ATP-hydrolysis, whereas right-side-out vesicles catalyzed Ca^{2+} uptake after alkalization of the outer medium.

Other studies in membrane vesicles supplied more direct support for the chemiosmotic concept of solute transport. Lactose is accumulated in relatively large amounts by membrane vesicles of *E. coli* ML308-225, and the uptake of lactose diminishes the accumulation of acetate[29,127] (accumulates in response to ΔpH) and the accumulation of $TPMP^+$[130] (accumulates in response to $\Delta\psi$). When the same experiments are carried out with ML30 membrane vesicles uninduced for lactose transport, the accumulation of acetate and $TPMP^+$ is unaffected by addition of lactose.[127,130] These results indicate that lactose is taken up by cotransport with protons, as suggestd by Mitchell.[8]

The relationship between $\Delta\psi$ and ΔpH and solute transport was studied by titration studies with valinomycin and nigericin in membrane vesicles from *E. coli* ML308-225.[10,22,29,127] At an external pH of 5.5, the $\Delta\tilde{\mu}_{H^+}$ is composed of a $\Delta\psi$ of −74 mV and a ΔpH of −115 mV. At pH 5.5, there are two general classes of transport systems: those that are primarily driven by $\Delta\tilde{\mu}_{H^+}$ (lactose, proline, serine, glycine, tyrosine, glutamate, leucine, lysine, cysteine, and succinate) and those that are primarily driven by ΔpH (glucose-6-P, lactate, glucuronate, and gluconate). For both systems, the steady-state levels of accumulation increase linearly with the driving force. At pH 7.5, the $\Delta\psi$ comprises the only component of $\Delta\tilde{\mu}_{H^+}$ (see Figure 3), and all of these transport systems are driven by $\Delta\psi$.

Of special interest is the observation that at external pH values exceeding 6.0 to 6.5, $\Delta\tilde{\mu}_{H^-}$ is insufficient thermodynamically to account for the concentration gradients observed for most solutes if it is assumed that the stoichiometry between protons and solute remains constant at 1:1 (Figure 16). These findings, and the observation that the accumulation of organic acids is coupled to ΔpH at relatively low external pH and to $\Delta\psi$ at relatively high external pH, led Ramos and Kaback[23] to the conclusion that the stoichiometry between protons and transport solutes increases from 1:1 at pH 5.5 to values of about 2:1 as the external pH is increased. Such a change in proton/solute stoichiometry is conceivably the result of a pH-dependent dissociation of a functional group of the carrier. A pK of approximately 6.8 was estimated for the functional group responsible for the increase in proton/proline stoichiometry.

Based on these considerations, Rottenberg[24] proposed a chemiosmotic model for solute transport which is shown in Figure 2. This model assumes that the inward moving complex of carrier, solute, and cotransported protons is always electroneutral. The stoichiometry between protons and solute thus depends on the charge of the solute and on the valency of the carrier which is thought to be determined by a pH-dependent dissociation of important functional groups. The overall transport cycle in this model implies the following steps: (1) binding of protons and solute to the carrier at the external membrane surface, (2) equilibration of an electroneutral complex of carrier, solute, and protons between the inner and outer surface of the membrane, (3) dissociation of protons and solute from the carrier at the internal membrane surface, and (4) equilibration of the unloaded carrier between the inner and outer surface. According to this model, solute transport is determined by the electrochemical proton gradient and the electrochemical solute gradient. In Table 1 are given the driving forces for solute transport by differently charged carriers with varying numbers of protons. Steady-state levels of solute transport will be reached when the total driving force is zero.

It is important to note that this model is based on the assumption that carrier-mediated transport is in essence a facilitated diffusion process which can translocate solutes and protons in both directions, depending on the direction of the electrochemical proton gradient and the electrochemical solute gradient ($\Delta\tilde{\mu}_A$). When $\Delta\tilde{\mu}_A$ exceeds and

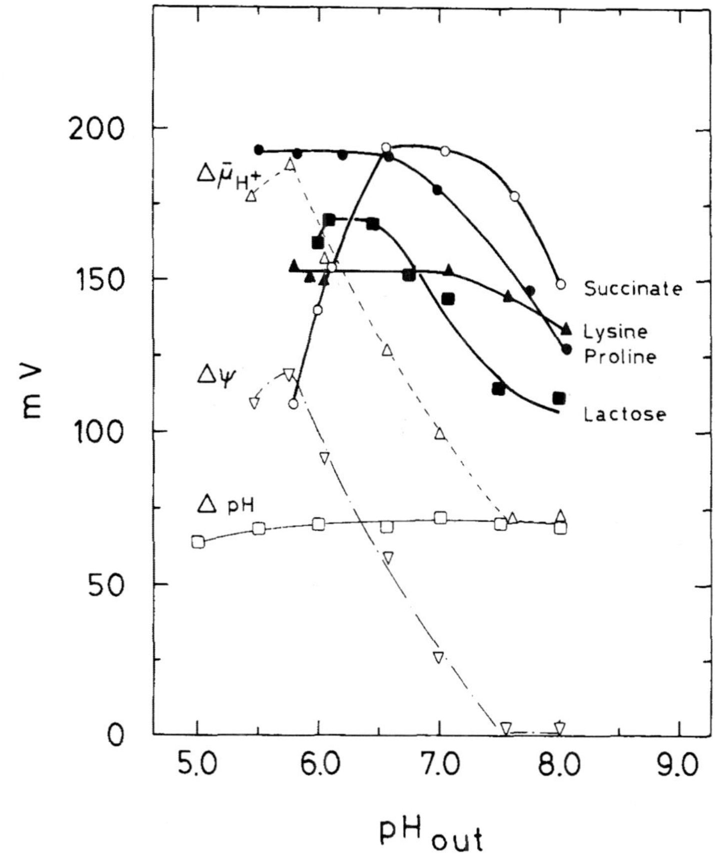

FIGURE 16. Effect of external pH on μ_H^+ (△), ΔpH (▽), and $\Delta\psi$ (□), and on steady-state levels of accumulations of lactose (■), proline (○), lysine ▲), and succinate (●) in membrane vesicles from *E. coli* ML 308-225. The electron donor was ascorbate (10 m*M*) + PMS (0.1 m*M*). The ΔpH and $\Delta\psi$ were determined as described in the legend to Figure 3. (Reprinted with permission from Ramos, S. and Kaback, H.R., *Biocheistry*, 16, 854, 1977. Copyright by the American Chemical Society.)

is opposite to $\Delta\tilde{\mu}_H^+$, solute transport will occur down the concentration gradient. This implies that influx of a solute driven by $\Delta\tilde{\mu}_A$ will result in the generation of $\Delta\tilde{\mu}_H^+$ until the total driving force is zero. Equilibration of solute will occur only when the electrochemical gradient is dissipated, for instance, by an uncoupler like FCCP. Evidence consistent with such a mechanism has been presented for β-galactoside transport in *E. coli*.[261,276]

As stated above, this model implies that solute translocation according to the electrochemical solute gradient can result in the generation of an electrochemical proton gradient. On the basis of these considerations, an "energy recycling" model has been postulated.[277] According to this model, efflux of fermentation products can result in the generation of an electrochemical proton gradient (Figure 17). This model visualizes that product excretion is not only a mechanism for the disposal of waste products, but also a form of energy supply for chemiosmotic membrane-bound processes. Such a mechanism will allow a cell to use ATP generated by substrate-level phosphorylation more effectively for biosynthetic purposes.

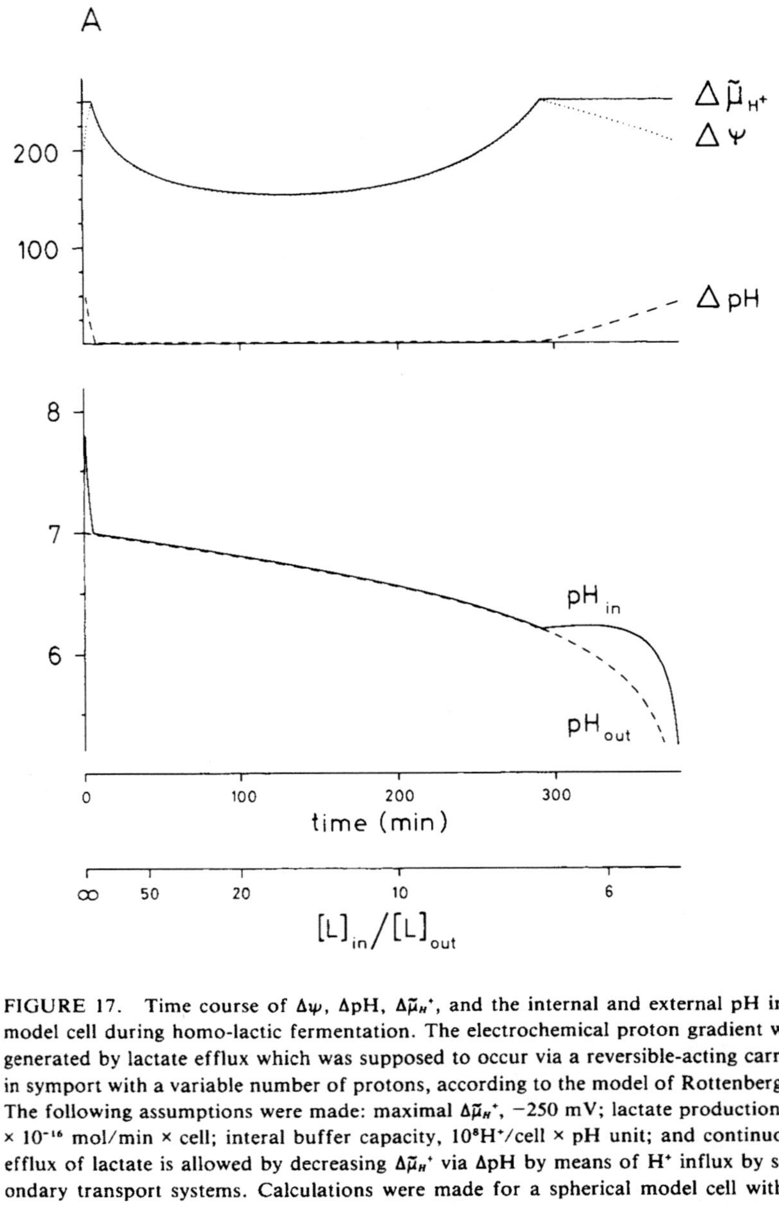

FIGURE 17. Time course of $\Delta\psi$, ΔpH, $\Delta\tilde{\mu}_{H^+}$, and the internal and external pH in a model cell during homo-lactic fermentation. The electrochemical proton gradient was generated by lactate efflux which was supposed to occur via a reversible-acting carrier in symport with a variable number of protons, according to the model of Rottenberg.[24] The following assumptions were made: maximal $\Delta\tilde{\mu}_{H^+}$, −250 mV; lactate production, 5 × 10^{-16} mol/min × cell; interal buffer capacity, $10^8 H^+$/cell × pH unit; and continuous efflux of lactate is allowed by decreasing $\Delta\tilde{\mu}_{H^+}$ via ΔpH by means of H^+ influx by secondary transport systems. Calculations were made for a spherical model cell with a diameter of 1 μm suspended to a density of 10^{11} ells/ℓ in a medium buffered with 50 m*M* phosphate. (From Michels, P.A.M., Michels J.P.J., Boonstra, J., and Konings, W.N., *FEMS Microbiol. Lett.*, 5, 357, 1979. With permission.)

B. Facilitated Secondary Transport Systems

Facilitated secondary transport of solutes across the cytoplasmic membranes of bacteria is mediated by specific proteins, the carrier proteins (previously termed permeases). Cytoplasmic membranes usually contain many of these carrier proteins, each having affinity for only one solute or a group of structurally related solutes. Evidence for the existence of distinct carriers for different solutes was presented by kinetic and competition studies (see, for instance, Lombardi and Kaback,[133] and Konings and Freese[62]). Strong evidence was also obtained from genetic studies. Many mutants have been isolated, mainly from *Escherichia coli*, which are defective in the transport of a specific solute (for a review see Halpern[51]). Carrier proteins perform secondary transport of a solute, often with a high affinity. K_s values range usually between 10^{-5} and

10^{-6} *M*, but also affinity constants as low as 10^{-8}*M* have been reported. It has been mentioned above that carriers have a specific affinity for a certain solute or a group of structurally related solutes. The transport systems can, therefore, be organized in groups which mediate transport of structurally related solutes:

1. Transport of amino acids
2. Transport of sugars
3. Transport of carboxylic acids
4. Transport of inorganic carbons
5. Transport of inorganic anions

Most bacteria perform secondary transport of amino acids, but the amino acids transported vary from organism to organism. Many bacteria, like *E. coli*, possess transport systems for essentially all amino acids.

In all bacteria studied, transport of amino acids is mediated by several distinct specific transport systems. The pattern of specificity appears to be surprisingly similar. For instance in *E. coli, Staphylococcus aureus*, and *B. subtilis*, nine amino acid transport systems have the same specificity.[13] Such a pattern of specificity might be common among bacteria. Each system mediates transport of only a few amino acids, for instance glycine and alanine or phenylalanine.

In addition to these general systems, other systems with a different pattern of specificity are found in some organisms. Multiple transport systems for a particular amino acid have also been demonstrated, each with its own specificity pattern and affinity constants.[51,133] At this moment, it is not clear whether these systems are functional under the same growth conditions. Also, information about the regulation of these transport systems is lacking.

Besides the secondary transport systems for amino acids, Gram-negative bacteria possess transport systems in which periplasmic binding proteins play an essential role. These systems will not be discussed here. The reader is directed for detailed information to Reference 289.

Bacteria possess several different transport mechanisms for sugars: group translocation, binding-protein-dependent systems, and secondary transport systems. In some organisms, more than one system is involved in the transport of a particular sugar. Thus, for instance, D-fructose and L-rhamnose are accumulated by *Arthrobacter pyridinolis* by a group translocation system and a secondary transport system.[290-292]

Secondary transport systems have been demonstrated in *E. coli* for β-galactosides such as lactose, arabinose, glucuronate, and glucose-6-phosphate;[293] in *Pseudomonas aeroginosa* for gluconate;[55,294] and in *Azotobacter vinelandii* for D-glucose.[105,295]

Studies in whole cells and membrane vesicles demonstrated the presence of secondary transport systems for dicarboxylic acids. Specific transport systems for the monocarboxylic acids D-lactate and L-lactate have been described for *B. subtilis* and *E. coli*,[115] and for pyruvate for *E. coli*.[115]

C_4-dicarboxylic acid transport systems for L-malate, fumarate, and succinate have been reported for *B. subtilis, E. coli*, and *Pseudomonas spp.*[115,296-301] Recently, an oxalate transport system has been described for *Pseudomonas oxalaticus*.[302]

Transport systems for inorganic cations have been demonstrated. Transport systems for sodium have been demonstrated in *S. faecalis*,[303] *S. typhimurium*,[285] *H. halobium*,[252,253,304] *E. coli*,[305] and *R. sphaeroides*.[131] Studies on whole cells and membrane vesicles from *E. coli*[305,306] and *H. haloium*[304] indicated that Na^+ is transported by a Na^+/H^+ antiport system. In *E. coli*, the system is highly specific and translocates only Na^+ and Li^+.[305] The stoichiometry of the exchange process is altered with changes in the external pH. At low pH values (6.6), the exchange iselectroneutral, at higher pH values (7.5), more than one proton is translocated per sodium ion. Many bacteria possess a highly specific transport system for the divalent cations Mn^{2+} [307] and Ca^{2+}.[308] These systems have been studied in intact cells and membrane vesicles. Evidence has also been presented for the existence of Mg^{2+} transport systems.[309] The Ca^{2+}-transport system has been studied in detail for *B. megaterium*,[310] *Azotobacter vinelandii*,[311,312] and *E. coli*.[84,275,313] Calcium appears to be actively extruded from the cells. Inside-out membrane vesicles accumulate calcium, most likely by an electroneutral calcium/proton antiport.[275,312]

Inorganic anion transport systems also exist. *E. coli*,[136,314] *S. faecalis*,[315] *Paracoccus denitrificans*,[273] and *Micrococcus lysodeiktus*[316] accumulate phosphate by a secondary transport system. In *P. denitrificans*[273] and *E. coli*,[136] phosphate transport in membrane vesicles could be driven by artificially generated pH gradients. It has been suggested that phosphate is transported via a proton symport. A proton symport mechanism has also been proposed for sulphate transport in *P. denitrificans*.[273]

C. Role of Cations in Facilitated Secondary Transport

For transport of various solutes across bacterial membranes, a requirement for specific ions has been demonstrated. Na^+ and K^+, especially, have been shown to be required for transport of a number of solutes.

The affinities of the uptake systems for melibiose in *Salmonella typhimurium*,[278] glutamate in *E. coli*,[279,280] α-aminoisobutyrate in a marine pseudomonas,[281] and succinate and glucose in *Micrococcus lysodeiktus*[282] were increased by sodium ions. The requirement for sodium for bacterial transport systems has been investigated in more detail. The main reason for this interest is that sodium gradients are involved in transport of metabolites across cytoplasmic membranes of mammalian cells by sodium-solute cotransport. Furthermore, transmembrane sodium and potassium gradients are responsible for fluctuations of the membrane potential during pulse conduction in nerve and muscle cells.

Gradients of sodium and potassium also exist across cytoplasmic membranes of bacteria. A number of solutes are transported by solute/Na^+ symport, and accumulation of solute in response to a Na^+ gradient was demonstrated for glutamate[283] and methyl-1-thio-β-D-galactopyranoside (TMG)[284] in *E. coli*, TMG in *S. typhimurium*,[285] amino acids in *H. halobium*,[286] and α-aminoisobutyrate in *B. alcalophilus*.[122]

A direct role of potassium in energy coupling has not been demonstrated. Potassium ions do not affect the affinity of transport, but increase the capacity of solute transport. It has been suggested that K^+ accelerates the cycling of the carriers due to binding of the cation at the internal surface of the membrane.[280,287] Information about the role of other cations in bacterial transport processes is limited. A Mg^{2+}-dependent cotransport of citrate has been reported for *B. subtilis*.[288]

D. Molecular Properties of Carrier Proteins

The first attempts to obtain information about the molecular properties of carrier proteins were made by Kennedy and co-workers.[317-319] These investigators estimated the amount of β-galactoside carrier (lactose permease, M- protein) in membranes of *E. coli* ML 30 by radioactive labeling with *N*-ethyl maleimide and by double labeling experiments with radioactive amino acids.[317-319] The amount of β-galactoside carrier was estimated to be 6% of the cytoplasmic protein. Similar values were obtained from binding studies with dansylgalactosides and azidophenylgalactosides in the presence of a $\Delta\tilde{\mu}_H{}^+$ across the membrane (for review see Reference 19).

Kennedy and co-workers determined a molecular weight 30,000 for the β-galactoside carrier. The same molecular weight was estimated by Overath et al.[320] These investigators showed by equilibrium binding experiments using dansylgalactosides and azidophenyl galactosides that the carrier binds 1 mole of substrate per mole of polypeptide. The β-galactoside carrier not only binds, but also transports dansylgalactosides and azidophenylgalactosides across the membrane in the presence f an electrochemical proton gradient. Too high amounts of lactose carriers are, therefore, estimated from binding studies in the presence of an electrochemical proton gradient. Overath et al.[320] estimated from binding studies in the absence of an electrochemical gradient and from double-labeling experiments that the lactose carrier comprises less than 1% of the cytoplasmic membrane protein.

Other studies on the lactose carrier were done by Lancaster and Hinkle.[321] In inverted membrane vesicles from *E. coli* ML 308-225, the same fluorescence changes as in right-side-out membrane vesicles were observed with dansylgalactosides when an electrochemical gradient (inside negative) was imposed or upon lactose efflux. These results provide evidence for a functionally symmetrical action of the lactose carrier, as was postulated in the chemiosmotic concept of solute transport.[8]

For more information about the molecular properties, isolation of carrier proteins is required. Several attempts have been made to solubilize and purify such proteins. Hirata et al.[96,100] solubilized an L-ala carrier from membranes of the thermophilic bacterium PS 3 with deoxycholate mixture and partially purified this protein by diethyl aminoethyl cellulose column chromatography and gel filtration. With this protein fraction, active transport could be restored to some extent in PS 3 phospholipid vesicles. Amanuma et al.[322] extracted a proline carrier from membranes of *E. coli* W1-1 with acetic *n*-butanol, partially purified this protein, and reconstituted with this protein fraction proline transport in *E. coli* phospholipid vesicles.

Altendorf et al.[323] used aprotic solvents such as hexamethyl phosphoric triamines, for the extraction of the lactose carrier from *E. coli* ML 308-225 and reconstituted lactose transport with this extract in transport negative *E. coli* ML 35 vesicles. In none of these investigations was more information about the molecular properties of the carriers supplied.

Such information, however, was obtained from studies on the purification and characterization of the alanine carrier of *B. subtilis* by Kusaka et al.[324] and Kusaka and Kanai.[97] A hydrophobic protein fraction was isolated by sucrose density gradient centrifugation of cytoplasmic membrane vesicles from *B. subtilis* grown on a complex medium with glucose. Treatment of this fraction with deoxycholate allowed the isolation of an alanine carrier which was purified to homogeneity. The carrier has a molecular weight of 7500, which seems to be suspiciously low when compared with the molecular weight of the lactose carrier.

Possibly, proteolytic activity present in the hydrophobic protein fraction might have degraded the original carrier protein. Each mole of carrier binds 1 mole of alanine with a dissociation constant of 0.2 μM. This binding is inhibited by *p*-chloromercuribenzoate, and this inhibition is reversed by dithiothreitol. This indicates an essential role of SH groups in the function of this carrier. Such evidence has been obtained for other carrier proteins.[317,325]

ACKNOWLEDGMENTS

The authors would like to thank Dr. K.J. Hellingwerf and Dr. J. Boonstra for their contributions to the text of this paper and for their constructive criticism. The help in the preparation of the manuscript from Mrs. M.Th. Broens-Erenstein and Mrs. M. Pras is greatly appreciated.

The studies performed in the laboratory of the authors were supported by the Netherlands Organization of Pure Scientific Research (Z.W.O.).

REFERENCES

1. **Konings, W.N. and Boonstra, J.**, Anaerobic electron transfer and active transport in bacteria, *Curr. Top Membr. Transp.*, 9, 177, 1977.
2. **Thauer, R.K., Jungermann, K., and Decker, K.**, Energy conservation in chemotrophic anaerobic bacteria, *Bacteriol. Rev.*, 41, 100, 1977.
3. **Kröger, A.**, Phosphorylative electron transport with fumarate and nitrate as terminal hydrogen acceptors, in *Microbial Energetics*, Haddock, B.A. and Hamilton, W.A., Eds., Cambridge University Press, Cambridge, 1977, 61.
4. **Haddock, B.A. and Jones, C.W.**, Bacterial respiration, *Bacteriol. Rev.*, 41, 47, 1977.
5. **Boonstra, J., Downie, A., and Konings, W.N.**, Energy supply for active transport in anaerobically grown *Escherichia coli*, *J. Bacteriol.*, 136, 844, 1978.
6. **Mitchell, P.**, Coupling of phosphorylation to electron and hydrogen transfer by a chemiosmotic type of mechanism, *Nature (London)*, 191, 144, 1961.
7. **Mitchell, P.**, Chemiosmotic coupling in oxidative and photosynthetic phosphorylation, *Biol. Rev. Cambridge Philos. Soc.*, 41, 445, 1966.
8. **Mitchell, P.**, *Chemiosmotic coupling and energy transduction*, Glynn Research Ltd., Bodmin, England, 1968.
9. **Simoni, R.D. and Postma, P.W.**, The energetics of bacterial active transport, *Annu. Rev. Biochem.*, 44, 523, 1975.
10. **Kaback, H.R.**, Molecular biology and energetics of membrane transport, *J. Cell. Physiol.* 89, 575, 1976.
11. **Hamilton, W.A.**, Energy coupling in substrate and group translocation, in *Microbial Energetics*, Haddock, B.A. and Hamilton, W.A., Eds., Cambridge University Press, Cambridge, 1977, 185.
12. **Harold, F.M.**, Membranes and energy transduction in bacteria, in *Current Topics in Bioenergetics*, Vol. 6, Rao Sanadi, D., Ed., Academic Press, New York, 1977, 84.
13. **Konings, W.N.**, Active transport of solutes in bacterial membrane vesicles, in *Advances in Microbial Physiology*, Vol. 15, Rose, A.H. and Tempest, D.W., Eds., Academic Press, London, 1977, 175.
14. **Rosen, B.P. and Kashket, E.R.**, Energetics of active transport, in *Bacterial Transport*, Rosen, B., Ed., Marcel Dekker, New York, 1978.
15. **Montal, M.**, Experimental membranes and mechanisms of bioenergy transductions, *Annu. Rev. Biophys. Bioeng.*, 5, 119, 1976.
16. **Rogers, H.J., Ward, J.B., and Burdett, I.D.J.**, Walls of Gram-positive bacteria, in *Relations between Structure and Function in the Prokaryotic Cell*, Stanier, R.Y., Rogers, H.J., and Ward, B.J., Eds., Cambridge University Press, Cambridge, 1978, 139.
17. **Decad, G.M. and Nikaido, H.**, Outer membrane of gram-negative bacteria. XII. Molecular-sieving function of cell wall, *J. Bacteriol.*, 128, 325, 1976.
18. **Braun, V.**, Structure-function relationships of the Gram-negative bacterial cell envelope, in *Relations Between Structure and Function in the Prokaryotic Cell*, Stanier, R.Y., Rogers, H.J., and Ward, B.J., Eds., Cambridge University Press, Cambridge, 1978, 111.
19. **Schuldiner, S. and Kaback, H.R.**, Fluorescent galactosides as probes for the lac carrier protein, *Biochim. Biophys. Acta*, 472, 399, 1977.
20. **Rudnick, G., Weil, R., and Kaback, H.R.**, Photoinactivation of the β-galactoside transport system in *Escherichia coli* membrane vesicles with an impermeant azidophenylgalactoside, *J. Biol. Chem.*, 250, 6847, 1975.
21. **Rudnick, G., Schuldiner, S., and Kaback, H.R.**, Equilibrium between two forms of the lac carrier protein in energized and non-energized membrane vesicles from *Escherichia coli*, *Biochemistry*, 15, 5126, 1976.
22. **Ramos, S. and Kaback, H.R.**, The relationship between the electrochemical proton gradient and active transport in *Escherichia coli* membrane vesicles, *Biochemistry*, 16, 854, 1977.
23. **Ramos, S. and Kaback, H.R.**, pH-Dependent changes in proton: substrate stoichiometries during active transport in *Escherichia coli* membrane vesicles, *Biochemistry*, 16, 4271, 1977.
24. **Rottenberg, H.**, The driving force for proton(s) metabolites cotransport in bacterial cells, *FEBS Lett.*, 66, 159, 1976.
25. **Skulachev, V.P.**, Membrane-linked energy buffering as the biological function of the Na^+/K^+ gradient, *FEBS Lett.*, 87, 171, 1978.
26. **Rottenberg, H.**, The measurement of transmembrane electrochemical proton gradients, *J. Bioenerg.*, 7, 61, 1975.
27. **Azzi, A. and Montecucco, C.**, Probes for energy transduction in membranes, *J. Bioenerg.*, 8, 257, 1976.

28. **Kell, D.B., John, P., Sorgato, M.C., and Ferguson, S.J.**, Continuous monitoring of the electrical potential across energy-transducing membranes using ion-selective electrodes: application to submitrochondrial particles and chromatophores, *FEBS Lett.*, 86, 294, 1978.
29. **Ramos, S., Schuldiner, S., and Kaback, H.R.**, The electrochemical gradient of protons and its relationship to active transport in *Escherichia coli* membrane vesicles, *Proc. Natl. Acad. Sci. U.S.A.*, 73, 1892, 1976.
30. **Grinius, L.L., Jasaitis, A.A., Kadziauskas, Y.P., Liberman, E.A., Skulachev, V.P., Topali, V.P., Tsofina, L.M., and Vladimorova, M.A.**, Conversion of biomembrane-produced energy into electric form. I. Submitochondrial particles, *Biochim. Biophys. Acta*, 216, 1, 1970.
31. **Bakeeva, L.E., Grinius, L.L., Jasaitis, A.A., Kuliene, V.V., Levitsky, D.O., Liberman, E.A., Severina, I.I., and Skulachev, V.P.**, Conversion of biomembrane-produced energy into electric form. II. Intact mitochondria, *Biochim. Biophys. Acta*, 216, 13, 1970.
32. **Jasaitis, A.A., Kuliene, V.V., and Skulachev, V.P.**, Anilinonaphtalenesulphonate fluorescence changes induced by non-enzymatic generation of membrane potential in mitochondrial and submitochrondrial particles, *Biochim. Biophys. Acta*, 234, 177, 1971.
33. **Jasaitis, A.A., van Chu, L., and Skulachev, V.P.**, Anilinonaphtalenesulphonate and other synthetic ions as mitochondrial membrane penetrants: an H^+ pulse technique study, *FEBS Lett.*, 31, 241, 1973.
34. **Waggoner, A.**, Optical probes of membrane potential, *J. Membr. Biol.*, 27, 317, 1976.
35. **Njus, D., Ferguson, S.J., Sorgato, M.C., and Radda, G.**, "The ANS Response". Eight years later, in *Structure and Function of Energy Transducing Membranes*, Vol. 14, van Dam, K. and van Gelder, B.F., Eds., Elsevier, Amsterdam, 1977, 237.
36. **Pick, U. and Avron, M.**, Measurement of transmembrane potentials in *Rhodospirillum rubrum* chromatophores with an oxacarbocyanine dye, *Biochim. Biophys. Acta*, 440, 189, 1976.
37. **Burckhardt, G.**, Non-linear relationship between fluorescence and membrane potential, *Biochim. Biophys. Acta*, 468, 227, 1977.
38. **Junge, W.**, Membrane potentials in photosynthesis, *Annu. Rev. Plant Physiol.*, 28, 503, 1977.
39. **Chance, B., Baltscheffsky, M., Vanderkooi, J., and Cheng, W.**, Localized and delocalized potentials in biological membranes, in *Perspectives in Membrane Biology*, Estrada-O, S. and Gitler, C., Eds., Academic Press, New York, 1974, 329.
40. **Jackson, J.B. and Crofts, A.R.**, The high energy state in chromatophores from *Rhodopseudomonas sphaeroides*, *FEBS Lett.*, 4, 185, 1969.
41. **Baccarini-Melandri, A., Casadio, R., and Melandri, B.A.**, Thermodynamics and kinetics of photophosphorylation in bacterial chromatophores and their relation with the transmembrane electrochemical potential difference of protons, *Eur. J. Biochem.*, 78, 389, 1977.
42. **Dijkema, C., Michels, P.A.M., and Konings, W.N.**, Light-induced spectral changes of carotenoids in chromatophores of *Rhodopseudomonas sphaeroides* Archives Biochem. Biophys., in press 1980.
43. **Mitchell, P. and Moyle, J.**, Estimation of membrane potential and pH difference across the cristae membrane of rat-liver mitochondria, *Eur. J. Biochem.*, 7, 471, 1969.
44. **Renthal, R. and Lanyi, J.K.**, Light-induced membrane potential and pH gradient in *Halobacterium halobium* envelope vesicles, *Biochemistry*, 15, 2136, 1976.
45. **Michels, P.A.M. and Konings, W.N.**, The electrochemical proton gradient generated by light in membrane vesicles and chromatophores from *Rhodopseudomonas sphaeroides*, *Eur. J. Biochem.*, 85, 147, 1978.
46. **Schuldiner, S., Rottenberg, H., and Avron, M.**, Determination of ΔpH in chloroplasts. II. Fluorescent amines as a probe for the determination of ΔpH in chloroplasts, *Eur. J. Biochem.*, 25, 64, 1972.
47. **Kraayenhof, R., Brocklehurst, J.R., and Lee, C.P.**, Fluorescent probes for the energized state in biological membranes, in *Biological fluorescence concepts*, Vol. 2, Chen, R.F. and Edelhoch, H., Eds., Marcel Dekker, New York, 1976, 767.
48. **Elema, R.P., Michels, P.A.M., and Konings, W.N.**, Response of 9-aminoacridine fluorescence to transmembrane pH-gradients in chromatophores from *Rhodopseudomonas sphaeroides*, *Eur. J. Biochem.*, 92, 381, 1978.
49. **Salhany, J.M., Yamane, T., Shulman, R.G., and Ogawa, S.**, High resolution P-31 nuclear magnetic resonance studies of intact yeast cells, *Proc. Natl. Acad. Sci. U.S.A.*, 72, 4966, 1975.
50. **Navon, G., Ogawa, S., Shulman, R.G., and Yamane, T.**, High-resolution ^{31}P nuclear magnetic resonance studies of metabolism in aerobic *Escherichia coli* cells, *Proc. Natl. Acad. Sci., U.S.A.*, 74, 888, 1977.
51. **Halpern, Y.S.**, Genetics of amino acid transport in bacteria, *Ann. Rev. Genet.*, 8, 103, 1974.
52. **Kaback, H.R. and Stadman, E.R.**, Proline uptake by an isolated cytoplasmic membrane preparation of *Escherichia coli*, *Proc. Natl. Acad. Sci. U.S.A.*, 55, 920, 1966.
53. **Kaback, H.R.**, Bacterial Membranes, in *Methods in Enzymology*, Vol. 22, Jakoby, W.B., Ed., Academic Press, New York, 1971, 99.

54. **Konings, W.N.**, Energization of solute transport in membrane vesicles from anaerobically grown bacteria., in *Methods in Enzymology*, Fleischer, S. and Packer, L., Eds., Academic Press, New York, Vol. 56, 370-388, 1979.
55. **Stinnet, J.D., Guymon, L.F., and Eagon, R.G.**, A novel technique for the preparation of transport-active membrane vesicles from *Pseudomonas aeruginosa:* observations on gluconate transport, *Biochem. Biophys. Res. Commun.*, 52, 284, 1973.
56. **Short, S.A., White, D.C., and Kaback, H.R.**, Active transport in isolated bacterial membrane vesicles. V. The transport of amino acids by membrane vesicles prepared from *Staphylococcus aureus, J. Biol. Chem.*, 247, 298, 1972.
57. **Short, S.A., White, D.C., and Kaback, H.R.**, Mechanisms of active transport in isolated bacterial membrane vesicles. IX. The kinetics and specificity of amino acid transport in *Staphylococcus aureus* membrane vesicles, *J. Biol. Chem.*, 247, 7452, 1972.
58. **Konings, W.N. and Kaback, H.R.**, Mechanisms of active transport in isolated bacterial membrane vesicles, XVII. Anaerobic transport in *Escherichia coli* membrane vesicles, *Proc. Natl. Acad. Sci. U.S.A.*, 70, 3376, 1973.
59. **Hellingwerf, K.J., Michels, P.A.M., Dorpema, J.W., and Konings, W.N.**, Transport of amino acids in membrane vesicles of *Rhodopseudomonas sphaeroides* energized by respiratory and cyclic electron flow, *Eur. J. Biochem.*, 55, 397, 1975.
60. **Konings, W.N., Bisschop, A., Veenhuis, M., and Vermeulen, C.A.**, New procedure for the isolation of membrane vesicles of *Bacillus subtilis* and an electron microscopy study of their ultrastructure, *J. Bacteriol.*, 116, 1456, 1973.
61. **Barnes, E.M. and Kaback, H.R.**, Mechanisms of active transport in isolated membrane vesicles. I. The site of energy coupling between D-lactic dehydrogenase and β-galactoside transport in *Escherichia coli* membrane vesicles, *J. Biol. Chem.*, 246, 5518, 1971.
62. **Konings, W.N. and Freese, E.**, Amino acid transport in membrane vesicles of *Bacillus subtilis, J. Biol. Chem.*, 247, 2408, 1972.
63. **Kaback, H.R.**, Transport across isolated bacterial cytoplasmic membranes, *Biochim. Biophys. Acta*, 265, 367, 1972.
64. **van Heerikhuizen, H., Boekhout, M., and Witholt, B.**, Proline transport activity in *Escherichia coli* membrane vesicles of different buoyant densities, *Biochim. Biophys. Acta*, 470, 453, 1977.
65. **Weissbach, H., Thomas, E., and Kaback, H.R.**, Studies on the metabolism of ATP by isolated bacterial membranes: formation and metabolism of membrane-bound phosphatidic acid, *Arch. Biochem. Biophys.*, 147, 249, 1972.
66. **Thomas, E.L., Weissbach, H., and Kaback, H.R.**, Further studies on metabolism and phosphatidic acid of isolated *Escherichia coli* membrane vesicles, *Arch. Biochem. Biophys.*, 150, 797, 1972.
67. **Thomas, E.L., Weissbach, H., and Kaback, H.R.**, Studies on the metabolism of ATP by isolated bacterial membranes: solubilization and phosphorylation of a protein component of the diglyceride kinase system, *Arch. Biochem. Biophys.*, 157, 327, 1973.
68. **Cox, G.S., Thomas, E., Kaback, H.R. and Weissbach, H.**, Synthesis of cyclopropane fatty acids in isolated bacterial membranes, *Arch. Biochem. Biophys.*, 158, 667, 1973.
69. **Joenje, H., Konings, W.N., and Venema, G.**, Interactions between exogenous deoxyribonucleic acid and membrane vesicles isolated from *Bacillus subtilis* 168, *J. Bacteriol.*, 119, 784, 1974.
70. **Joenje, H., Konings, W.N., and Venema, G.**, Interactions between exogenous deoxyribonucleic acid and membrane vesicles isolated from competent and noncompetent *Bacillus subtilis, J. Bacteriol.*, 121, 771, 1975.
71. **Costerton, J.W., Ingram, J.M., and Cheng, K.-J.**, Structure and function of the cell envelope of Gram-negative bacteria, *Bacteriol. Rev.*, 38, 87, 1974.
72. **Hare, J.F., Olden, K., and Kennedy, E.P.**, Heterogeneity of membrane vesicles from *E. coli* and their subfractionation with antibody to ATPase, *Proc. Natl. Acad. Sci. U.S.A.*, 71, 4843, 1974.
73. **Futai, M.**, Orientation of membrane vesicles from *Escherichia coli* prepared by different procedures, *J. Membr. Biol.*, 15, 15, 1974.
74. **Short, S.A. and Kaback, H.R.**, Localization of D-lactate dehydrogenase in native and reconstituted *E. coli* membrane vesicles, *J. Biol. Chem.*, 250, 4291, 1975.
75. **Kaback, H.R.**, Transport studies in bacterial membrane vesicles. Cytoplasmic membrane vesicles devoid of soluble constituents catalyze the transport of many metabolites, *Science*, 186, 882, 1974.
76. **Altendorf, K.H. and Staehelin, L.A.**, Orientation of membrane vesicles from *E. coli* as detected by freeze-cleave electron microscopy, *J. Bacteriol.*, 117, 888, 1974.
77. **Futai, M. and Tanaka, Y.**, Localization of D-lactate dehydrogenase in membrane vesicles prepared by using a French press or ethylenediamine-tetraacetate-lysozyme from *Escherichia coli, J. Bacteriol.*, 124, 470, 1975.
78. **Short, S.A., Kaback, H.R., and Kohn, L.D.**, Localization of D-lactate dehydrogenase in native and reconstituted *Escherichia coli* membrane vesicles, *J. Biol. Chem.*, 250, 4291, 1975.

79. **Owen, P. and Kaback, H.R.**, Molecular structure of membrane vesicles from *Escherichia coli, Proc. Natl. Acad. Sci. U.S.A.*, 75, 3148, 1978.
80. **Jones, O.T.G.**, Photosynthetic bacteria, in *Microbial Energetics*, Haddock, B.A. and Hamilton, W.A., Eds., Cambridge University Press, Cambridge, 1977, 151.
81. **Oelze, J. and Drews, G.**, Membranes of photosynthetic bacteria, *Biochim. Biophys. Acta*, 265, 209, 1972.
82. **Michels, P.A.M. and Konings, W.N.**, Structural and functional properties of chromatophores and membrane vesicles from *Rhodopseudomonas sphaeroides, Biochim. Biophys. Acta*, 507, 353, 1978.
83. **Herzberg, E.L. and Hinkle, P.C.**, Oxidative phosphorylation and proton translocation in membrane vesicles prepared from *Escherichia coli, Biochem. Biophys. Res. Commun.*, 58, 178, 1974.
84. **Rosen, B.P. and McClees, J.S.**, Active transport of calcium in inverted membrane vesicles of *E. coli, Proc. Natl. Acad. Sci. U.S.A.*, 71, 5042, 1974.
85. **Bangham, A.D., Hill, M.W., and Miller, N.G.A.**, Preparation and use of liposomes as model of biological membrane, in *Methods in Membrane Biology*, Vol. 1, Karn, E.D., Ed., Plenum Press, New York, 1974, 1.
86. **Kagawa, Y. and Racker, E.**, Partial resolution of the enzymes catalyzing oxidative phosphorylation. XXV. Reconstitution of vesicles catalyzing ^{32}Pi-adenosine triphosphate exchange, *J. Biol. Chem.*, 246, 5477, 1971.
87. **Racker, E.**, A new procedure for the reconstitution of biological active phospholipid vesicles, *Biochem. Biophys. Res. Commun.* 55, 224, 1973.
88. **Racker, E., Chien, T.F., and Kandrach, A.**, A cholate dilution procedure for the reconstitution of the Ca^{++}-pump, ^{32}Pi-ATP exchange and oxidative phosphorylation, *FEBS Lett.*, 57, 14, 1975.
89. **Sone, N., Yoshida, M., Hirata, H., and Kagawa, Y.**, Adenosine triphosphate synthesis by electrochemical proton gradient in vesicles reconstituted from purified adenosine triphosphatase and phospholipids of thermophilic bacterium, *J. Biol. Chem.*, 252, 2956, 1977.
90. **Racker, E. and Kandrach, A.**, Reconstitution of the third site of oxidative phosphorylation, *J. Biol. Chem.*, 246, 7069, 1971.
91. **Shertzer, H.G. and Racker, E.**, Reconstitution and characterization of the adenine nucleotide transporter derived from bovine heart mitochondria, *J. Biol. Chem.*, 251, 2446, 1976.
92. **Kagawa, Y.**, Reconstitution of the energy transformer, gate and channel subunit reassembly, crystalline ATPase and ATP synthesis, *Biochim. Biophys. Acta*, 505, 45, 1978.
93. **Hwang, S.-B. and Stoeckenius, W.**, Purple membrane vesicles: morphology and proton translocation, *J. Membr. Biol.*, 33, 325, 1977.
94. **Happe, M., Teather, R.M., Overath, P., Knobling, A., and Oesterhelt, D.**, Direction of proton translocation in proteoliposomes formed from purple membrane and acidic lipids depends on the pH during reconstitution, *Biochim. Biophys. Acta*, 465, 415, 1977.
95. **Hellingwerf, K.J., Scholte, B.J., and van Dam, K.**, Bacteriorhodopsin vesicles. An outline of the requirements for light-dependent H^+-pumping, *Biochim. Biophys. Acta*, 513, 66, 1978.
96. **Hirata, H., Sone, N., Yoshida, M., and Kagawa, Y.**, Solubilization and partial purification of alanine carrier from membranes of a thermophilic bacterium and its reconstitution into functional vesicles, *Biochem. Biophys. Res. Commun.*, 69, 665, 1976.
97. **Kusaka, I. and Kanai, K.**, Purification and characterization of alanine carrier isolated from H-protein of *Bacillus subtilis, Eur. J. Biochem.*, 83, 307, 1978.
98. **Racker, E. and Stoeckenius, W.**, Reconstitution of purple membrane vesicles catalyzing light-driven proton uptake and adenosine triphosphate formation, *J. Biol. Chem.*, 249, 662, 1974.
99. **Boyer, P.D., Chance, B., Ernster, L., Mitchell, P., Racker, E., and Slater, E.C.**, Oxidative phosphorylation and photophosphorylation, *Annu. Rev. Biochem.*, 46, 955, 1977.
100. **Hirata, H., Sone, N., Yoshida, M., and Kagawa, Y.**, Isolation of the alanine carrier from the membranes of a thermophilic bacterium and its reconstitution into vesicles capable of transport, *J. Supramol. Struct.*, 6, 77, 1977.
101. **Konings, W.N. and Freese, E.**, L-serine transport in membrane vesicles of *Bacillus subtilis* energized by NADH or reduced phenazine methosulfate, *FEBS Lett.*, 14, 65, 1971.
102. **Dietz, G.W.**, Dehydrogenase activity involved in the uptake of glucose-6-phosphate by a bacterial membrane system, *J. Biol. Chem.*, 247, 4561, 1972.
103. **Short, S.A. and Kaback, H.R.**, Mechanisms of active transport in isolated bacterial membrane vesicles. XIX. Further studies on amino acid transport in *Staphylococcus aureus* membrane vesicles, *J. Biol. Chem.*, 249, 4275, 1974.
104. **Kaczorowski, G., Shaw, L., Fuentes, M., and Walsh, C.**, Coupling of alanine racemase and D-alanine dehydrogenase to active transport of amino acids in *E. coli* B membrane vesicles, *J. Biol. Chem.*, 250, 2855, 1975.
105. **Barnes, E.M.**, Respiration-coupled glucose transport in membrane vesicles from *Azotobacter vinelandii, Arch. Biochem. Biophys.*, 152, 795, 1972.

106. **Kaback, H.R. and Milner, L.S.**, Relationship of a membrane-bound D-(−)-lactic dehydrogenase to amino acid transport in isolated bacterial membrane preparations *Proc. Natl. Acad. Sci. U.S.A.*, 66, 1008, 1970.
107. **Barnes, E.M. and Kaback, H.R.**, β-Galactoside transport in bacterial membrane preparations: energy coupling via membrane-bound D-lactic dehydrogenase, *Proc. Natl. Acad. Sci. U.S.A.*, 66, 1190, 1970.
108. **Smith, L.**, Cytochrome systems in aerobic electron transport, in *The Bacteria*, Vol. 2, Gunsalus, I. and Stanier, R.Y., Eds., Academic Press, New York, 1961, 365.
109. **Cox, G.B., Newton, N.A., Gibson, F., Snoswell, A.M., and Hamilton, J.A.**, The function of ubiquinone in *Escherichia coli*, *Biochem. J.*, 117, 551, 1970.
110. **Miki, K., Sekuzu, I., and Okunuki, K.**, Cytochromes of *Bacillus subtilis*. I. Cytochrome system in the particulate preparation, *Annu. Rep. Sci. Works Fac. Sci. Osaka Univ.*, 15, 33, 1967.
111. **Konings, W.N.**, Localization of membrane proteins in membrane vesicles of *Bacillus subtilis*, *Arch. Biochem. Biophys.*, 167, 570, 1975.
112. **Hirata, H., Asano, A., and Brodie, A.F.**, Respiration dependent transport of proline by electron transport from *Mycobacterium phlei*, *Biochem. Biophys. Res. Commun.*, 44, 368, 1971.
113. **Bisschop, A., Bergsma, J., and Konings, W.N.**, Site of interaction between phenazine methosulphate and the respiratory chain of *Bacillus subtilis*, *Eur. J. Biochem.*, 93, 369, 1979.
114. **Singh, A.P. and Bragg, P.D.**, Ascorbate-phenazine methosulfate dependent membrane energization in respiratory chain mutants of *Escherichia coli*, *Biochem. Biophys. Res. Commun.*, 72, 195, 1976.
115. **Matin, A. and Konings, W.N.**, Transport of lactate and succinate by membrane vesicles of *E. coli*, *Bacillus subtilis* and a *Pseudomonas* species, *Eur. J. Biochem.*, 34, 58, 1973.
116. **Futai, M.**, Stimulation of transport into *Escherichia coli* membrane vesicles by internally generated reduced nicotinamide adenine dinucleotide, *J. Bacteriol.*, 120, 861, 1974.
117. **Stroobant, P. and Kaback, H.R.**, Ubiquinone-mediated coupling of NADH dehydrogenase to active transport in membrane vesicles from *Escherichia coli*, *Proc. Natl. Acad. Sci. U.S.A.*, 72, 3970, 1975.
118. **Bisschop, A., Boonstra, J., Sips, H.J., and Konings, W.N.**, Respiratory chain linked ferricyanide reduction drives active transport in membrane vesicles from *Bacillus subtilis*, *FEBS Lett.*, 60, 11, 1975.
119. **Prince, R.C., Baccarini-Melandri, A., Hauska, G.A., Melandri, B.A., and Crofts, A.R.**, Assymmetry of an energy transducing membrane. The location of cytochrome c_2 in *Rhodopseudomonas sphaeroides* and *Rhodopseudomonas capsulata*, *Biochim. Biophys. Acta*, 387, 212, 1975.
120. **Dutton, P.L., Petty, K.M., Bonner, H.S., and Morse, S.D.**, Cytochrome c_2 and reaction centre of *Rhodopseudomonas sphaeroides* Ga membranes. Extinction coefficients, content, halfreduction potentials, kinetics and electric field alterations, *Biochim. Biophys. Acta*, 387, 536, 1975.
121. **Padan, E., Zilberstein, D., and Rottenberg, H.**, The proton electrochemical gradient in *Escherichia coli* cells, *Eur. J. Biochem.*, 63, 533, 1976.
122. **Guffanti, A.A., Susman, P., Blanco, R., and Krulwich, T.A.**, The protonmotive force and α-aminoisobutyric acid transport in an obligately alkalophilic bacterium, *J. Biol. Chem.*, 253, 708, 1978.
123. **Hsung, J.C. and Haug, A.**, Membrane potential of *Thermoplasma acidophila*, *FEBS Lett.*, 73, 47, 1977.
124. **Hsung, J.C. and Haug, A.**, Intracellular pH of *Thermoplasma acidophila*, *Biochim. Biophys. Acta*, 389, 477, 1975.
125. **Krulwich, T.A., Davidson, L.F., Filip, S.J. Jr., Zuckerman, R.S., and Guffanti, A.A.**, The protonmotive force and β-galactoside transport in *Bacillus acidocaldarius*, *J. Biol. Chem.*, 253, 4599, 1978.
126. **Collins, S.H. and Hamilton, W.A.**, Magnitude of the protonmotive force in respiring *Staphylococcus aureus* and *Escherichia coli*, *J. Bacteriol.*, 126, 1224, 1976.
127. **Ramos, S. and Kaback, H.R.**, The electrochemical proton gradient in *Escherichia coli* membrane vesicles, *Biochemistry*, 16, 848, 1977.
128. **Hirata, H., Altendorf, K.H., and Harold, F.M.**, Role of an electrical potential in the coupling of metabolic energy to active transport by membrane vesicles of *Escherichia coli*, *Proc. Natl. Acad. Sci. U.S.A.*, 70, 1804, 1973.
129. **Altendorf, K.H., Harold, F.M., and Simoni, R.D.**, Impairment and restoration of the energized state in membrane vesicles of a mutant of *Escherichia coli* lacking adenosine triphosphatase, *J. Biol. Chem.*, 249, 4587, 1974.
130. **Schuldiner, S. and Kaback, H.R.**, Membrane potentials and active transport in membrane vesicles from *Escherichia coli*, *Biochemistry*, 14, 5451, 1975.
131. **Michels, P.A.M.**, Light-Induced Transport Processes in Isolated Membranes of *Rhodopseudomonas sphaeroides*, Ph.D. thesis, University of Groningen, The Netherlands, 1978.
132. **Kaback, H.R. and Barnes, E.M.**, Mechanisms of active transport in isolated membrane vesicles. II. The mechanism of energy coupling between D-lactic dehydrogenase and β-galactosidase transport in membrane preparations from *Escherichia coli*, *J. Biol. Chem.*, 246, 5523, 1971.

133. **Lombardi, F.J. and Kaback, H.R.,** Mechanisms of active transport in isolated bacterial membrane vesicles. VIII. The transport of amino acids by membrane preparations from *Escherichia coli, J. Biol. Chem.*, 247, 7844, 1972.

134. **Lombardi, F.J., Reeves, J.P., and Kaback, H.R.,** Mechanisms of active transport in isolated bacterial membrane vesicles. XIII. Valinomycin-induced *Rubidium* transport, *J. Biol. Chem.*, 248, 3551, 1973.

135. **Konings, W.N., Barnes, E.M., and Kaback, H.R.,** Mechanisms of active transport in isolated membrane vesicles. III. The coupling of reduced phenazine methosulfate to the concentrative uptake of galactosides and amino acids, *J. Biol. Chem.*, 246, 5857, 1971.

136. **Konings, W.N. and Rosenberg, H.,** Phosphate transport in membrane vesicles from *Escherichia coli, Biochim. Biophys. Acta*, 508, 370, 1978.

137. **Harold, F.M.,** Ion currents and physiological functions in microorganisms, *Annu. Rev. Microbiol.*, 31, 181, 1977.

138. **Bisschop, A., and Konings, W.N.,** Reconstitution of reduced nicotinamide adenine dinucleotide oxidase activity with menadione in membrane vesicles from the menaquinone deficient *Bacillus subtilis* aroD, *Eur. J. Biochem.*, 67, 357, 1976.

139. **Farrand, S.K. and Taber, H.W.,** Pleiotropic menaquinone-deficient mutant of *Bacillus subtilis, J. Bacteriol.*, 115, 1021, 1973.

140. **Robertson, D.E., Tokuda, H., and Kaback, H.R.,** Influence of the electrochemical proton gradient on the lac carrier protein in *Escherichia coli, Fed. Proc. Fed. Am. Soc. Exp. Biol.*, 37, 1294, 1978.

141. **Postgate, J.R.,** Recent advances in the study of the sulfate-reducing bacteria, *Bacteriol. Rev.*, 29, 425, 1965.

142. **Barton, L.L., le Gall, J. and Peck, H.D.,** Oxidative phosphorylation in the obligate anaerobe *Desulfovibrio gigas*, in *Horizons of Bioenergetics*, Pietro, A.S. and Gest, H., Eds., Academic Press, New York, 1972, 33.

143. **le Gall, J. and Postgate, J.R.,** The physiology of sulphate reducing bacteria, in *Advances in Microbial Physiology*, Vol. 10, Rose, A.H. and Tempest, D.W., Eds., Academic Press, London, 1973, 81.

144. **de Groot, G.N. and Stouthamer, A.H.,** Regulation of reductase formation in *Proteus mirabilis*. III. Influence of oxygen, nitrate and azide on thiosulfate reductase and tetrathionate reductase formation, *Arch. Mikrobiol.*, 74, 326, 1970.

145. **Faust, P.J. and Vandemark, P.J.,** Phosphorylation coupled to NADH oxidation with fumarate in *Streptococcus faecalis* 10Cl., *Arch. Biochem. Biophys.*, 137, 392, 1970.

146. **Stouthamer, A.H.,** *Yield Studies in Microorganisms*, Meadowfield Press, Durham, England, 1976.

147. **Stouthamer, A.H.,** Energetic aspects of growth of microorganisms, in *Microbial Energetics*, Haddock, B.A. and Hamilton, W.A., Eds., Cambridge University Press, Cambridge, 1977, 285.

148. **Kröger, A.,** Fumarate as terminal acceptor of phosphorylative electron transport, *Biochim. Biophys. Acta*, 505, 129, 1978.

149. **Taniguchi, S., Sato, R., and Egami, F.,** The enzymatic mechanism of nitrate and nitrite metabolism in bacteria, in *Inorganic Nitrogen Metabolism*, McElroy, W.D. and Glass, B., Eds., Johns Hopkins Press, Baltimore, 1956, 87.

150. **Garland, P.B., Downie, J.A., and Haddock, B.A.,** Proton translocation and the respiratory nitrate reductase of *Escherichia coli, Biochem. J.*, 152, 547, 1975.

151. **Ruiz-Herrera, J. and de Moss, J.A.,** Nitrate reductase complex of *Escherichia coli* K12; Participation of specific formate dehydrogenase and cytochrome b_1 components in nitrate reduction, *J. Bacteriol.*, 99, 720, 1969.

152. **Enoch, H.G. and Lester, R.L.,** The purification and properties of formate dehydrogenase and nitrate reductase from *Escherichia coli, J. Biol. Chem.*, 250, 6693, 1975.

153. **de Moss, J.A.,** Limited proteolysis of nitrate reductase purified from membranes of *Escherichia coli, J. Biol. Chem.*, 252, 1696, 1977.

154. **Stiefel, E.I.,** Proposed molecular mechanism for the action of molybdenum in enzymes: coupled proton and electron transfer, *Proc. Natl. Acad. Sci. U.S.A.*, 70, 988, 1973.

155. **Bray, R.C.,** Molybdenum iron-sulfur flavin hydroxylases and related enzymes, *Enzymes*, 12, 299, 1976.

156. **Ruiz-Herrera, J., Showe, M.K. and DeMoss, J.A.,** Nitrate reductase complex of *Escherichia coli* K12; isolation and characterization of mutants unable to reduce nitrate, *J. Bacteriol.*, 97, 1291, 1969.

157. **Haddock, B.A., Downie, J.A., and Garland, P.B.,** Kinetic characterization of the membrane bound cytochromes of *Escherichia coli* under a variety of conditions by using a stopped-flow dual-wavelength spectrophotometer, *Biochem. J.*, 154, 285, 1976.

158. **Kemp, M.B., Haddock, B.A., and Garland, P.B.,** Synthesis and sidedness of membrane-bound respiratory nitrate reductase (EC 1.7.99.4) in *Escherichia coli* lacking cytochromes, *Biochem. J.*, 148, 329, 1975.

159. **Mac Gregor, C.H.**, Biosynthesis of membrane-bound nitrate reductase in *Escherichia coli*: evidence for a soluble precursor, *J. Bacteriol.*, 126, 122, 1976.
160. **Clegg, R.A.**, Purification and some properties of nitrate reductase (EC 1.7.99.4) from *Escherichia coli* K12, *Biochem. J.*, 153, 533, 1976.
161. **Azoulay, E., Riviere, C., Giardano, G., Pommier, J., Denis, M., and Ducet, G.**, Participation of cytochrome b to the in vitro reconstitution of the membrane-bound formate-nitrate of *Escherichia coli* K12 and the possible role of sulfhydryl groups and temperature in the reconstitution process, *FEBS Lett.*, 79, 321, 1977.
162. **Enoch, H.G. and Lester, R.L.**, The role of a novel cytochrome b-containing nitrate reductase and quinone in the in vitro reconstitution of formate-nitrate reductase activity of *Escherichia coli*, *Biochem. Biophys. Res. Commun.*, 61, 1234, 1974.
163. **Mac Gregor, C.H.**, Anaerobic cytochrome b_1 in *Escherichia coli*: association with and regulation of nitrate reductase, *J. Bacteriol.*, 121, 1111, 1975.
164. **Jones, R.W. and Garland, P.B.**, Sites and specificity of the reaction of bipyridylium compounds with anaerobic respiratory enzymes of *Escherichia coli*. Effects of permeability barriers imposed by the cytoplasmic membrane, *Biochem. J.*, 164, 199, 1977.
165. **Boxer, D.H. and Clegg, R.A.**, A transmembrane-location for the proton-translocating reduced ubiquinone-nitrate reductase segment of the respiratory chain of *Escherichia coli*, *FEBS Lett.*, 60, 54, 1975.
166. **Mac Gregor, C.H. and Christofer, A.R.**, Asymmetric distribution of nitrate reductase subunits in the cytoplasmic membrane of *Escherichia coli*. Evidence derived from surface labelling studies with transglutaminase, *Arch. Biochem. Biophys.*, 185, 204, 1978.
167. **Jones R.W.**, Proton-Translocating Nitrate Reductase of *Escherichia coli*, Ph. D. thesis, University of Dundee, Scotland, 1978.
168. **Garland, P.B., Clegg, R.A., Boxer, D.H., Downie, J.A., and Haddock, B.A.**, Proton-translocating nitrate reductase of *Escherichia coli*, in *Electron transfer chains and oxidative phosphorylation*, Quagliariello, E., Papa, S., Palmieri, F., Slater, E.C., and Siliprandi, N., Eds., North-Holland — American Elsevier, Amsterdam, 1975, 351.
169. **Boonstra, J., Sips, H.J., and Konings, W.N.**, Active transport by membrane vesicles from anaerobically grown *Escherichia coli* energized by electron transfer to ferricyanide and chlorate, *Eur. J. Biochem.*, 69, 35, 1976.
170. **Yamamoto, I. and Ishimoto, M.**, Anaerobic growth of *Escherichia coli* on formate by reduction of nitrate, fumarate, and trimethylamine *N*-oxide, *Z. Allg. Mikrobiol.*, 17, 235, 1977.
171. **Ota, A., Yamanaka, T., and Okuniki, K.**, Oxidative phosphorylation coupled with nitrate respiration. II. Phosphorylation coupled with anaerobic nitrate reduction in a cell-free extract of *Escherichia coli*, *Biochem. J.*, 55, 131, 1964.
172. **Boonstra, J. and Konings, W.N.**, Generation of an electrochemical proton gradient by nitrate respiration in membrane vesicles from anaerobically grown *Escherichia coli*, *Eur. J. Biochem.*, 78, 361, 1977.
173. **Haddock, B.A. and Kendall-Tobias, M.W.**, Functional anaerobic electron transport linked to the reduction of nitrate and fumarate in membranes from *Escherichia coli* as demonstrated by quenching of atebrin fluorescence, *Biochem. J.*, 152, 655, 1975.
174. **Boonstra, J. Huttunen, T., Kaback, H.R., and Konings, W.N.**, Anaerobic transport in *Escherichia coli* membrane vesicles, *J. Biol. Chem.*, 250, 6792, 1975.
175. **Konings, W.N., Boonstra, J., and de Vries, W.**, Amino acid transport in membrane vesicles of obligately anaerobic *Veillonella alcalescens*, *J. Bacteriol.*, 122, 245, 1975.
176. **Laanbroek, H.J., Lambers, J.T., de Vos, W.M., and Veldkamp, H.**, L-Aspartate fermentation by a free-living *Campylobacter* species, *Arch. Microbiol.*, 117, 109, 1978.
177. **Miki, K. and Lin, E.C.C.**, Enzyme complex which couples glycerol-3-phosphate dehydrogenation to fumarate reduction in *Escherichia coli*, *J. Bacteriol.*, 114, 767, 1973.
178. **Peck, H.D., Smith, O.H., and Gest, H.**, Comparitive biochemistry of the biological reduction of fumaric acid, *Biochim. Biophys. Acta*, 25, 142, 1957.
179. **Hirsch, C.A., Raminsky, M., Davis, B.D., and Lin, E.C.C.**, A fumarate reductase in *Escherichia coli* distinct from succinate dehydrogenase, *J. Biol. Chem.*, 238, 3770, 1963.
180. **Spencer, M.E. and Guest, J.R.**, Isolation and properties of fumarate reductase mutants of *Escherichia coli*, *J. Bacteriol.*, 114, 563, 1973.
181. **Amarasingham, C.R. and Davis, B.D.**, Regulation of α-ketoglutarate dehydrogenase formation in *Escherichia coli*, *J. Biol. Chem.*, 240, 3554, 1965.
182. **Gray, C.T., Wimpenny, J.W.T., Hudges, D.E., and Mossman, M.R.**, Regulation of metabolism in facultative bacteria. I. Structural and functional changes in *Escherichia coli* associated with shifts between the aerobic and anaerobic states, *Biochim. Biophys. Acta*, 117, 22, 1966.

183. **Singh, A.P. and Bragg, P.D.**, Reduced nicotinamide adenine dinucleotide-dependent reduction of fumarate coupled to membrane energization in a cytochrome deficient mutant of *Escherichia coli* K12, *Biochim. Biophys. Acta*, 396, 229, 1975.
184. **Miki, L. and Lin, E.C.C.**, Anaerobic energy-yielding reaction associated with transhydrogenation from glycerol-3-phosphate to fumarate by an *Escherichia coli* system, *J. Bacteriol.*, 124, 1282, 1975.
185. **Gutowski, S.J. and Rosenberg, H.**, Effects of dicyclohexylcarbodi-imide on proton translocation coupled to fumarate reduction in anaerobically grown cells of *Escherichia coli* K12, *Biochem. J.*, 160, 813, 1976.
186. **Gutowski, S.J. and Rosenberg, H.**, Proton translocation coupled to electron flow from endogenous substrates to fumarate in anaerobically grown *Escherichia coli* K12, *Biochem. J.*, 164, 265, 1977.
187. **Singh, A.P. and Bragg, P.D.**, Anaerobic transport of amino acids coupled to the glycerol-3-phosphate-fumarate oxidoreductase system in a cytochrome-deficient mutant of *Escherichia coli*, *Biochim. Biophys. Acta*, 423, 450, 1976.
188. **Macy, J., Kulla, H., and Gottschalk, G.**, H_2-dependent anaerobic growth of *Escherichia coli* on L-malate: succinate formation, *J. Bacteriol.* 125, 423, 1976.
189. **Berger, E.**, Different mechanisms of energy coupling for the active transport of proline and glutamine in *Escherichia coli*, *Proc. Natl. Acad. Sci. U.S.A.*, 70, 1514, 1973.
190. **Berger, E.A. and Heppel, L. A..**, Different mechanisms of energy coupling for shock sensitive and shock resistant amino acid permease of *Escherichia coli*, *J. Biol. Chem.*, 249, 7747, 1974.
191. **Brice, J.M., Law, J.F., Meyer, D.J., and Jones, C.W.**, Energy conservation in *Escherichia coli* and *Klebsiella pneumoniae*, *Biochem. Soc. Trans.*, 2, 523, 1974.
192. **Butlin, J.D.**, Ph.D. thesis, Australian University, Canberra City, Australia, 1973.
193. **Rosenberg, H., Cox, G.B., Butlin, J.D., and Gutowski, S.J.**, Metabolite transport in mutants of *Escherichia coli* K12 defective in electron transport and coupled phosphorylation, *Biochem. J.*, 146, 417, 1975.
194. **Parson, W.W.**, Bacterial Photosynthesis, *Annu. Rev. Microbiol.*, 28, 41, 1974.
195. **Cogdell, R.J., Parson, W.W., and Kerr, M.**, The type, amount, location and energy transfer properties of the carotenoid in reaction centers from *Rhodopseudomonas sphaeroides*, *Biochim. Biophys. Acta*, 430, 83, 1976.
196. **Parson, W.W. and Cogdell, R.J.**, The primary photochemical reaction of bacterial photosynthesis, *Biochim. Biophys. Acta*, 416, 105, 1975.
197. **Dutton, P.L. and Wilson, D.F.**, Redox potentiometry in mitochondrial and photosynthetic bioenergetics, *Biochim. Biophys. Acta*, 346, 165, 1974.
198. **Prince, R.C. and Dutton, P.L.**, The primary acceptor of bacterial photosynthesis: its operating midpoint potential?, *Arch. Biochem. Biophys.*, 172, 329, 1976.
199. **Rockley, M.G., Windsor, M.W., Cogdell, R.J., and Parson, W.W.**, Picosecond detection of an intermediate in the photochemical reaction of photosynthesis, *Proc. Natl. Acad. Sci. U.S.A.*, 72, 2251, 1975.
200. **Fajer, J., Brune, D.C., Davis, M.S., Forman, A., and Spaulding, L.D.**, Primary charge separation in bacterial photosynthesis — oxidized chlorophylls and reduced pheophytin, *Proc. Natl. Acad. Sci. U.S.A.*, 72, 4956, 1975.
201. **Kaufman, K.J., Petty, K.M., Dutton, P.L., and Rentzepis, P.M.**, Picosecond kinetics in reaction centers of *Rhodopseudomonas sphaeroides*, and the effects of ubiquinone extraction and reconstitution, *Biochem. Biophys. Res. Commun.*, 70, 839, 1976.
202. **Petty, K.M. and Dutton, P.L.**, Properties of the flash induced proton binding encountered in membranes of *Rhodopseudomonas sphaeroides*: a functional pK on the ubisemiquinone?, *Arch. Biochem. Biophys.*, 172, 335, 1976.
203. **Petty, K.M., Jackson, J.B., and Dutton, P.L.**, Kinetics and stoichiometry of proton binding in *Rhodopseudomonas sphaeroides* chromatophores, *FEBS Lett.*, 84, 299, 1977.
204. **Prince, R.C. and Dutton, P.L.**, Single and multiple turnover reactions in the ubiquinone-cytochrome b-c_2 oxidoreductase of *Rhodopseudomonas sphaeroides*, *Biochim. Biophys. Acta*, 462, 731, 1977.
205. **Crofts, A.R., Crowther, D., Bowyer, J., and Tierney, G.V.**, Electron transport through the antimycin sensitive site in *Rhodopseudomonas capsulata*, in *Structure and Function of Energy-Transducing Membranes*, Vol. 14, van Dam, K. and van Gelder, B.F., Eds., Elsevier, Amsterdam, 1977, 139.
206. **Baccarini-Melandri, A. and Melandri, B.A.**, A role for ubiquinone-10 in the b-c_2 segment of the photosynthetic bacterial electron transport chain, *FEBS Lett.*, 80, 459, 1977.
207. **Prince, R.C., Bashford, C.L., Takamiya, K., van den Berg, W.H., and Dutton, P.L.**, Second order kinetics of the reduction of cytochrome c_2 by the ubiquinone cytochrome b-c_2 oxidoreductase of *Rhodopseudomonas sphaeroides*, *J. Biol. Chem.*, 253, 4137, 1978.
208. **Dutton, P.L. and Prince, R.C.**, Equilibrium and disequilibrium in the ubiquinone-cytochrome b-c_2 oxidoreductase of *Rhodopseudomonas sphaeroides*, *FEBS Lett.*, 91, 15, 1978.

209. **Mitchell, P.**, Protonmotive redoxmechanism of the cytochrome b-c_1 complex in the respiratory chain: protonmotive ubiquinone cycle, *FEBS Lett.*, 56, 1, 1975.
210. **Mitchell, P.**, The protonmotive Q cycle: a general formulation, *FEBS Lett.*, 59, 137, 1975.
211. **Feher, G. and Okamura, M.Y.**, Reaction centers from *Rhodopseudomonas sphaeroides, Brookhaven Symp. Biol.*, 28, 183, 1976.
212. **Dutton, P.L., Petty, K.M., Prince, R.C., and Cogdell, R.J.**, Mechanisms of membrane electron and proton transfer, in *Molecular Aspects of Membrane Phenomena*, Kaback, H.R., Neurath, H., Radda, G.K., Schwyzer, R., and Wiley, W.R., Eds., Springer-Verlag, Berlin, 1975, 278.
213. **Duysens, L.N.M., van Grondelle, R., and del Valle-Tascón, S.**, Electron transport and photophosphorylation associated with primary reactions in purple bacteria, in *Photosynthesis '77*, Proc. 4th Int. Congr. Photosynthesis, Hall, D.O., Coombs, J. and Goodwin, T.W., Eds., The Biochemical Society, London, 1977, 173.
214. **Chance, B., Crofts, A.R., Nishimura, M., and Price, B.**, Fast membrane H^+ binding in the light-activated state of *Chromatium* chromatophores, *Eur. J. Biochem.*, 13, 364, 1970.
215. **Barouch, I. and Clayton, R.K.**, Ubiquinone reduction and proton uptake by chromatophores of *Rhodopseudomonas sphaeroides* R-26. Periodicity of two in consecutive light flashes, *Biochim. Biophys. Acta*, 462, 785, 1977.
216. **Jackson, J.B. and Dutton, P.L.**, The kinetic and redox potentiometric resolution of the carotenoid shift in *Rhodopseudomonas sphaeroides* chromatophores: their relationship to electric field alterations in electron transport and energy coupling, *Biochim. Biophys. Acta*, 325, 102, 1973.
217. **Barsky, E.L. and Samuilov, V.D.**, Absorption changes of carotenoids and bacteriochlorophyll in energized chromatophores of *Rhodospirillum rubrum*, *Biochim. Biophys. Acta*, 325, 454, 1973.
218. **Barsky, E.L., Bonch-Osmolovskaya, E.A., Ostroumov, S.A., Samuilov, V.D. and Skulachev, V.P.**, A study on the membrane potential and pH gradient in chromatophores and intact cells of photosynthetic bacteria, *Biochim. Biophys. Acta*, 387, 388, 1975.
219. **Rottenberg, H.**, The proton electrochemical potential and active transport in bacterial cells, in *The Proton and Calcium Pumps*, Azzone, G.F., Avron, M., Metcalfe, J.C., Quagliarello, E., and Siliprandi, N., Eds., Elsevier/North-Holland Biomedical Press, Amsterdam, 1978, 125.
220. **Schuldiner, S., Padan, E., Rottenberg, H., Gromet-Elhanan, Z., and Avron, M.**, ΔpH and membrane potential in bacterial chromatophores, *FEBS Lett.*, 49, 174, 1974.
221. **Leiser, M. and Gromet-Elhanan, Z.**, Comparison of the electrochemical proton gradient and phosphate potential maintained by *Rhodospirillum rubrum* chromatophores in the steady state, *Arch. Biochem. Biophys.*, 178, 79, 1977.
222. **Kell, D.B., Ferguson, S.J., and John, P.**, Measurement by a flow dialysis technique of the steady state protonmotive force in chromatophores from *Rhodospirillum rubrum*. Comparison with phosphorylation potential, *Biochim. Biophys. Acta*, 502, 111, 1978.
223. **Casadio, R., Baccarini-Melandri, A., Zannoni, D., and Melandri, A.B.**, Electrochemical proton gradient and phosphate potential in bacterial chromatophores, *FEBS Lett.*, 49, 203, 1974.
224. **Baccarini-Melandri, A., Casadio, R., and Melandri, B.A.**, Thermodynamics and kinetics of photophosphorylation in bacterial chromatophores and their relation with the transmembrane electrochemical potential difference of protons, *Eur. J. Biochem.*, 78, 389, 1977.
225. **Casadio, R., Baccarini-Melandri, A., and Melandri, B.A.**, On the determination of the transmembrane pH difference in bacterial chromatophores using 9-aminoacridine, *Eur. J. Biochem.*, 47, 121, 1974.
226. **Gibson, J.**, Uptake of C_4 dicarboxylates and pyruvate by *Rhodopseudomonas sphaeroides*, *J. Bacteriol.*, 123, 471, 1975.
227. **Jasper, P.**, Potassium transport system of *Rhodopseudomonas capsulata*, *J. Bacteriol.*, 133, 1314, 1978.
228. **Jasper, P. and Silver, S.**, Divalent cation transport systems of *Rhodopseudomonas capsulata*, *J. Bacteriol.*, 133, 1323, 1978.
229. **Rinehart, C.A. and Hubbard, J.S.**, Energy coupling in the active transport of proline and glutamate by the photosynthetic halophile, *Ectothiorhodospira halophila*, *J. Bacteriol.*, 127, 1255, 1976.
230. **Knaff, D.B.**, Active transport in the photosynthetic bacterium *Chromatium vinosum*, *Arch. Biochem. Biophys.*, 189, 225, 1978.
231. **Gromet-Elhanan, Z. and Leiser, M.**, Interchangeability of the membrane potential with the pH gradient in *Rhodospirillum rubrum* chromatopnores, *Arch. Biochem. Biophys.*, 159, 583, 1973.
232. **Oesterhelt, D. and Stoeckenius, W.**, Rhodopsin-like protein from the purple membrane of *Halobacterium halobium*, *Nature (London) New Biol.*, 233, 149, 1971.
233. **Blaurock, A.E. and Stoeckenius, W.**, Structure of the purple membrane, *Nature (London) New Biol.*, 233, 152, 1971.

234. **Oesterhelt, D. and Stoeckenius, W.**, Isolation of the cell membrane of *Halobacterium halobium* and its fractionation into red and purple membrane, in *Methods in Enzymology*, Vol. 31A, Fleischer, S. and Packer, L., Eds., Academic Press, New York, 1974, 667.
235. **Ovchinnikov, Y.A., Abdulaev, N.G., Feigina, M.Y., Kiselev, A.V., and Lobanov, N.A.**, Recent findings in the structure — functional characteristics of bacteriorhodopsin, *FEBS Lett.*, 84, 1, 1977.
236. **Long, M.M., Urry, D.W., and Stoeckenius, W.**, Circular dichroism of biological membranes: purple membrane of *Halobacterium halobium*, *Biochem. Biophys. Res. Commun.*, 75, 725, 1977.
237. **Henderson, R. and Unwin, P.N.T.**, Three-dimensional model of purple membrane obtained by electron microscopy, *Nature (London)*, 257, 28, 1975.
238. **Stoeckenius, W. and Lozier, R.H.**, Light energy conversion in *Halobacterium halobium*, *J. Supramol. Struct.*, 2, 769, 1974.
239. **Sherman, W.V. and Caplan, S.R.**, Chromophore mobility in bacteriorhodopsin, *Nature (London)*, 265, 273, 1977.
240. **Danon, A. and Stoeckenius, W.**, Photophosphorylation in *Halobacterium halobium*, *Proc. Natl. Acad. Sci. U.S.A.*, 71, 1234, 1974.
241. **Oesterhelt, D. and Stoeckenius, W.**, Functions of a new photoreceptor membrane, *Proc. Natl. Acad. Sci. U.S.A.*, 70, 2853, 1973.
242. **Hellingwerf, K.J., Arents, J.C., Scholte, B.J., and Westerhoff, H.V.**, Bacteriorhodopsin in liposomes. II. Experimental evidence in support of a theoretical model, *Biochim. Biophys. Acta*, 1979, 547, 561-582.
243. **Hellingwerf, K.J., Arents, J.C., and van Dam, K.**, Light-stimulated oxygen uptake by vesicles containing cytochrome c oxidase and bacteriorhodopsin, *FEBS Lett.*, 67, 164, 1976.
244. **Baker, E.P., Rottenberg, H., and Caplan, S.R.**, An estimation of the light-induced electrochemical potential difference of protons across the membrane of *Halobacterium halobium*, *Biochim. Biophys. Acta*, 440, 557, 1976.
245. **Michel, H. and Oesterhelt, D.**, Light-induced changes of the pH gradient and the membrane potential in *Halobacterium halobium*, *FEBS Lett.*, 65, 175, 1976.
246. **Kanner, B.I. and Racker, E.**, Light-dependent proton and rubidium translocation in membrane vesicles from *Halobacterium halobium*, *Biochem. Biophys. Res. Commun.*, 64, 1054, 1975.
247. **Mac Donald, R.E. and Lanyi, J.K.**, Light-induced leucine transport in *Halobacterium halobium* envelope vesicles: a chemi-osmotic system, *Biochemistry*, 14, 2882, 1975.
248. **Drachev, L.A., Jasaitis, A.A., Kaulen, A.D., Kondrashin, A.A., Liberman, E.A., Nemecek, I.B., Ostroumov, S.A., Semenov, A. Yu., and Skulachev, V.P.**, Direct measurement of electric current generation by cytochrome oxidase, H^+-ATPase and bacteriorhodopsin, *Nature (London)*, 249, 321, 1974.
249. **Shieh, P. and Packer, L.**, Photo-induced potentials across a polymer stabilized planar membrane, in the presence of bacteriorhodopsin, *Biochem. Biophys. Res. Commun.*, 71, 603, 1976.
250. **Kagawa, Y., Ohno, K., Yoshida, M., Takeuchi, Y., and Sone, N.**, Proton translocation by ATPase and bacteriorhodopsin, *Fed. Proc. Fed. Am. Soc. Exp. Biol.*, 36, 1815, 1977.
251. **Blok, M.C., Hellingwerf, K.J., Kaptein, R., and de Kruijff, B.**, Light-induced changes inside bacteriorhodopsin vesicles as measured by ^{31}P-N.M.R., *Biochim. Biophys. Acta*, 514, 178, 1978.
252. **Lanyi, J.K. and Mac Donald, R.E.**, Existence of electrogenic hydrogen ion/sodium ion antiport in *Halobacterium halobium* cell envelope vesicles, *Biochemistry*, 15, 4608, 1976.
253. **Eisenbach, M., Cooper, S., Garty, H., Johnstone, R.M., Rottenberg, H., and Caplan, S.R.**, Light-driven sodium transport in sub-bacterial particles of *Halobacterium halobium*, *Biochim. Biophys. Acta*, 465, 599, 1977.
254. **Belliveau, J.W. and Lanyi, J.K.**, Calcium transport in *Halobacterium halobium* envelope vesicles, *Arch. Biochem. Biophys.*, 186, 98, 1978.
255. **Wagner, G., Hartmann, R. and Oesterhelt, D.**, Potassium uniport and ATP synthesis in *Halobacterium halobium*, *Eur. J. Biochem.*, 89, 169, 1978.
256. **Mac Donald, R.E. and Lanyi, J.K.**, Light-activated amino acid transport in *Halobacterium halobium* envelope vesicles, *Fed. Proc. Fed. Am. Soc. Exp. Biol.*, 36, 1828, 1977.
257. **Helgerson, S.L. and Lanyi, J.K.**, Methionine transport in *Halobacterium halobium* vesicles: noncompetitive asymmetric inhibition by L-cysteine, *Biochemistry*, 17, 1042, 1978.
258. **Belliveau, J.W. and Lanyi, J.K.**, Analogies between respiration and a light-driven proton pump as sources of energy for active glutamate transport in *Halobacterium halobium*, *Arch. Biochem. Biophys.*, 178, 308, 1977.
259. **Bayley, S.T. and Morton, R.A.**, Recent developments in the molecular biology of extremely halophilic bacteria, *CRC Crit. Rev. Microbiol.*, 6, 151, 1978.
260. **Hubbard, J.S., Rinehart, C.A., and Baker, R.A.**, Energy coupling in the active transport of amino acids by bacteriorhodopsin containing cells of *Halobacterium halobium*, *J. Bacteriol.*, 125, 181, 1976.

261. **West, I.C. and Mitchell, P.**, Proton-coupled β-galactoside translocation in non-metabolizing *Escherichia coli*, *J. Bioenerg.*, 3, 445, 1972.
262. **West, I.C. and Mitchell, P.**, Stoicheiometry of lactose-H^+ symport across the plasma membrane of *Escherichia coli*, *Biochem. J.*, 132, 587, 1973.
263. **Harold, F.M. and Baarda, J.R.**, Effects of nigericin and monactin on cation permeability of *Streptococcus faecalis* and metabolic capacities of potassium-depleted cells, *J. Bacteriol.*, 96, 2025, 1968.
264. **Pavlasova, E. and Harold, F.M.**, Energy coupling in the transport of β-galactosides by *Escherichia coli*: effect of proton conductors, *J. Bacteriol.*, 98, 198, 1969.
265. **Asghar, S.S., Levin, E., and Harold, F.M.**, Accumulation of neutral amino acids by *Streptococcus faecalis*, *J. Biol. Chem.*, 248, 5225, 1973.
266. **Kashket, E.R. and Wilson, T.H.**, Galactoside accumulation associated with ion movements in *Streptococcus lactis*, *Biochem. Biophys. Res. Commun.*, 49, 615, 1972.
267. **Kashket, E.R. and Wilson, T.H.**, Proton-coupled accumulation of galactoside in *Streptococcus lactis* 7962, *Proc. Natl. Acad. Sci. U.S.A.*, 70, 2866, 1973.
268. **Niven, D.F. and Hamilton, W.A.**, Valinomycin-induced amino acid uptake by *Staphylococcus aureus*, *FEBS Lett.*, 37, 244, 1973.
269. **Booth, I.R. and Morris, J.G.**, Protonmotive force in the obligately anaerobic bacterium *Clostridium pasteurianum*: a role in galactose and gluconate uptake, *FEBS Lett.*, 59, 153, 1975.
270. **Flagg, J.L. and Wilson, T.H.**, Galactoside accumulation by *Escherichia coli*, driven by a pH gradient, *J. Bacteriol.*, 125, 1235, 1976.
271. **Niven, D.F. and Hamilton, W.A.**, Mechanisms of energy coupling to the transport of amino acids by *Staphylococcus aureus*, *Eur. J. Biochem.*, 44, 517, 1974.
272. **Burnell, J.N., John, P., and Whatley, F.R.**, The reversibility of active sulphate transport in membrane vesicles of *Paracoccus denitrificans*, *Biochem. J.*, 150, 527, 1975.
273. **Burnell, J.N., John, P., and Whatley, F.R.**, Phosphate transport in membrane vesicles of *Paracoccus denitrificans*, *FEBS Lett.*, 58, 215, 1975.
274. **Lancaster, J.R. and Hinkle, P.C.**, Studies of the β-galactoside transporter in inverted membrane vesicles of *Escherichia coli*. I. Symmetrical facilitated diffusion and proton gradient-coupled transport, *J. Biol. Chem.*, 252, 7657, 1977.
275. **Tsuchiya, T. and Rosen, B.P.**, Calcium transport driven by a proton gradient in inverted membrane vesicles of *Escherichia coli*, *J. Biol. Chem.*, 251, 962, 1976.
276. **Cecchini, G. and Koch, A.L.**, Energy coupling to transport of β-galactosides in *Escherichia coli*, *Abstr. Annu. Meet. Am. Soc. Microbiol.*, 193, 1974.
277. **Michels, P.A.M., Michels, J.P.J., Boonstra, J., and Konings, W.N.**, Generation of an electrochemical proton gradient in bacteria by the excretion of metabolic endproducts, *FEMS Microbiol. Lett.*, 5, 357, 1979.
278. **Stock, J. and Roseman, S.**, A sodium-dependent sugar co-transport system in bacteria, *Biochem. Biophys. Res. Commun.*, 44, 132, 1971.
279. **Frank, L. and Hopkins, I.**, Sodium stimulated transport of glutamate in *Escherichia coli*, *J. Bacteriol.*, 100, 329, 1969.
280. **Halpern, Y.S., Barash, H., Dover, S., and Druck, K.**, Sodium and potassium requirements for active transport of glutamate by *Escherichia coli* K12, *J. Bacteriol.*, 114, 53, 1973.
281. **Wong, P.T.S., Thompson, J., and MacLeod, R.A.**, Nutrition and metabolism of marine bacteria. XVII. Ion-dependent retention of α-aminoisobutyric acid and its relation to Na^+-dependent transport in a marine pseudomonad, *J. Biol. Chem.*, 244, 1016, 1969.
282. **Ariel, M. and Grossowicz, N.**, Enhancement of transport in *Micrococcus lysodeikticus* by sodium ions, *Biochim. Biophys. Acta*, 352, 122, 1974.
283. **Mac Donald, R.E., Lanyi, J.K., and Greene, R.V.**, Sodium-stimulated glutamate uptake in membrane vesicles of *Escherichia coli*: The role of ion gradients, *Proc. Natl. Acad. Sci. U.S.A.*, 74, 3167, 1977.
284. **Tsuchiya, T., Raven, J., and Wilson, T.H.**, Co-transport of Na^+ and methyl-β-D-thiogalactopyranoside mediated by the melibiose transport system of *Escherichia coli*, *Biochem. Biophys. Res. Commun.*, 76, 26, 1977.
285. **Tokuda, H. and Kaback, H.R.**, Sodium-dependent methyl 1-thio-β-D-galactopyranoside transport in membrane vesicles isolated from *Salmonella typhimurium*, *Biochemistry*, 17, 2130, 1977.
286. **Mac Donald, R.E., Greene, R.V., and Lanyi, J.K.**, Light-activated amino acid transport systems in *Halobacterium halobium* envelope vesicles: role of chemical and electrical gradients, *Biochemistry*, 16, 3227, 1977.
287. **Thompson, J. and MacLeod, R.A.**, Functions of Na^+ and K^+ in the active transport of α-aminoisobutyric acid in a marine pseudomonad, *J. Biol. Chem.*, 246, 4066, 1971.
288. **Willecke, K., Gries, E.M., and Oehr, P.**, Coupled transport of citrate and magnesium in *Bacillus subtilis*, *J. Biol. Chem.*, 248, 807, 1973.

289. **Boos, W.**, Pro and contra carrier protein; sugar transport via periplasmic galactose-binding protein, in *Curr. Top. Membr. Transp.*, 5, 51, 1975.
290. **Wolfson, E.B. and Krulwich, T.A.**, Requirement for a functional respiration-coupled D-fructose transport system for induction of phosphoenolpyruvate: D-fructose phosphotransferase activity, *Proc. Natl. Acad. Sci. U.S.A.*, 71, 1739, 1974.
291. **Wolfson, E.B., Sobel, M.E., Blanco, R., and Krulwich, T.A.**, Pathways of D-fructose transport in *Arthrobacter pyridinolis, Arch. Biochem. Biophys.*, 160, 440, 1974.
292. **Levinson, S.L. and Krulwich, T.A.**, Alternate pathways of L-ramnose transport in *Arthrobacter pyridinolis, Arch. Biochem. Biophys.*, 160, 445, 1974.
293. **Kaback, H.R. and Hong, J.S.**, Membranes and transport, *CRC Crit. Rev. Microbiol.*, 2, 333, 1973.
294. **Guymon, L.F. and Eagon, R.G.**, Transport of glucose, gluconate and methyl-α-D-glucoside by *Pseudomonas aeruginosa, J. Bacteriol.*, 117, 1261, 1974.
295. **Barnes, E.M.**, Multiple sites for coupling of glucose transport to the respiratory chain of membrane vesicles from *Azotobacter vinelandii, J. Biol. Chem.*, 248, 8120, 1973.
296. **Bisschop, A., Doddema, H., and Konings, W.N.**, Dicarboxylic acid transport in membrane vesicles from *Bacillus subtilis, J. Bacteriol.*, 124, 613, 1975.
297. **Fournier, R.E., McKillen, M.N., Pardee, A.B., and Willecke, K.**, Transport of dicarboxylic acids in *Bacillus subtilis*. Inducible uptake of L-malate, *B. Biol. Chem.*, 247, 5587, 1972.
298. **Ghei, O. K. and Kay, W. W.**, A dicarboxylic acid transport system in *Bacillus subtilis, FEBS Lett.*, 20, 137, 1972.
299. **Willecke, K. and Lange, R.**, C_4-dicarboxylate transport in *Bacillus subtilis* studied with 3-fluoro-L-erythro-malate as a substrate, *J. Bacteriol.*, 117, 373, 1974.
300. **Rayman, M.K., Lo, T.C.Y., and Sanwell, B.D.**, Transport of succinate in *Escherichia coli*. II. Characterization of uptake and energy coupling with transport in membrane preparations, *J. Biol. Chem.*, 247, 6332, 1972.
301. **Murakawa, S., Izaki, K., and Takabashi, H.**, A transport system for C_4-dicarboxylic acids in isolated membrane preparations from *Escherichia coli, Agric. Biol. Chem.*, 37, 1905, 1973.
302. **Dijkhuizen, L., Groen, L., Harder, W., and Konings, W.N.**, Active transport of oxalate by *Pseudomonas oxalaticus* OX1, *Arch. Microbiol.*, 115, 223, 1977.
303. **Harold, F.M. and Papineau, D.**, Cation transport and electrogenesis by *Streptococcus faecalis, J. Membr. Biol.*, 8, 27, 1972.
304. **Lanyi, J.K., Renthal, R. and Mac Donald, R.E.**, Light-induced glutamate transport in *Halobacterium halobium*. II. Evidence that the driving force is a light-dependent sodium gradient, *Biochemistry*, 15, 1603, 1976.
305. **Schuldiner, S. and Fishkes, H.**, Sodium-proton antiport in isolated membrane vesicles of *Escherichia coli, Biochemistry*, 17, 706, 1978.
306. **West, I.C. and Mitchell, P.**, Proton/sodium antiport in *Escherichia coli, Biochem. J.*, 144, 87, 1974.
307. **Silver, S. and Jasper, P.**, Manganese transport in microorganisms, in *Microorganisms and Minerals*, Weinberg, E.D., Ed., Marcel Dekker, New York, 1977, 105.
308. **Silver, S.**, Calcium transport in microorganisms, in *Microorganisms and Minerals*, Weinberg, E.D., Ed., Marcel Dekker, New York, 1977, 49.
309. **Jasper, P. and Silver, S.**, Magnesium transport in microorganisms, in *Microorganisms and Minerals*, Weinberg, E.D., Ed., Marcel Dekker, New York, 1977, 7.
310. **Bronner, F., Nash, W.E., and Golub, E.E.**, in *Spores* Vol. 6, Gerhardt, P., Sadoff, H.L., and Costillow, R.N., Eds., American Society for Microbiology, Washington, D.C., 1975, 356.
311. **Barnes, E.M.**, Respiration-coupled calcium transport by membrane vesicles from *Azotobacter vinelandii, Fed. Proc. Fed. Am. Soc. Exp. Biol.*, 33, 1475, 1974.
312. **Barnes, E.M., Jr., Roberts, P.R., and Bhattacharyya, P.**, Respiration-coupled calcium transport by membrane vesicles from *Azotobacter vinelandii, Membr. Biochem.*, 1, 73, 1978.
313. **Tsuchiya, T. and Rosen, B.P.**, Characterization of an active transport system for calcium in inverted membrane vesicles of *Escherichia coli, J. Biol. Chem.*, 250, 7687, 1975.
314. **Rosenberg, H., Gerdes, R.G., and Chegwidden, K.**, Two systems for the uptake of phosphate in *Escherichia coli, J. Bacteriol.*, 131, 505, 1977.
315. **Harold, F.M. and Spitz, E.**, Accumulation of arsenate, phosphate and aspartate by *Streptococcus faecalis, J. Bacteriol.*, 122, 266, 1975.
316. **Friedberg, I.**, Phosphate transport in *Micrococcus lysodeikticus, Biochim. Biophys. Acta*, 466, 451, 1977.
317. **Fox, C.F. and Kennedy, E.P.**, Specific labeling and partial purification of the M protein, a component of the β-galactoside transport systems of *Escherichia coli, Proc. Natl. Acad. Sci. U.S.A.*, 54, 891, 1965.
318. **Fox, C.F., Carter, J.R., and Kennedy, E.P.**, Genetic control of the membrane protein component of the lactose transport system of *Escherichia coli, Proc. Natl. Acad. Sci. U.S.A.*, 57, 698, 1967.

319. **Jones, T.H.D. and Kennedy, E.P.,** Characterization of the membrane protein component of the lactose transport system of *Escherichia coli, J. Biol. Chem.*, 244, 5981, 1969.
320. **Overath, P., Teather, R.M., Simoni, R.D., Aichele, G., and Wilhelm, U.,** Lactose carrier protein of *Escherichia coli.* Transport and binding of 2′-(*N*-dansyl) aminoethyl-β-D-thiogalactopyranoside and *p*-nitrophenyl-α-D-galactopyranoside, *Biochemistry*, 18, 1, 1979.
321. **Lancaster, J.R. and Hinkle, P.C.,** Studies of the β-galactoside transporter in inverted membrane vesicles of *Escherichia coli.* II. Symmetrical binding of a dansylgalactoside induced by an electrochemical proton gradient and by lactose efflux, *J. Biol. Chem.*, 252, 7662, 1977.
322. **Amanuma, H., Motojima, K., Yamaguchi, A., and Anraku, Y.,** Solubilization of a functionally active proline carrier from membranes of *Escherichia coli* with an organic solvent, *Biochem. Biophys. Res. Commun.*, 74, 366, 1977.
323. **Altendorf, K., Müller, C.R. and Sanderman, H. Jr.,** β-D-Galactoside transport in *Escherichia coli.* Reversible inhibition by aprotic solvents and its reconstitution in transport-negative membrane vesicles, *Eur. J. Biochem.*, 73, 545, 1977.
324. **Kusaka, I., Hayakawa, K., Kanai, K., and Fukui, S.,** Isolation and characterization of hydrophobic proteins (H proteins) in the membrane fraction of *Bacillus subtilis.* Involvement in membrane biosynthesis and the formation of biochemically active membrane vesicles by combining H proteins with lipids, *Eur. J. Biochem.*, 71, 451, 1976.
325. **Kaback, H.R. and Patel, L.,** The role of functional sulfhydryl groups in active transport in *Escherichia coli* membrane vesicles, *Biochemistry*, 17, 1640, 1978.
326. **Kaback, H.R., Rudnick, G., Schuldiner, S., Short, S.A., and Stroobant, P.,** Molecular aspects of active transport, in *The Structural Basis of Membrane Function*, Academic Press, New York, 1976, 107.
327. **Boonstra, J.,** Energy Supply for Active Transport in Anaerobically-Grown *Escherichia coli*, Ph. D. thesis, University of Groningen, The Netherlands, 1978.
328. **Boonstra, J. and Konings, W.N.,** unpublished results.

Chapter 3

TEMPORAL DIVERSITY OF BACTERIAL RESPIRATORY SYSTEMS: MEMBRANE AND RESPIRATORY DEVELOPMENT DURING THE CELL CYCLE

R. K. Poole

TABLE OF CONTENTS

ABSTRACT

Studies of the structure and function of energy-conserving membranes during the bacterial cell cycle are frustrated by methods of cell cycle analysis. Such methods are reviewed and their suitability for such studies discussed. Respiration rates of synchronous cultures generally exhibit marked discontinuities increasing in an oscillatory or stepwise fashion. The pattern of respiratory development may, in part, be influenced by the nature of the growth medium, but the molecular mechanism(s) of control remain to be identified. The possibility of control by adenylate pools is discussed. Proteins and lipids may be synthesized and incorporated into the membrane asynchronously so that fluctuations in the protein to lipid ratio of membranes occur during the cycle. Such fluctuations result in changing membrane density and fluidity, and may thus modulate the activity of membrane-bound enzymes. Activities of succinate, NADH dehydrogenases, and of adenosine triphosphate, and the amount of cytochrome b, all exhibit discontinuous increases in the cell cycles of various bacteria. Again, underlying control mechanisms remain obscure. Alternative models for the spatiotemporal growth of membranes, and their consequences for the segregation at cell division of respiratory chain components, are described. Localization of membrane growth, together with conservation of newly and previously synthesized membrane zones, is proposed to account for the observed unequal segregation of nitrate reductase and cytochromes at cell division.

ABBREVIATIONS

δ-ALA, δ-aminolaevulinic acid; ATCase, aspartate transcarbamylase (EC 2.1.3.2); ATPase, adenosine triphosphatase (EC 3.6.1.3); CCCP, carbonylcyanide *m*-chlorophenylhydrazone; DCCD, *NN*-dicyclohexylcarbodi-imide; and Nbf-Cl, 4-chloro-7-nitrobenzofurazan.

I. INTRODUCTION

"We are no longer satisfied with a good measurement or a good observation or the clever elucidation of some mechanism in the cell unless we can locate it on a time axis, and the time I refer to is not our time, but that of the cell itself, as expressed in its life history."[1]

That respiratory events are ordered in time as well as in space is implicit in other contributions in these volumes. The explicit aim of this chapter is to survey the change in respiratory activity and in the composition and organization of energy-conserving membranes that occur in a special period of time, the division cycle of the bacterial cell. The conceptual and experimental problems that this area of investigation pose appear to bridge two areas of microbial physiology that have evolved with only little interaction, namely bioenergetics and biogenesis.

I have chosen to view the growth of a bacterial membrane and the development of its respiratory functions as the result of an ordered temporal and spatial expression of genetic information in which four types of assembly reactions occur (Figure 1). Anderson[2] has used this scheme to define the locations for the input of genetic information during assembly of a supramolecular structure, but the scheme serves equally well to suggest an *order* of events that occurs in a temporally structured manner during the cell cycle. The synthesis of a membrane requires that genetic information be expressed during the synthesis (and perhaps modification, "activation") of the constituent proteins and lipids and during assembly. Attainment of a morphological "end-point" by these primary, secondary, and tertiary processes may be followed by a quartenary modification of the membrane so that it acquires its full functional activity in the cell.[3] Fluctuations in respiratory activity during the cell cycle may be due to this kind of modulation of membrane function to meet the changing energy demands of the cell, rather than to an expression of the synthesis of a component of the respiratory apparatus. Some studies, to be reported later, suggest that synthesis and assembly (i.e., incorporation of the component into the membrane) are virtually simultaneous. Nevertheless, the formal distinction is valuable since it may prompt a search for pools of synthesized, but unassembled, membrane components.

The basic unit of time, within which these reactions must be ordered, is the cell cycle. Therefore, the reactions must be completed within a cycle time if the cell's membrane complement is to be undiminished in successive cell cycles. Time "domains" within the cycle can be visualized[4,5] when the durations of reactions or other processes are shown as constituting a hierarchy on a scale which employs, as an expression of time magnitude, the common logarithm of time.[6] Thus, within the time "domain" of the cell cycle are domains of ever-decreasing time constants, namely genome replication, membrane assembly, and the reactions of intermediary metabolism, energy conservation, and electron transport. This chapter explores the current state of our knowledge and ignorance of the interrelationships between these domains.

It will become apparent that our knowledge is only rudimentary, particularly when compared with the spectacular advances that have been made in other areas covered in this volume. Recent progress in our understanding of bacterial respiration, coupled with a careful use of the techniques of cell-cycle research, should enable significant advances to be made in this area. We will start by examining the techniques used in cell-cycle research, partly because they are probably unfamiliar to many readers, and partly because many of the techniques appear to be incompatible with those used in bioenergetics.

II. METHODS OF CELL-CYCLE ANALYSIS

A. Preamble and Nomenclature

The methodology that distinguishes the experimental work described in this chapter from that reviewed by other contributors is the analysis of cell cycles. These techniques are the subject of much debate among workers in this field, and so, this section presents an overview of the methods available and a critical, if somewhat personal, evaluation of their potentials and limitations.

Much can be deduced about the cell cycle from measurements on cells growing asynchronously in a mixture whose age distribution is known,[7] most conveniently an exponential culture. However, most of our knowledge has arisen from studies of temporal organization in single cells or, much more commonly, of the amplification of these events in (1) cultures in which the growth and division of cells is synchronized, or (2) fractions separated from an exponentially growing culture, each fraction representing cells of a particular age or size class.

Despite the diversity of labels that the latter approach has acquired (e.g., culture fractionation,[8] fractionation by cell age,[9] zonal gradient sizing[10]), the centrifugal separations of bacteria reported so far all rely on fractionation of cells into size classes. The terminology used for synchronous cultures is (probably deceptively) more descriptive. James[11] classified the methods used for preparation of synchronous cultures into two, those employing "selection" and "induction". In the former, cells in a growing culture that are at a similar stage in their division cycle are selected from the remainder of the culture and grown up as a synchronous culture. In the latter method, all cells in a culture are induced to divide together by relieving the culture from some treatment that had previously blocked the progress of the cell cycle, thereby bringing all cells to the same position in the cycle. Earlier, Abbo and Pardee[12] had proposed the terms "synchronous" and "synchronized" to describe the cultures resulting, respectively, from selection and induction methods. This classification has not been rigidly adhered to in the literature and may be confusing. In the present review, the terms of James[11] are used.

Some classification appears necessary because of the continuing debate as to the relative merits of the two systems, and in particular, whether cultures made by selection rather than induction methods reflect more accurately the events occurring during the cell cycle in "normal, balanced growth".

B. The Quest for Balanced Growth

This concept was defined by Campbell[13] as follows: "Growth is balanced over a time interval if, during that interval, every extensive property* of the growing system increases by the same factor." This condition may be restated[14] as:

$$\frac{dx_1}{x_1 dt} = \frac{dx_2}{x_2 dt} = \ldots\ldots = \frac{dx_j}{x_j dt} = \alpha \qquad (1)$$

where the x_is are the extensive properties of the population of which there are a total of j. When α is a constant, the population is in a state of balanced exponential growth, although this is a special case of the Campbell definition. Experimentally, growth is said to be balanced when extensive properties of the population all increase at the same specific rate. In synchronous cultures, the temporal complexity of the cell cycle is revealed, and thus, in these cases, growth may be balanced only over intervals which are integral multiples of the doubling time. Commonly, this is assessed by measuring an extensive property (such as respiration rate) at successive mid-points of the doublings in cell numbers. An approximate doubling of the extensive property is then interpreted as representing at least partial fulfillment of the requirements for balanced growth. Inspection of the often complex patterns of change in such extensive properties, however, (Section III and Figures 2 to 4) show how difficult such an assessment may be. An alternative approach to assessing balanced growth exploits the conclusions of Painter and Marr[15] who showed that, if the distribution of intensive properties of a culture (e.g., cell size) is in a steady state, the system must be in a state of balanced exponential growth. In a synchronous culture, mean or modal cell volume is not invariant, but reflects the growth and division of the partially synchronized population. Again, balanced growth will be reflected in the constancy of mean volume measured at multiple integrals of the cycle time.

It has been argued[16] that balanced growth is not a requirement for providing an insight into the essential activities for division (e.g., respiration), since cells continue to divide in populations whose growth is patently unbalanced. An understanding of the distortions introduced in a developmental sequence by, say, induction synchronization can increase our knowledge of that sequence. Helmstetter also points out[16] that cells undergoing balanced exponential growth are in only one physiological state of interest. One may, with undiminished scientific merit, wish to study the cell cycle during a shift in growth conditions such as those that are implicit in certain induction procedures. However, what is clearly essential is that a synchronizing procedure, whether it be induction or selection, does not introduce unknown alterations in the

* "Extensive" properties refer to overall measures of population size, such as numbers, and biomass. "Intensive" properties refer to frequency distributions of age, size and chemical components within the population.

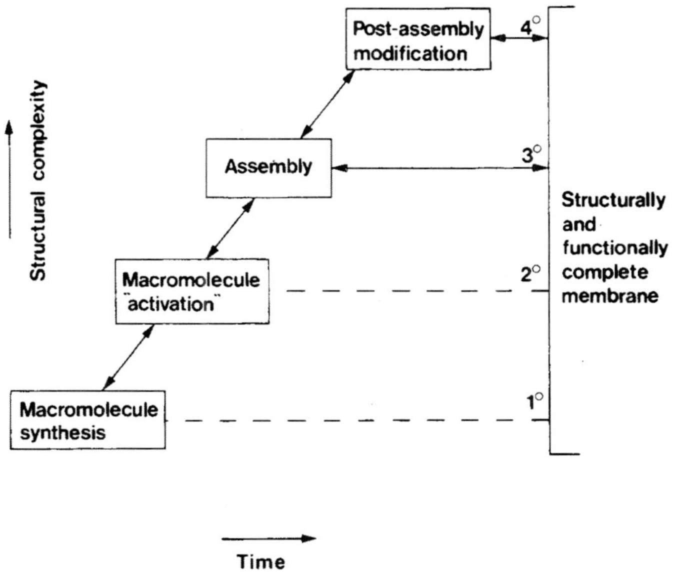

FIGURE 1. Spatiotemporal organization of membrane biogenesis. Membrane formation is depicted as the result of the assembly of previously synthesized and activated membrane components (proteins and lipids) and of further post-assembly modification. Dashed lines indicate that spontaneous assembly from macromolecules is unlikely and that expression of genetic information is required for assembly. Reversibility of the reactions are indicative only of membrane turnover and do not imply a mechanism (Modified from Anderson, R.G.W., *J. Theor. Biol.*, 67, 535, 1977. With permission.)

growth or physiology of the culture. Since it is far from clear what the side effects of "careful" selection methods are, the problems in interpreting data from induction synchronization must be formidable. Regrettably, few papers allow the reader to judge to what extent the growth of the culture is removed from "balanced growth", and even fewer describe adequate control experiments to test the effect of the synchronizing procedure on subsequent growth. Some synchronization methods preclude the possibility of performing such controls.

C. Synchrony Indices

Several indices[17] have been developed to quantitate the similarity of experimental "synchronous" cultures to the theoretical perfectly synchronized culture. This ideal theoretical culture would be one in which, at any given time, every cell was doing exactly the same thing.

Certain precautions in the use of synchrony indices have been referred to by Mitchison,[7] but two further points may be made. First, application of different indices to the same culture can result in quite different numerical values. Secondly, attention should be paid to those features of culture growth that an index can conceal. For example, the index of Blumenthal and Zahler[18] (which is mentioned here because it is used in Figures 2 to 4 and 7) pools two important characteristics of the culture. The index, F, is given by:

$$F = N/N_0 - 2^{t/g} \qquad (2)$$

where N and N_0 are, respectively, the numbers of cells $m\ell^{-1}$ after and before a "dou-

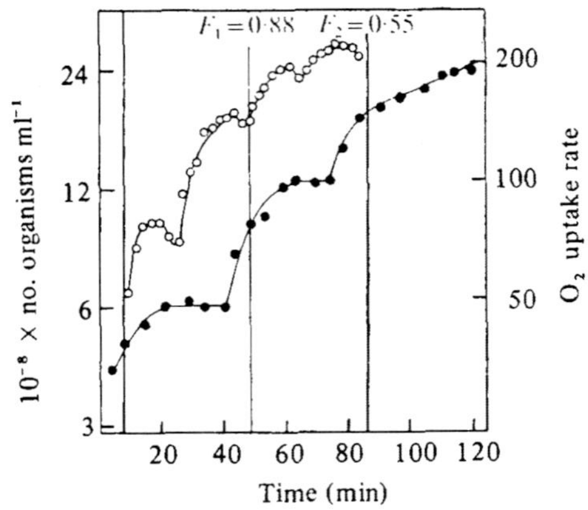

FIGURE 2. Oxygen uptake in a synchronous culture of a wild-type strain of *E. coli* K12 grown with glucose and casein hydrolysate. An exponential culture was passed through a continuous action rotor at 161 m*l* min^{-1}. Rotor speed was 15.9×10^3 r/min. Integrated force-time was 2.2×10^4 g-min at half-maximal radius in the rotor. The rotor effluent, containing 6.9% of the cell population in the original culture, was concentrated 13-fold, and 2 m*l* samples were cultured in open O_2 electrode vessels. ●, cell numbers; ○, O_2 uptake rates calculated at intervals of 2.5 min from the polarographic trace and expressed as ng-atom O/min/m*l*. F_1 and F_2 denote the synchrony indices calculated according to Blumenthal and Zahler of the first and second increases of cell numbers, respectively. Vertical lines indicate the midpoints of divisions. (From Poole, R.K., *J. Gen. Microbiol.*, 99, 369, 1977. With permission.)

bling instant'' of duration t. The mean generation time or cycle time is taken to be g. Clearly, identical values of F can result from two cultures in which the values of N/N_o and t/g are dissimilar. A culture in which, for example, only 70% of the population divides in each cycle, but which does so in a time that is very short relative to the cycle length, may give a satisfactory synchrony index. An identical index might be calculated for a culture in which all cells divide in each cycle, but over a longer period. Indeed, some published figures show division to occur in a time that is a very small fraction of the cycle time and incompatible with the known heterogeneity of generation times[10,15,19,20] observed for bacteria growing in exponential culture. This distribution of generation times would be expected to result in a progressive decay of synchrony in successive cycles,[21] and the use of synchrony indices can be valuable in checking this. Induction synchrony (Section II.D.2) often results in anomalous growth.[19] In one example,[22] t/g appears incompatible with accepted estimates of the variation in generation times; $N/N_o = 1.7$, and the degree of synchrony in the second cycle is considerably better than that in the first.

D. Preparation of Synchronous Cultures

1. Selection Synchrony

Excellent reviews already exist on the multitude of ingenious methods devised for selecting synchronous cultures of bacteria[7,16], so that Table 1 and what follows here

TABLE I

A Survey of Selection Methods for Preparing Synchronous Cultures of Bacteria

Principle of method	Application	Duration of selection	Yield	Perturbation?	Comments
Size selection by rate centrifugation through a stabilizing gradient, e.g., of sucrose[23-29], Ficoll[30], dextran[30], Ludox®[31]	*Escherichia coli,*[23,25,27,28,30-32] *Staphylococcus aureus*[24], *Bacillus subtilis,*[24] *Rhodopseudomonas palustris,*[29] *Streptococcus faecium*[26]	Typically 15[32] to 30 min[24]	Dependent on gradient capacity. Zonal rotors[27,28,31] give greatly increased yield	Probably little, at least in gradients of high mol wt	
Size selection by filtration, e.g., through filter paper[33], fiber filters[34] or glasswool and Ballotini beads[35]	*E. coli,*[35] *B. subtilis,*[34] *Rhodomicrobium vannielii,*[35] *Alcaligenes faecalis,*[36] *B. megaterium,*[16]	Yields of up to 10 ℓ in 15 min.[35] Other methods give lower yields and cannot readily be scaled up.[16]		?	
Selection by passage through continuous-flow centrifuge rotor, approximately 10% of culture allowed to escape harvest[37] [a]	*E. coli,*[38-42] *Alcaligenes eutrophus*[43,44]	Typically, 160 to 300 mℓ min^{-1} [40,44]		Assumed to be minimal for larger eukaryotic cells,[37] but much greater *g* required for bacteria	Controversy over age of selected cells[39,41,42]
Density selection by isopyonic centrifugation in a density gradient, e.g., of Ludox® polyvinylpyrrolidone[39] or Ficoll®[46]	*E. coli,*[39] *Lineola longa*[46]	60[39] to 90 min[46] centrifugation	Theoretically almost unlimited	Selected cells divide after lag and with increased cycle time.[39]	Not all cells show density fluctuations[10,42,47]
Age selection by elution of newly born cells from a culture bound to a membrane[16]	*E. coli K12,*[48] *E. coli* B/r[16]	Typically, yields 4 to 5 × 10^8 cells min^{-1}.[16] Methods of scale-up have been devised.[16]		?	
Differential centrifugation in growth medium (no gradient)	*Caulobacter crescentus*[49]	Three successive centrifugations[49] of 5 min each	Up to 6 ℓ[49]	?	

[a] Koch[45] has devised a flow-through rotor in which a sucrose cushion traps the harvested cells. The low speed of the rotor limits its use to preparing small-scale synchronous cultures.

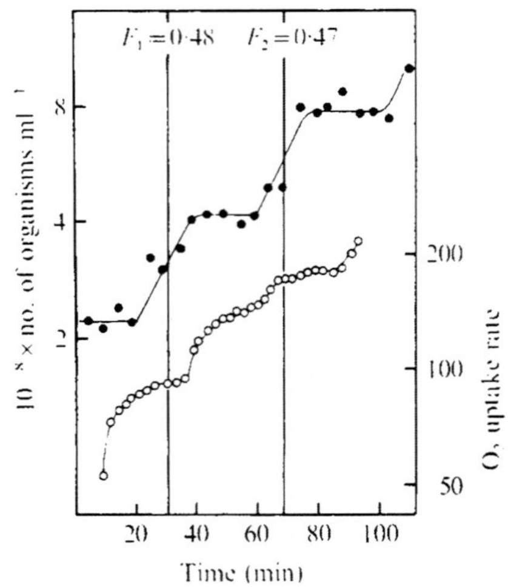

FIGURE 3. Oxygen uptake in a synchronous culture of a wild-type strain of *E. coli* K12 grown with glycerol and casein hydrolysate. An exponential culture was passed through a continuous action rotor at 170 m*l*/min. Rotor speed was 15×10^3 r/min. Integral force time was 1.85×10^4 g-min at half-maximl radius in the rotor. The rotor effluent, containing 8.6% of the cell population in the original culture, was concentrated 10-fold and 2 m*l* samples were cultured in open O_2 electrode vessels. ●, cell numbers; ○, O_2 uptake rates calculated at intervals of 2.4 min from the polarographic trace and expressed as ng atom O/min/m*l*. F_1 and F_2 denote the synchrony indices calculated according to Blumenthal and Zahler of the first and second increases of cell numbers, respectively. Vertical lines indicate the midpoints of division. (From Poole, R.K., *J. Gen. Microbiol.*, 99, 369, 1977. With permission.)

is restricted mainly to newer methods or to a discussion of their suitability for investigations of respiratory metabolism in the cell cycle.

Methods that use rate centrifugation (i.e., separation on the basis of differences in S - values) are widely used. An advantage of these methods is the ease with which control experiments can be performed to study the effects of exposure to potentially perturbing conditions such as the presence of gradient media, centrifugation and, perhaps, anaerobiosis. The resultant gradient fractions are mixed, and their growth after inoculation into medium is studied. Division synchrony has not been reported in controls of this type,[25] but recent disturbing results indicate that, for yeast, fluctuations in enzyme activity[50] or adenylate pools[51] occur even in the absence of division synchrony, after gradient selection,[50] or merely after short centrifugation and resuspension in growth medium.[51] Bellino[52] suggests that earlier results showing discontinuous increases in activity of ATCase during the bacterial cell cycle were artifacts of the synchronizing method. Clearly, similar further work is required.

Continuous flow centrifugation appears to overcome many of the disadvantages of gradient methods.[37] However, it suffers from the serious drawback that the design of meaningful controls is problematic. The selected cells do not experience the prolonged

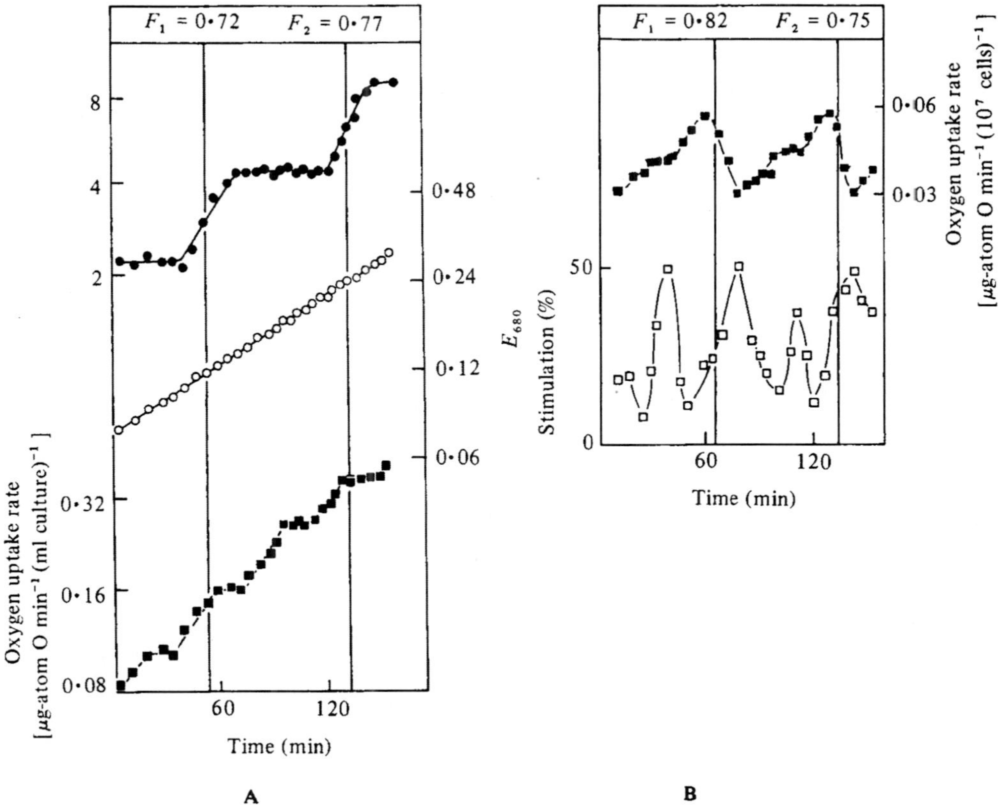

FIGURE 4. Growth and respiration of *Alcaligenes eutrophus* in synchronous culture. An exponential culture was passed through a continuous action rotor at 300 mℓ/min. Rotor speed was 17 × 10^3 r/min. The rotor effluent contained 3 to 4% of the population in the original culture. Oxygen uptake was measured by conventional polarographic methods on 2.4 mℓ samples removed from the culture. CCCP was added as a methanolic solution to the electrode vessel. In experiment (A), ●, cell numbers; ○, E_{680}; ■, oxygen uptake rate. In experiment (B), ■, specific oxygen uptake (expressed with respect to cell number); □, percentage specific stimulation of specific oxygen uptake following addition of 16 μ*M*-CCCP. F_1 and F_2 denote the synchrony indices calculated according to Blumenthal and Zahler of the first and second increases in cell numbers, respectively. Vertical lines indicate the midpoints of divisions. (From Edwards, C. and Jones, C.W., *J. Gen. Microbiol.*, 99, 383, 1977. With permission.)

centrifugation or pelleting that the other 90% do, so that mixing the two populations is not a truly relevant control. Allowing increasingly larger proportions of cells into the effluent, monitoring the degree of division synchrony, and checking for perturbation of the growth pattern that prevailed in the culture before centrifugation is an alternative approach.

There have been few studies of the physiological consequences of manipulations commonly used in synchronization procedures. Exposure of *E. coli* to sudden increases in osmotic pressure results in extrusion of putrescine (a polyamine) and K^+ uptake.[53] Chilling of cells eluted from a membrane results in reduced rates of acetate and leucine uptake at the beginning of synchronous growth.[54] In the absence of further evidence, experimental practice must be dictated largely by intuition.

2. Induction Synchrony

Less attention is paid here to induction methods, mainly because almost all lend themselves to scale-up and, thus, suitability for bioenergetic investigations. References to work published before 1971 are to be found in earlier reviews.[7,16] Techniques amenable to exploitation include amino acid starvation,[55] temperature shifts combined with centrifugation,[56] temperature shifts of a temperature-sensitive mutant,[57] repeated dilution of stationary phase cultures,[58] and "periodic"[59] or "phased"[60-62] addition of nutrient to a continuous[59] or semicontinuous[60-62] culture system.

E. Culture Fractionation by Centrifugation

The aim of culture fractionation is the direct separation of cells at different stages in the cell cycle, obviating the preparation of a synchronous culture. This author sympathizes with the view of Koch and Blumberg[10]: "We feel that the method of choice is zonal gradient sizing. When experiments are performed such that all critical events occur during undisturbed balanced growth, and then fractionation is achieved rapidly in a way that does not depend on the subsequent physiological behaviour of a perturbed cell population, then it seems to the authors that artefacts are minimal and experimental results amenable to clear and direct interpretation." Since it is believed[63] that the determinants in the normal bacterial cell cycle that control key events are more directly related to cell size than to age, then an experimental method that generates size classes will be *a priori* more tightly coupled to the event under study than a method based on age separation.

Adequate separations of this type have been achieved for *E. coli* on sucrose gradients[64,27] and in a zonal rotor prefilled with a uniform concentration (10%, w/v) of sucrose.[28] The maximum loading of cells in the latter case could probably be increased by using a gradient, thus avoiding local instability arising from a zone containing cells that exceeds the density of the underlying zone.[65] The separation of *E. coli* into size classes on cesium chloride gradients[66] is not suitable where cell viability after fractionation is required.[16] Exponential cultures of *Myxobacter* strain AL-1, but not certain other strains, can be fractionated on sucrose gradients without loss of viability.[67]

The main disadvantage of these methods, namely that successive cell cycles cannot be studied, has previously been discussed together with practical aspects of the approach.[9] In fact, the frequently rapid decay of synchrony in selection-synchronized bacterial cultures[19] makes the analysis of successive cycles in these cultures precarious. The outstanding advantage of the method is that almost all cells in the exponential culture, and not just a selected minority (commonly about 10%), are available for analysis. The method is, therefore, particularly suitable for analyses requiring large quantities of cells. Maximal resolution of cells from an exponentially growing culture does present a real challenge to separation technology, since the S-value of a rod-shaped cell such as *E. coli* varies less than twofold during a cell cycle.[10,41]

F. Culture Fractionation on the Basis of Cell Age

In addition to its use in preparing synchronous cultures, membrane elution can be used to study continuously the cells in a single age class (the youngest) in a population.[16] Cells from a culture, previously pulse-labeled with a precursor of the macromolecule under study, are bound to the filter. The radioactivity in eluted new-born cells then reflects the amount of label incorporated into their ancestors during the pulse. Further information is given by the method's innovator,[16] and an application of the technique is presented later (Section IV.A and Figure 6).

The relatively new instruments that allow sorting of populations on a cell-by-cell basis open up exciting possibilities for the analysis of bacterial cell cycles.[68]

III. DIVERSITY IN THE RATES OF RESPIRATION IN SYNCHRONOUS CULTURES

A. Continuous, Oscillatory, and Stepwise Increases in Respiration

The history of measurements of respiration rates as a function of the cell cycle is a rather long, if uneventful, one. In 1956, Maruyama used Warburg manometry to measure both O_2 uptake and CO_2 evolution in cultures of *E. coli* synchronized by filtration.[69] The rate of O_2 uptake increased smoothly during the first cell cycle with no evidence of discontinuity so as to approximately double during the cycle. Total cell nitrogen increased in a fashion indistinguishable from that of O_2 uptake. A transient, but reproducible, decrease in the rate of CO_2 evolution at about 0.5 of the cycle* resulted in an oscillation of the respiratory quotient.

In contrast to this simple pattern of respiratory development, recent results from two laboratories show that the respiration rates of *E. coli*, in cultures synchronized by continuous-flow selection, oscillate during the cell cycle. *E. coli* W1485 growing in a defined medium with alanine as carbon source exhibited respiratory oscillations with a periodicity of 23 to 40 min in a cycle time of about 100 min.[38]** The amplitude of the oscillations decreased over two successive cycles, even though the synchrony index remained high and constant over the first two division periods.

In strain K12, growing in a medium containing glucose and casein hydrolysate as carbon source, respiration rates oscillated with a periodicity of half a cycle (Figure 2).[40] Enhanced resolution of the oscillations was achieved by continuous monitoring of the O_2 tension in the culture and subsequent calculation of O_2 uptake rates at intervals of 2.5 min. The oscillations were damped, i.e., decreased in amplitude with time, and the synchrony index also decayed significantly in two successive divisions. Selection was performed when the mean cell volume in the exponential culture was constant,[40] indicating that, for a short period at least, growth of the culture was balanced. In the synchronous culture, the "overall" rate of O_2 uptake (i.e., after smoothing the oscillations) doubled, in two successive cycles, in a period very similar to the average cycle time.

Synchronous cultures of an ATPase-deficient mutant (A103c) also showed oscillating respiration rates with a periodicity of half a cycle, even though the length of the cycle was about 50% greater than that of the wild type.[40] This mutant is defective in the Mg^{2+}-Ca^{2+}-activated membrane-bound ATPase and, thus, oxidative phosphorylation. The persistence of oscillations was, therefore, interpreted as evidence that the oscillations of similar periodicity in cultures of the wild-type were not a reflection of in vivo respiratory control. Respiration rates of mitochondria depend on the level of ADP. State 4 is an aerobic ADP-limited state characterized by a low "resting" respiration rate, and state 3 is a state of active respiration and phosphorylation with adequate supplies of substrate and phosphate acceptor (ADP).[70] Such a means of regulation of respiratory rate, i.e., a cycle-dependent switch between state 3 and state 4, operates during the cell cycle of certain eukaryotic microorganisms.[51,71,72] Further evidence for the involvement, or otherwise, of in vivo respiratory control in synchronous bacterial cultures could come from measurements of intracellular adenine nucleotide pools during growth. Previous measurements on selection-synchronized *E. coli* B/r

* The timing of events during a cell cycle is expressed as that fraction of the cycle elapsed at the time of the event. The cell cycle is defined as the period between successive mid-points in the synchronous increase of cell numbers.

** Unfortunately the published figure[38] is erroneous. These data are those measured by the author from corrected results kindly supplied by Dr. Evans.

demonstrated oscillations of ATP during the cell cycle,[73] but ADP was not measured.

The pattern of respiration during the cell cycle appears to be influenced by the nature of the carbon and energy source. When *E. coli* K12 was grown in synchronous culture under conditions identical to those used in Figure 2, but with glycerol rather than glucose as carbon source, the results shown in Figure 3 were obtained. Respiratory activity increased in a number of "steps".[40] To measure the periodicity of the steps, the timing of the inflections between each plateau and the subsequent step was measured. With this convention, the periodicity of the steps observed in two separate experiments (one of which is shown in Figure 3) was 0.76 of a cell cycle (S.D. ± 0.11). These synchronous cultures were prepared from exponential cultures in which growth was not balanced, reflected in the continuously decreasing mean cell volume.[40] The specific growth rate in volume, and presumably mass, was lower than that in number density. Likewise, in the synchronous culture, the overall rate of increase in respiration rate was somewhat slower than that of cell numbers. It is thus tempting to speculate that the periodicity of steps would be close to 0.5 of a cycle (i.e., two steps per cycle) in a culture exhibiting balanced growth. A similar phenomenon in induction-synchronized cultures of yeast[74] has been interpreted as a dissociation of two cellular "clocks", those of "growth" (and respiration rate) and of cell division.

Periodic stepwise increases in respiration rates have also been observed in synchronous cultures of *Alcaligenes eutrophus* prepared by continuous-flow selection (Figure 4).[43] Synchrony indices were high, in one case higher in the second cell cycle than in the first. Respiration rates of untreated culture samples increased discontinuously during each of two successive cycles. Two steps per cycle were observed. The mid-point of the first occurred at about 0.4 of a cycle, that of the second at approx. 0.9 (Figure 4A). The effect on the specific respiration rate (O_2 uptake per min per cell) of adding the uncoupler CCCP to culture samples varied during the cycle. At those times when respiration was maximal, the degree of stimulation by uncoupler was minimal. When respiration rates were minimal, the converse was true (Figure 4B). Similar fluctuations in sensitivity to uncoupler occur in synchronous cultures of yeast and amoeba.[71,72]

There is scope for much improvement in the design and execution of these experiments. Rarely have appropriate control experiments been performed to demonstrate that the respiratory fluctuations are not a consequence of the synchronizing procedure. Mixing pelleted cells in the continuous flow rotor with the effluent is probably a poor control (Section II.D.1.). In the one reported attempt at this control experiment,[40] the resulting culture did not show division synchrony or oscillations in respiration rates. However, respiratory oscillations have been reported after inoculation of stationary phase cells into fresh medium even though cell division was not synchronized.[40] Had division been synchronized, these results might have been interpreted as demonstrating cell-cycle-specific discontinuities in respiration rate. Measurements of respiration rates do not demand large quantities of cells, and so the whole gamut of selection and induction methods awaits exploitation. Culture fractionation methods are probably unsuitable for measurements of those events, like respiratory rates, which are under rapid control because of the inevitable delay between harvesting and analysis of the fractionated population.

B. High Frequency Oscillations in Continuous Culture

The results of Harrison[75] are mentioned here because, although not explicity performed on synchronous cultures, they reveal a different domain of temporal organization of bacterial respiration. Damped, high-frequency oscillations were observed in pyridine nucleotide fluorescence after a chemostat culture of *Klebsiella aerogenes* had been subjected to an anaerobic shock (Figure 5A). After repeated anaerobic shocks, undamped oscillations arose which persisted for several days if the culture was left

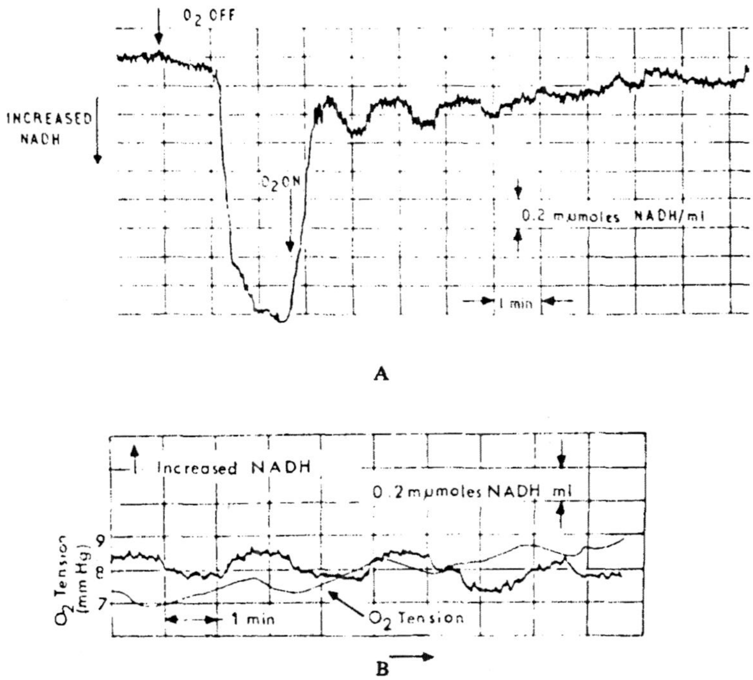

FIGURE 5. (A) Damped oscillations in pyriine nucleotide fluorescence following an anaerobic shock to an aerobic chemostat culture of *Klebsiella aerogenes.* The delay between turning off the oxygen supply and the increase in fluorescence represents the time required for the culture to become anaerobic. Steady-state oxygen tension, 43 mm Hg. (B) Oscillations in pyridine nucleotide fluorescence and corresponding oscillations in dissolved oxygen tension. In both (A) and (B), calibration of fluorescence is in units of NADH (m*l* culture)$^{-1}$, based on enzyme assay. Growth was glucose-limited at a rate of 0.2 hr^{-1}. (From Harrison, D.E.F., *J. Cell Biol.*, 45, 514, 1970. With permission.)

undisturbed. The oscillations could be transiently stopped by interrupting the medium flow. They were lost if the flow was switched off for more than 30 min and could not be obtained on recommencing the medium flow until another anaerobic shock was applied. Oxygen tension of the culture oscillated in phase with pyridine nucleotide fluorescence (Figure 5B), but oscillations of fluorescence persisted anaerobically, suggesting that the respiratory oscillations were a consequence, rather than the cause, of the oscillations in pyridine nucleotide. The regime of regular anaerobic shocks is reminiscent of the phasing and pulsing methods of synchronization referred to in Section II. The oscillations may, therefore, have resulted from the induction of synchronous growth in the chemostat, although this explanation was considered unlikely[75] because no simple correlation was found between growth rate and frequency of oscillation. Of particular interest is the finding that the frequency of the oscillations could be varied simply by changing the oxygen tension in the culture at a constant dilution rate. Oscillations of high frequency such as these (2 to 3 min period) would not have been detected in the experimental designs applied so far to synchronous cultures and described above. Their study awaits the application of on-line, real-time monitoring of respiratory metabolism, as with the fluorometric methods[76] that have been used recently to detect high-frequency metabolic oscillations in synchronous cultures of eukaryotic microorganisms.[77]

Few steps have been taken to elucidate the regulation of respiration rates during the cell cycle. However, it is worth considering the levels at which regulation of respiration rates may occur in growing bacteria[78] so that they may be considered in the design and interpretation of future experiments.

C. Regulation of Respiration in Synchronous Cultures

1. Regulation by Substrate Uptake

Where rates of solute uptake have been measured as a function of the bacterial cell cycle, they show rather simple patterns that do not appear to be directly related to changes in cell respiration rate. Kubitschek has suggested that the constant rates of uptake of all major growth factors in the cell cycle of *E. coli* result in linear growth in volume during the cycle.[79] Uptake is limited by the presence of a constant number of functional binding or accumulation sites which double near the end of the cycle.[19,79] Similar results have been described by Ohki[57] for α-glycerophosphate. In this case, the increase in uptake rate was coincident with cytochrome *b* synthesis and turnover of phosphatidylglycerol. In *Alcaligenes faecalis*, the rate of methionine uptake is constant during the cycle and doubles at the time of cell division.[36] Thus, the net rate of transport of solutes appears to be constant, or nearly so, during the cell cycle and cannot alone control the discontinuities in respiration rates described. In no experiments have rates of uptake of both solute and O_2 been measured. A simple model for control of respiration by uptake alone would not allow for balanced energy metabolism and growth in substrate-limited environments. Superimposed on a mechanism that regulates respiration by uptake, there must, therefore, be other regulatory mechanisms that would allow for further modulation of respiration rates and for balanced anabolic and catabolic use of substrate.

2. Regulation by Adenylate Pools

Harrison has concluded[78] that the ADP levels of growing bacterial cells do not "appear to fall low enough to give the equivalent of the 'state 4' of isolated mitochondria" and, therefore, has disfavored the control of respiration rates in bacteria by a respiratory control mechanism.[70] However, the measurements cited relate to cells grown in asynchronous cultures and thus describe the adenylate pools only in cells averaged over the cell cycle. In the absence of careful measurements of adenylate pools during the bacterial cell cycle, these conclusions should not deter us from testing the hypothesis that respiration rates may be modulated by fluctuating adenylate pools[51,72] or the adenylate "energy charge".[80]

3. Regulation by Composition and Electron Flux of the Electron Transport Chain

Our scant knowledge of the changing activities and amounts of the respiratory chain components during the cell cycle is reviewed in the next section. It is probably premature to make correlations between these and the observed respiration rates. By analogy with the control of mitochondrial respiration in state 3 (conditions of adequate substrate, O_2, and ADP), it might be supposed that the respiration rates of actively growing bacteria could be controlled by the cytochrome content. With few exceptions, the potential respiration rate of bacterial cultures is unrelated to the amount of the terminal oxidase.[78] A further argument against limitation of respiration rate by the terminal oxidase reaction is that bacteria usually have a vast excess of terminal oxidase over their respiratory requirements. Harrison[78] has calculated that, using conservative estimates for the turnover number of a bacterial oxidase, a spectroscopically undetectable amount of the enzyme could support a significant respiration rate. Therefore, inferences regarding the quantitative importance of a cytochrome for respiration may be unsound. Other possible levels of control over respiration rates are the activities of the substrate dehydrogenases, alterations in the rate of electron flux through alternative respiratory chains, and the loss or acquisition of energy coupling sites along these chains.

IV. TEMPORAL ASPECTS OF MEMBRANE SYNTHESIS IN THE BACTERIAL CELL CYCLE

A. General Patterns of Lipid and Protein Synthesis

Recognition of the diversity of functions performed by the bacterial cytoplasmic membrane has led to many studies of the changing composition of the membrane through the cycle. Since the membrane is the site of the reactions of electron-transfer and oxidative phosphorylation, and these functions are modulated by the membrane make-up,[3,81] investigations of the patterns of total membrane biogenesis are presented in Table 2. As yet, it is not clear to what extent these patterns are reflected in the syntheses of individual components. The few respiratory enzymes that have been studied are described later. There is disagreement as to the continuity or otherwise of the rates of overall membrane synthesis, but when both total (or soluble) protein and membrane protein have been carefully studied in the same experiment,[84,86] the latter, but not the former, has been found to be synthesized discontinuously (Figure 6). In *E. coli,* bulk inner membrane polypeptides are synthesized at an exponential rate and do not contribute to the sharp doubling in the rate of synthesis of overall membrane protein.[90] The basis of the underlying control is obscure, but is not dependent on DNA synthesis. Proteins of the outer, but not inner, membrane of *E. coli* are specified by mRNA species of unusually long half-lives.[91] Control may occur via a factor involved in the initiation of translation, which is itself subject to a stepwise change in the rate of synthesis or activation.

It cannot be concluded whether the diverse patterns of synthesis are in any way reflections of the various methods of cell-cycle analysis employed. The synthesis of a membrane protein during only a short period of the cycle in membrane elution experiments[86] was later found to be induced by merely filtering the bacterial culture and resuspending the cells in identical fresh medium.[92]

A causal relationship between increased membrane synthesis and other marker events of the cycle remains to be established. Increased synthesis has been reported to occur at division,[82] which may be coincident with termination of chromosome replication,[88] at septation.[57,87] and at the initiation of DNA replication.[93] Sargent grew synchronous cultures of *B. subtilis* in media that supported different growth rates.[94] In succinate- and glucose-containing cultures, the rate of membrane synthesis doubled coincidentally with the termination of chromosome replication and nuclear segregation, rather than at a fixed fraction of the cycle.

Changes in the composition of the cytoplasmic membrane probably have important consequences for membrane function. The most clearly documented case of fluctuations in gross membrane composition is that which occurs in cultures of *Rhodopseudomonas sphaeroides* synchronized by dilutions of stationary phase cultures. Kaplan and his group have shown that, while membrane-associated proteins are made at a constant rate and continuously inserted into the growing membrane,[87] phospholipid content increases discontinuously[89] as a result of discontinuous synthesis in the absence of detectable turnover.[88] The fluctuating protein to phospholipid ratio is reflected in the intrinsic buoyant density of the isolated membranes[88] and oscillations in membrane fluidity revealed by fluorescence polarization techniques.[95] Increases in the protein to phospholipid ratio of the intracytoplasmic membrane result in changes in the relative mobility of the fluorophore, α-parinaric acid.[96] The magnitude of the polarization changes that occur during the cell cycle is equivalent to those induced by temperature changes of $\sim$20°C. It is tempting to speculate that such changes are in part responsible for the modulation of activity of membrane-bound enzymes during the cell cycle, may alter the permeability properties of the membrane, and thus, influence the availability of substrate for such enzymes.

TABLE 2

Patterns of Synthesis of Major Membrane Components During the Cell Cycles of Bacteria

Organism	Method of cell cycle analysis	Membrane component(s)	Synthesis	Reference
Bacillus megaterium and *Escherichia coli*	Synchronization by amino acid starvation (both); culture fractionation *(E. coli)*	Phospholipid	Discontinuous rates of synthesis, maximal at division	82
Bacillus licheniformis	Synchronous outgrowth from spores	Phospholipids	Discontinuous rates of synthesis	83
E. coli K12	Synchronization by heat treatment of thermosensitive strain	Phospholipids	Continuous synthesis	57
Bacillus subtilis	Synchronization by filtration	(1) Tightly membrane-bound proteins, (2) loosely membrane-bound proteins, and (3) phospholipids	Continuous synthesis of (2) and (3), discontinuous synthesis of (1)[a]	84[a]
Caulobacter crescentus	Synchronization by differential centrifugation	Protein, phospholipid	Rate of synthesis declines during cycle	85
E. coli B/r	Membrane elution	(1) Total protein, (2) envelope protein, and (3) phospholipid	Rate of synthesis of (1) and (3) increases exponentially; rate of synthesis of (2) increases step-wise	86
E. coli K12	Repeated dilution of stationary phase cultures	Phospholipids	Synthesis is continuous and occurs at a constant rate	58
E. coli B/r	Synchronization by membrane elution	Proteins and phospholipid	Rates of synthesis oscillate	54
Rhodopseudomonas sphaeroides	Repeated division of stationary phase cultures	(1) Total cell and soluble protein, (2) total particulate protein, and (3) phospholipid	Continuous accumulation of (1) and (2). Rate of synthesis of (1) increases exponentially, that of (2) doubles at division. Discontinuous synthesis and accumulation of (3)	87, 88 89

[a] Pulse-labeling suggests addition of proteins to membrane from cytoplasm and reverse flux during periods of zero net synthesis.

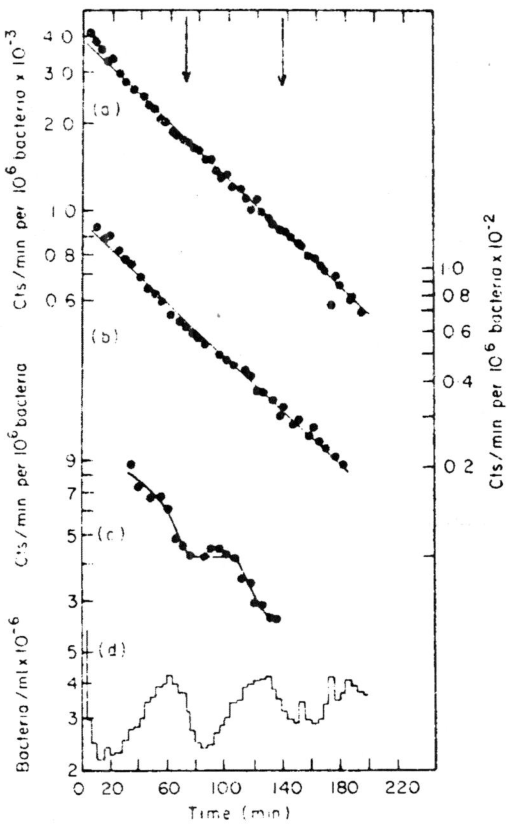

FIGURE 6. Rate of total protein, phospholipid, and envelope protein synthesis in *E. coli* B/r determined by membrane elution. An exponential culture of *E. coli* B/r was pulse-labeled for 5 min with [^{14}C]-leucine (for labeling protein) or [2 - ^{3}H]-glycerol (for labeling phospholipid). The pulse was promptly terminated and the washed bacteria collected on a membrane filter. The following were determined in successively eluted fractions: (a) total protein, (b) phospholipid, (c) envelope protein, and (d) total bacterial number. The age of bacteria with respect to the original pulse-labeling increases from right to left, and the arrows indicate the approximate timing of division. (From Churchward, G.G. and Holland, I.B., *J. Mol. Biol.*, 105, 245, 1976. With permission.)

B. Activities and Syntheses of Specific Components of the Respiratory System

1. Succinate Dehydrogenase

This enzyme complex has been used in many studies as a representative membrane-bound component of the respiratory chain. In a variety of bacteria synchronized by diverse methods, enzyme activity has been shown to increase in a stepwise fashion through the cell cycle. Such patterns have been observed in *Bacillus subtilis*[84] (synchronized by filtration), in *Myxobacter* AL-1[97] (after culture fractionation), and in phototrophic growth of *Rhodopseudomonas sphaeroides* (synchronized by repeated dilution of stationary phase cultures).[98]

Ohki and Mitsui[99] have used a mutant of *E. coli* which is temperature-sensitive for the formation of membrane components and cell division. After a shift to a temperature restrictive for growth (42°C), no increase in succinate dehydrogenase activity occurred. When cells were transferred to the permissive temperature (30°C) after 35 min incubation at 42°C, enzyme activity increased immediately and, after a period of little further increase, rose again. Cell division was also synchronized after the shift to 30°C. It was proposed that, because of a defect in the regulation of membrane biogenesis, the cell population was reversibly aligned at a specific stage in the cycle during incubation at 42°C. Thus, a tight coupling between cycle progress and the development of respiratory function is implied.

2. Other Substrate Dehydrogenases

Total NADH dehydrogenase activity, reported to be somewhat loosely bound to the membrane, increased continuously in selection-synchronized *Bacillus subtilis*.[84] In contrast, the activity of the membrane-bound enzyme increased in a stepwise fashion in induction-synchronized cultures of *Rhodopseudomonas sphaeroides*.[98] D-Lactate dehydrogenase[100] is synthesized concurrently with cytochrome b[57] (see below) and the permeases for lactose[100] and L-α-glycerophosphate[57] in the cell cycle of *E. coli.*

3. Adenosine Triphosphatase

Marked oscillations in ATPase activity have been observed by Edwards et al.[44] during growth of *Alcaligenes eutrophus* synchronized by a continuous-flow selection technique (Figure 7). Maximal activities were observed at about 0.4 and 0.9 of a cycle concurrently with maximum susceptibility of the enzyme to inhibition by Nbf-Cl, minimum inhibition by DCCD, and maximal rates of oxygen uptake (Figure 4). These compounds inhibit, respectively, the BF_1 (loosely membrane bound) and the BF_o (tightly membrane bound) components of the bacterial ATPase complex. It was proposed that the variations in ATPase activity could be due either to the differential periodic synthesis of a catalytic and/or regulatory subunit of component BF_1 or, and less likely, to variations in the affinity of BF_1 for ATP. That variation in the binding of BF_1 to BF_o was responsible for the fluctuating sensitivity to inhibitors was suggested by similar cell-cycle-dependent oscillations in the proton conductance of whole cells. Maximal proton conductance was proposed to result from increased dissociation of the BF_1 component from the membrane. Quotients ($\rightarrow H^+/O$) for the oxidation of endogenous substrates remained relatively constant throughout the cycle (~8) and were assumed to reflect the presence of four proton-translocating segments of the respiratory chain.

4. Cytochromes

The spectroscopic study of cellular cytochrome content has the distinct advantage that conclusions may be drawn regarding the net synthesis and/or degradation of the hemoprotein molecules. In contrast, studies of enzyme activity such as those described above generally yield little information about enzyme synthesis, although in cell-cycle studies "activity" and "synthesis" are frequently equated. The paucity of measurements of cytochrome content during the bacterial cell cycle can be explained, at least in part, by the large quantities of cells or membranes required for spectral analysis. Both studies described here use induction-synchronized cultures, which are more easily modified to produce such yields.

Difference spectra recorded on samples withdrawn during synchronous growth of *E. coli*[57] (Figure 8) reveal a stepwise pattern of net synthesis of *b*-type cytochromes. In fact, interpretation of these results is probably oversimplified since at least two *b*-type cytochromes which differ in their ability to bind CO,[101] contribute to this absorption band, and Shipp has shown that multiple absorption bands can be resolved using

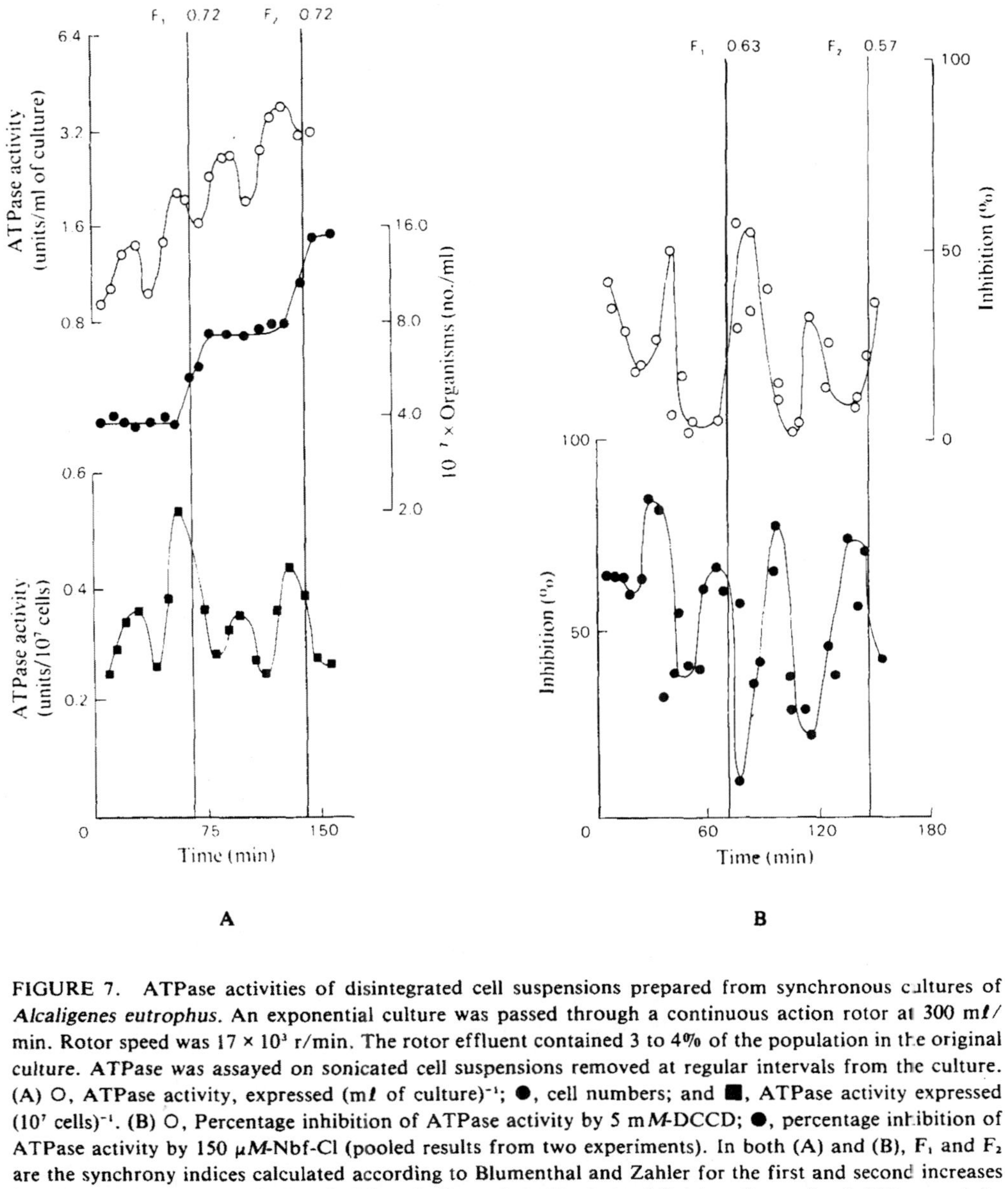

FIGURE 7. ATPase activities of disintegrated cell suspensions prepared from synchronous cultures of *Alcaligenes eutrophus*. An exponential culture was passed through a continuous action rotor at 300 m*l*/min. Rotor speed was 17×10^3 r/min. The rotor effluent contained 3 to 4% of the population in the original culture. ATPase was assayed on sonicated cell suspensions removed at regular intervals from the culture. (A) O, ATPase activity, expressed (m*l* of culture)$^{-1}$; ●, cell numbers; and ■, ATPase activity expressed (10^7 cells)$^{-1}$. (B) O, Percentage inhibition of ATPase activity by 5 m*M*-DCCD; ●, percentage inhibition of ATPase activity by 150 μ*M*-Nbf-Cl (pooled results from two experiments). In both (A) and (B), F_1 and F_2 are the synchrony indices calculated according to Blumenthal and Zahler for the first and second increases in cell numbers, respectively. Vertical lines indicate the midpoints of division. One unit of ATPase activity is 1 nmol of ATP hydrolyzed/min. (From Edwards, C., Spode, J.A., and Jones, C.W., *Biochem. J.*, 172, 253, 1978. With permission.)

higher derivative spectral analysis.[102] Description of the periods of synthesis in relation to the division cycle is complicated by the protraction of the first cycle. Steps occur at about 0.7 in the first cycle and 0.5 in the second concurrently with turnover of labeled phosphatidylglycerol. Furthermore, the steps in both cell numbers and, more markedly, in cytochrome amounts are considerably less than doublings. Experiments with the temperature-sensitive mutant of *E. coli* described earlier[99] (Section IV.B.1.) also show that cytochrome *b* synthesis, like the activities of succinate dehydrogenase, α-glycerophosphate dehydrogenase, and the transport system for L-α-glycerophosphate, is under the control of a gene product whose function is expressed 20 to 30 min before cell division.

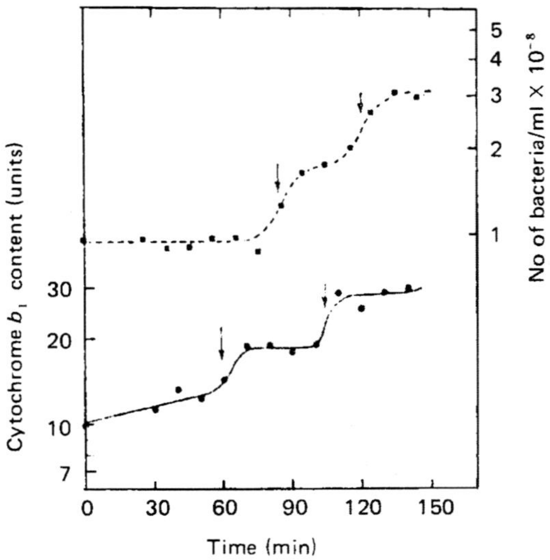

FIGURE 8. Cytochrome *b* synthesis during synchronous growth of a thermosensitive strain of *E. coli* K12. An exponential culture was incubated at the restrictive temperature (42°C) for 80 min. Synchronous growth and division was started by the addition of two volumes of medium, bringing the temperature to 30°C. Cytochrome *b* concentrations ● in samples removed from the culture were determined from dithionite-reduced *minus* ferricyanide-oxidized difference spectra at 560 to 575 nm. One unit corresponds to 2 × 10^{-4} $E_{560-575}$. Viable cell counts are also shown (■). Arrows indicate the midpointsof cell division or cytochrome synthesis. (From Ohki, M., *J. Mol. Biol.*, 68, 249, 1972. With permission.)

In contrast, the b- and c-type cytochromes of *Rhodopseudomonas sphaeroides* are synthesized and incorporated into the membrane continuously during synchronized phototrophic growth.[98]

5. Other Enzymes

Kurz et al.[62] have used the continuous phased culture technique to study nitrogenase activity during the cell cycle of *Azotobacter vinelandii.* Activity rose gradually during most of the cycle, doubling during one cycle. Activities of several enzymes involved in the synthesis of the tetrapyrrole of bacteriochlorophyll (succinyl CoA thiokinase, δ-aminolaevulinic acid dehydrase, and δ-aminolaevulinic acid synthetase) increased discontinuously in synchronous cultures of *Rhodopseudomonas sphaeroides*[103] while bacteriochlorophyll synthesis was continuous.[98]

V. SPATIAL ASPECTS OF MEMBRANE SYNTHESIS IN THE BACTERIAL CELL CYCLE

A. Localized vs. Delocalized Membrane Growth

A complete understanding of membrane assembly during the cell cycle requires information on its topography, i.e., its spatial organization. A review of this field discloses that there is considerable disagreement over the key issues. Does growth of the cell envelope (specifically the cytoplasmic membrane) occur by insertion of newly synthesized components into localized growing zones that are conserved in subsequent generations? Alternatively, are such zones undetectable because of (1) diffusely organized assembly or (2) post-assembly randomizing processes such as diffusion of lipids and proteins within the plane of the membrane? Few studies have been made of the topographical growth of the bacterial cytoplasmic membrane compared with those on cell wall growth,[104] and even fewer studies have been directed to components of the respiratory chain. Jacob and co-workers[105] used tellurite reduction by the respiratory chain, resulting in deposition of tellurium crystals in the membrane, to show that the segregation of preexisting respiratory membrane was uneven among daughter cells. Other studies support nonrandom distribution of preexisting and newly synthesized membrane components. A suitable identifiable marker component of the *B. subtilis* membrane is the basal body of the flagellum. In a mutant unable to develop flagellae at high temperatures, basal pieces of parental flagellae were distributed preferentially toward the polar parts of cell chains formed by cell multiplication.[106] The labeling of membrane phospholipids of *Bacillus megaterium* with radioactive palmitic acid led Morrison and Morowtiz[107] to conclude that lipid incorporation into membranes was localized at the cell poles and that these regions were conserved during growth and division.

The technique developed by Kepes and Autissier is described in some detail here because it has been used by them to demonstrate nonrandom segregation of various permeases, of respiratory nitrate reductase, of phospholipid and DNA,[108-110] and in our laboratory, to study the segregation of cytochromes at cell division in *E. coli*.

A culture of *E. coli* is induced for nitrate reductase by anaerobic growth on glucose with nitrate as nitrogen source and terminal electron acceptor. Deinduction is performed in glucose-containing medium with ammonia as nitrogen source for one, two, or three generations. Population samples removed from the deinduced culture (in which further synthesis of nitrate reductase is prevented) are tested for the presence of reductase-containing and reductase-less bacteria as follows. The first test involves incubation of the cells in medium that contains nitrate (as the sole nitrogen source) and penicillin. Only cells which contain the reductase will grow and be lysed by penicillin. The second test exploits the ability of nitrate reductase to reduce chlorate to chlorite, a highly toxic product. In this case, cells are incubated with ammonium as nitrogen source together with chlorate and penicillin. Those that contain nitrate reductase reduce chlorate to chlorite are killed and, therefore, not lysed by penicillin. Those cells that are fully induced and those in samples taken one and two generations later are killed by chlorite and, therefore, not lysed by penicillin. After three generations, about half of the cells in the population lack nitrate reductase and the ability to produce chlorite, and because they grow, are lysed by penicillin. Further evidence that, in this and related experiments, the heterogeneity of the population is due to the membrane marker is presented by Kepes and Autissier.

B. Cytochrome Segregation in *E. coli*

Our own experiments[111] have exploited a mutant of *E. coli* K12, A1004a, that is deficient in heme biosynthesis and thus, cytochrome biosynthesis. It can grow on a nonfermentable carbon source such as glycerol only when provided with the heme precursor, δ-aminolaevulinic acid.[112] After transfer of exponentially growing cells from a medium containing glycerol and δ-ALA to one lacking only δ-ALA, further cytochrome synthesis is stopped, but cells continue to grow exponentially for about 2.5 generation times. Growth of the culture then follows more closely linear, rather than exponential, kinetics. Cells withdrawn from the culture up to 2.2 generation times after resuspension in δ-ALA-deficient medium are readily lysed by ampicillin in medium

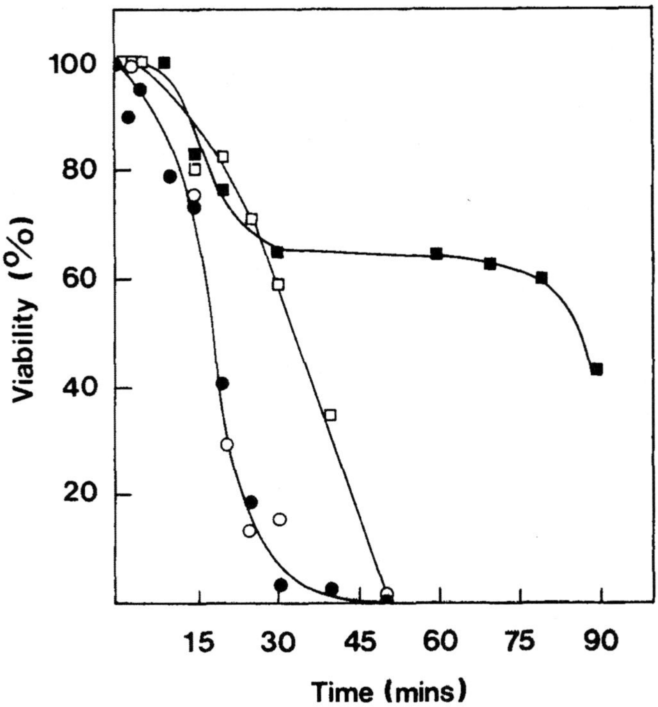

FIGURE 9. Time course of killing of a heme-deficient mutant of *E. coli* K12 after transfer to δ-ALA-deficient medium. See text for principles of experiment. The mutant was grown in a medium containing δ-ALA and glycerol to the exponential phase of growth. Cells were harvested, washed, and resuspended in the same medium lacking only δ-ALA. At zero (○), 1.1 (●), 2.2 (□), or 3.3 (■) generation times after resuspension, cells were removed and incubated with ampicillin (1 mg/m*l*) in a medium containing glycerol, but lacking δ-ALA. Killing by ampicillin was monitored by viable counts. Cells inheriing cytochromes grow in the absence of δ-ALA and are killed by ampicillin. After 3.3 generation times, a significant proportion of the population remain viable, indicating the presence of cells that have not inherited parental cytochromes. (Unpublished figure of R.K. Poole, R.I. Scott, and Christine H. Britnell. See also Reference 109).

that contains glycerol (Figure 9) and, thus, allows only growth of cytochrome-containing cells. However, after 3.3 generations, only about 40% of the population lyse under these conditions, presumably those cells that inherited parental cytochromes.

One model for membrane growth and segregation, of several that are compatible[108-110] with these data, is presented in Figure 10. The essential feature is that an equatorial growing zone separates the two poles of the cell, each of which contains a conserved region of marker-containing membrane that was synthesized in previous cell cycles. Towards the end of the cycle, a septum is formed within the growing zone and two new growing zones appear in the equatorial plane of daughter cells. If deinduction starts with a newly divided cell, it will give rise to a population, half of which lacks the parental marker membrane, after two divisions (Figure 10A). However, a population that is growing exponentially at the time of deinduction consists of cells at all stages of the cycle. The distribution of cell ages in such a population when all cells have the same doubling time has been extensively documented.[10] In Figure 10B, a cell at 0.4 of the cycle is taken as the "average" cell. This will give rise to descendants

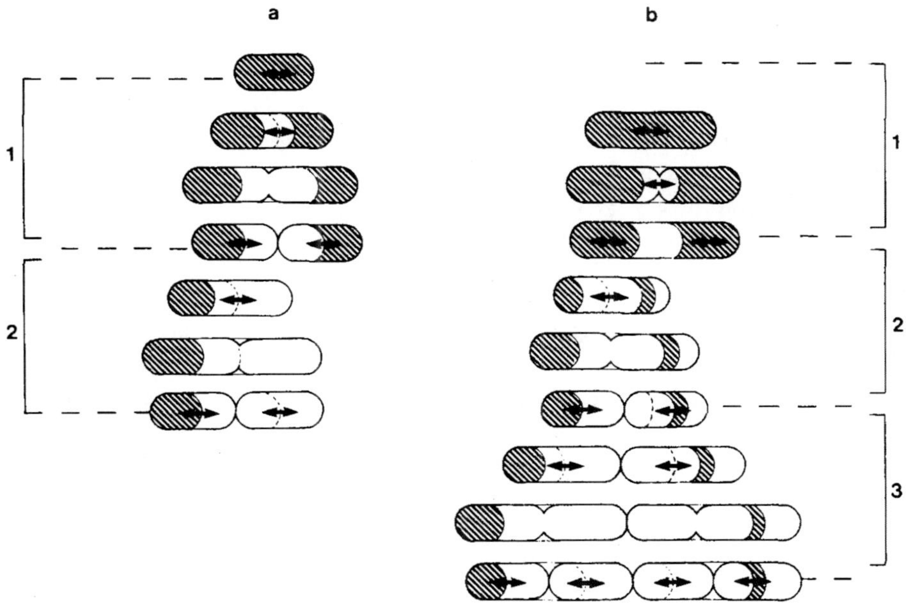

FIGURE 10. Model for distribution of parental membrane among descendants in bacteria with a single median growing zone. Parental marker (a cytochrome) is the hatched area. Growing zone is shown as a dashed line, when different from the border between parental and new membrane. Arrows indicate the elongation of new membrane. (A) A cell transferred to δ-ALA-deficient medium just after the last division. (B) A cell transferred to δ-ALA-deficient medium about midway between two cell divisions. If x is the time elapsed since the last cell division, and t the generation time ($O < x < t$), the cell will yield four cytochrome-containing and four cytochrome-deficient segregants after a time of $3t - x$. All cell clones will yield 50% cytochrome-deficient segregants after 3t. The numerals indicate complete generations. A generation is taken as starting and finishing at the time of fission. (From Kepes, A. and Autissier, F., *Biochim. Biophys. Acta*, 265, 443, 1972. With permission.)

having no parental membrane only after its third division, which will occur 2.6 generation times after deinduction. After three generations, all members of the deinduced population will have undergone a third division, and the resulting population will contain equal numbers of marker-containing and marker-lacking cells. The ratio of viable to nonviable (40:60) cells obtained in our experiment is to be expected after more than three generations (3.3).

Contradictory conclusions as to the topography of cytochrome synthesis and/or segregation have been reached by Green and Schaechter.[113] They used a minicell-forming mutant of *E. coli* to analyze membrane at the cell poles. Cytochrome-containing membrane was first labeled with [^{3}H]-δ-aminolaevulinic acid. During growth of this parental-labeled membrane, it was diluted at the same rate with new membrane in both normal cells and minicells, thus indicating that the cell poles were not different from the interpolar regions with respect to the distribution of parental cytochromes. Further evidence for the lack of conserved regions of newly synthesized membranes has been obtained in a number of laboratories.[85,114-116] Clearly, further work is required to resolve this controversy.

VI. CONCLUDING REMARKS

More questions have been raised in this survey than have been answered. Many of these, however, are amenable to direct experimental study. The importance of considering bacterial respiration as being temporally diverse is twofold. Firstly, the elucidation of the synthesis of the cytoplasmic membrane, which has such diverse and indistion of the cell cycle and cell division. Secondly, a complete understanding of the composition of the respiratory chain and of the mechanisms of electron transfer and associated reactions of energy conversion in the single cell is unlikely to be obtained by continued emphasis on exponentially growing cultures. Such populations reveal only properties of an average membrane in an average cell and conceal the temporal diversity of bacterial respiratory systems.

ACKNOWLEDGMENTS

It is a pleasure to acknowledge the co-operation of those colleagues who have allowed me to view and cite their unpublished findings. Work described here from my own laboratory was supported in part by the Science Research Council through grant GR/A/2252.5.

REFERENCES

1. **Mazia, D.**, The Cell Cycle, *Sci. Am.*, 230, 54, 1974.
2. **Anderson, R. G. W.**, The biogenesis of cell structures and the expression of assembly information, *J. Theor. Biol.*, 67, 535, 1977.
3. **George-Nascimento, C., Zehner, Z. E., and Wakil, S. J.**, Assembly of lipids and proteins in *Escherichia coli* membranes, *J. Supramol. Struct.*, 2, 646, 1974.
4. **Lloyd, D.**, *The Mitochondria of Micro-organisms*, Academic Press, London, 1974, vi.
5. **Poole, R. K.**, Structure and Function of the Fission Yeast *Schizosaccharomyces pombe*, Ph.D. thesis, University of Wales, Cardiff, 1973.
6. **Kamen, M. D.**, *Primary Processes in Photosynthesis*, Academic Press, New York, 1963.
7. **Mitchison, J. M.**, *The Biology of the Cell Cycle*, Cambridge University Press, Cambridge, 1971.
8. **Wells, J. R. and James, T. W.**, Cell cycle analysis by culture fractionation, *Exp. Cell Res.*, 75, 465, 1972.
9. **Mitchison, J. M. and Carter, B. L. A.**, Cell cycle analysis, in *Methods in Cell Biology*, Vol. 11, Prescott, D. M., Ed., Academic Press, New York, 1975, chap. 11.
10. **Koch, A. L. and Blumberg, G.**, Distribution of bacteria in the velocity gradient centrifuge, *Biophys. J.*, 16, 389, 1976.
11. **James, T. W.**, Cell synchrony, a prologue to discovery, in *Cell Synchrony*, Cameron, I. L. and Padilla, G. M., Eds., Academic Press, New York, 1966, 1.
12. **Abbo, F. E. and Pardee, A. B.**, Synthesis of macromolecules in synchronously-dividing bacteria, *Biochim. Biophys. Acta*, 39, 478, 1960.
13. **Campbell, A.**, Synchronization of cell division, *Bacteriol. Rev.*, 21, 263, 1957.
14. **Bazin, M. J., Richards, L., and Saunders, P. T.**, Automated data processing of microbial size distributions, in *Haematological and Biological Applications of the Coulter Counter*, Coulter Electronics, Harpenden, England, 1977, 426.
15. **Painter, P. R. and Marr, A. G.**, Mathematics of microbial populations, *Annu. Rev. Microbiol.*, 22, 519, 1968.
16. **Helmstetter, C. E.**, Methods for studying the microbial division cycle, in *Methods in Microbiology*, Vol. 1, Norris, J. R. and Ribbons, D. W., Eds., Academic Press, New York, 1969, 327.
17. **Burnett-Hall, D. G. and Waugh, W. A. O'N.**, Indices of synchrony in cellular cultures, *Biometrics*, 23, 693, 1967.
18. **Blumenthal, L. K. and Zahler, S. A.**, Index for measurement of synchronization of cell populations, *Science*, 135, 724, 1962.
19. **Kubitschek, H. E., Freeman, M. L., and Silver, S.**, Potassium uptake in synchronous and synchronized cultures of *Escherichia coli*, *Biophys. J.*, 11, 787, 1971.
20. **Schaechter, M., Williamson, J. P., Hood, J. R., and Koch, A. L.**, Growth, cell and nuclear division in some bacteria, *J. Gen. Microbiol.*, 29, 421, 1962.

21. **Engelberg, J.**, The decay of synchronization of cell division, *Exp. Cell Res.*, 36, 647, 1964.
22. **Nishi, A., Okamara, S., and Yanagita, T.**, Shift of cell age distribution in the later phases of *Escherichia coli* culture, *J. Gen. Appl. Microbiol.*, 13, 103, 1967.
23. **Mitchison, J. M. and Vincent, W. S.**, Preparation of synchronous cell cultures by sedimentation, *Nature (London)*, 205, 987, 1965.
24. **Chatterjee, A. N., Taber, H., and Young, F. E.**, A rapid method for synchronization of *Staphylococcus aureus* and *Bacillus subtilis*, *Biochem. Biophys. Res. Commun.*, 44, 1125, 1971.
25. **Kubitschek, H. E.**, Linear cell growth in *Escherichia coli*, *Biophys. J.*, 8, 792, 1968.
26. **Hinks, R. P., Dunco-Moore, L., and Shockman, G. D.**, Cellular autolytic activity in synchronized populations of *Streptococcus faecium*, *J. Bacteriol.*, 133, 822, 1978.
27. **Beck, B. D. and Park, J. T.**, Activity of three murein hydrolases during the cell division cycle of *Escherichia coli* as measured in toluene-treated cells, *J. Bacteriol.*, 126, 1250, 1976.
28. **Kung, F.-C. and Glaser, D. A.**, Synchronization of *Escherichia coli* by zonal centrifugation, *Appl. Environ. Microbiol.*, 34, 328, 1977.
29. **Westmacott, D. and Primrose, S. B.**, Synchronous growth of *Rhodopseudomonas palustris* from the swarmer phase, *J. Gen. Microbiol.*, 94, 117, 1976.
30. **Burdett, I. D. J. and Murray, R. G. E.**, Electron microscope study of septum formation in *Escherichia coli* strains B and B/r during synchronous growth, *J. Bacteriol.*, 119, 1039, 1974.
31. **Scott, R. I. and Poole, R. K.**, unpublished data, 1978.
32. **Gudas, L. T. and Pardee, A. B.**, Deoxyribonucleic acid synthesis during the division cycle of *Escherichia coli*: a comparison of strains B/r, K-12, 15 and 15 T^- under conditions of slow growth, *J. Bacteriol.*, 117, 1216, 1974.
33. **Maruyama, Y. and Tanagita, T.**, Physical methods for obtaining synchronous culture of *Escherichia coli*, *J. Bacteriol.*, 71, 542, 1956.
34. **Sargent, M. G.**, Synchronous cultures of *Bacillus subtilis* obtained by filtration with glass fiber filters, *J. Bacteriol.*, 116, 736, 1973.
35. **Whittenbury, R. and Dow, C. S.**, Morphogenesis and differentiation in *Rhodomicrobium vannielii* and other budding and prosthecate bacteria, *Bacteriol. Rev.*, 41, 754, 1977.
36. **Lark, K. G. and Lark, C.**, Changes during the division cycle in bacterial cell wall synthesis, volume and ability to concentrate free amino acids, *Biochim. Biophys. Acta*, 43, 520, 1960.
37. **Lloyd, D., John, L., Edwards, C., and Chagla, A. H.**, Synchronous cultures of micro-organisms: large-scale preparation by continuous-flow size selection, *J. Gen. Microbiol.*, 88, 153, 1975.
38. **Evans, J. B.**, Preparation of synchronous cultures of *Escherichia coli* by continuous flow size selection, *J. Gen. Microbiol.*, 91, 188, 1975.
39. **Poole, R. K.**, Fluctuations in buoyant density during the cell cycle of *Escherichia coli* K12: significance for the preparation of synchronous cultures by age selection, *J. Gen. Microbiol.*, 98, 177, 1977.
40. **Poole, R. K.**, The influence of growth substrate and capacity for oxidative phosphorylation on respiratory oscillations in synchronous cultures of *Escherichia coli* K12, *J. Gen. Microbiol.*, 99, 369, 1977.
41. **Poole, R. K.**, Preparation of synchronous cultures of micro-organisms by continuous-flow selection: which cells are selected? *FEMS Microbiol. Lett.*, 1, 305, 1977.
42. **Poole, R. K. and Pickett, A. M.**, Which cells are selected from exponential cultures by continuous-flow centrifugation? The selection of small cells from cultures of *Escherichia coli* and *Schizosaccharomyces pombe* that exhibit minimal density fluctuations during their cell cycles, *J. Gen. Microbiol.*, 107, 399, 1978.
43. **Edwards, C. and Jones, C. W.**, Respiratory properties of synchronous cultures of *Alcaligenes eutrophus* H16 prepared by a continuous-flow size selection method, *J. Gen. Microbiol.*, 99, 383, 1977.
44. **Edwards, C., Spode, J. A., and Jones, C. W.**, The properties of adenosine triphosphatase from exponential and synchronous cultures of *Alcaligenes eutrophus* H16, *Biochem. J.*, 172, 253, 1978.
45. **Koch, A. L.**, personal communication, *1978*.
46. **Baldwin, W. W. and Wegener, W. S.**, Selection of synchronous bacterial cultures by density sedimentation, *Can. J. Microbiol.*, 22, 390, 1976.
47. **Kubitschek, H. E.**, Constancy of the ratio of DNA to cell volume in steady-state cultures of *Escherichia coli* B/r, *Biophys. J.*, 14, 119, 1974.
48. **Cummings, D. J.**, Synchronization of *E. coli* K12 by membrane selection, *Biochem. Biophys. Res. Commun.*, 41, 471, 1970.
49. **Shapiro, L. and Agabian-Keshishian, N.**, Specific assay for differentiation in the stalked bacterium *Caulobacter crescentus*, *Proc. Natl. Acad. Sci. U.S.A.*, 67, 200, 1970.
50. **Mitchison, J. M.**, Enzyme synthesis during the cell cycle, in *Cell Differentiation in Micro-organisms, Plants and Animals*, Nover, L. and Mothes, K., Eds., North-Holland, Amsterdam, 1977, 377.

51. **Poole, R. K. and Salmon, I.**, The pool sizes of adenine nucleotides in exponentially-growing, stationary-phase and 2′-deoxyadenosine-synchronized cultures of *Schizosaccharomyces pombe* 972h⁻, *J. Gen. Microbiol.*, 106, 153, 1978.
52. **Bellino, F. L.**, Continuous synthesis of partially derepressed aspartate transcarbamylase during the division cycle of *Escherichia coli* B/r, *J. Mol. Biol.*, 74, 223, 1973.
53. **Munro, G. F., Hercules, K., Morgan, J., and Sauerbier, W.**, Dependence of the putrescine content of *Escherichia coli* on the osmotic strength of the medium, *J. Biol. Chem.*, 247, 1272, 1972.
54. **Hakenbeck, R., and Messer, W.**, Oscillations in the synthesis of cell wall components in synchronized cultures of *Escherichia coli, J. Bacteriol.*, 129, 1234, 1977.
55. **Ron, E. Z., Grossman, N., and Helmstetter, C. E.**, Control of cell division in *Escherichia coli*: effect of amino acid starvation, *J. Bacteriol.*, 129, 569, 1977.
56. **Koníčková-Radochová, M. and Koníček, J.**, Mutagenesis by *N*-methyl-*N*-nitroso-*N*-nitroguanidine in synchronous cultures of *Mycobacterium phlei, Folia Microbiol. (Prague)*, 19, 16, 1974.
57. **Ohki, M.**, Correlation between metabolism of phosphatidylglycerol and membrane synthesis in *Escherichia coli, J. Mol. Biol.*, 68, 249, 1972.
58. **Bauza, M. T., de Loach, J. R., Aguanno, J. J., and Larrabee, A. R.**, Acyl carrier protein prosthetic group exchange and phospholipid synthesis in synchronized cultures of a pantothenate auxotroph of *Escherichia coli, Arch. Biochem. Biophys.*, 174, 344, 1976.
59. **Goodwin, B. C.**, Synchronization of *Escherichia coli* in a chemostat by periodic phosphate feeding, *Eur. J. Biochem.*, 10, 511, 1969.
60. **Anagnostopoulos, G. D.**, Unbalanced growth in a semi-continuous culture system designed for the synchronization of cell division, *J. Gen. Microbiol.*, 65, 23, 1971.
61. **Anagnostopoulos, G. D.**, Induction and stabilization of synchrony in the cell division of *Escherichia coli, Arch. Microbiol.*, 107, 199, 1976.
62. **Kurz, W. G. W., la Rue, T. A., and Chatson, K. B.**, Nitrogenase in synchronized *Azotobacter vinelandii* OP, *Can. J. Microbiol.*, 21, 984, 1975.
63. **Koch, A. L.**, Does the initiation of chromosome replication regulate cell division? in *Advances in Microbiol Physiology*, Vol. 16, Rose, A. H. and Tempest, D. W., Eds., Academic Press, London, 1977, 49.
64. **Kubitschek, H. E., Bendigkeit, H. E., and Loken, M. R.**, Onset of DNA synthesis during the cell cycle in chemostat cultures, *Proc. Natl. Acad. Sci. U.S.A.*, 57, 1611, 1967.
65. **Lloyd, D. and Poole, R. K.**, Subcellular fractionation: isolation and characterization of organelles, in *Techniques in the Life Sciences*, Vol. B2/1, Kornberg, H. L., Ed., Elsevier/North-Holland Scientific Publishers, Ltd., Amsterdam, 1979, 1.
66. **Manor, H. and Haselkorn, R.**, Size fractionation of exponentially growing *Escherichia coli, Nature (London)*, 214, 983, 1967.
67. **Tan, I., Hartman, W., Guntermann, U., Hüttermann, A., and Kühlwein, H.**, Studies on the cell cycle of Myxobacter AL-1. I. Size fractionation of exponentially-growing cells by zonal centrifugation, *Arch. Microbiol.*, 100, 389, 1974.
68. **Horan, P. K. and Wheeless, L. L.**, Quantitative single cell analysis and sorting, *Science*, 198, 149, 1977.
69. **Maruyama, Y.**, Biochemical aspects of the cell growth of *Escherichia coli* as studied by the method of synchronous culture, *J. Bacteriol.*, 72, 821, 1956.
70. **Chance, B. and Williams, G. R.**, The respiratory chain and oxidative phosphorylation, *Adv. Enzymol.*, 17, 65, 1956.
71. **Poole, R. K.**, Development of respiratory activity during the cell cycle of *Schizosaccharomyces pombe* 972h⁻: respiratory oscillations and heat dissipation in cultures synchronized with 2′-deoxyadenosine, *J. Gen. Microbiol.*, 103, 19, 1977.
72. **Edwards, S. W. and Lloyd, D.**, Oscillations of respiration and adenine nucleotides in synchronous cultures of *Acanthaoeba castellanii*: mitochondrial respiratory control *in vivo, J. Gen. Microbiol.* , 108, 197, 1978.
73. **Huzyk, L. and Clark, D. J.**, Nucleoside triphosphate pools in synchronous cultures of *Escherichia coli, J. Bacteriol.*, 108, 74, 1971.
74. **Scopes, A. W. and Williamson, D. H.**, The growth and oxygen uptake of synchronously dividing cultures of *Saccharomyces cerevisiae, Exp. Cell Res.*, 35, 361, 1964.
75. **Harrison, D. E. F.**, Undamped oscillations of pyridine nucleotide and oxygen tension in chemostat cultures of *Klebsiella aerogenes, J. Cell Biol.*, 45, 514, 1970.
76. **Chance, B., Barlow, C., Nakase, Y., Takeda, H., Mayevsky, A., Fischetti, R., Graham, N., and Sorge, J.**, Heterogeneity of oxygen delivery in normoxic and hypoxic states: a fluorometer study, *Am. J. Physiol.*, 235, H809, 1978.

77. **Bashford, C. L., Chance, B., Lloyd, D., and Poole, R. K.,** Oscillations of redox states in synchronously-dividing cultures of *Acanthamoeba castellanii* and *Schizosaccharomyces pombe, Biophys. J.*, 28, in press, 1980.
78. **Harrison, D. E. F.,** The regulation of respiration rate in growing bacteria, in *Advances in Microbiol Physiology*, Vol. 14, Rose, A. H., and Tempest, D. W., Eds., Academic Press, New York, 1976, 243.
79. **Kubitschek, H. E.,** Constancy of uptake during the cell cycle in *Escherichia coli, Biophys. J.*, 8, 1401, 1968.
80. **Lloyd, D., Edwards, C., Edwards, S. W., El'Khayat, G., Jenkins, S. J., John, L., Phillips, C. A., and Statham, M.,** The stability of adenylate energy charge values, *Trends Biochem. Sci.*, 3, N138, 1978.
81. **Cronan, J. E.,** Molecular biology of bacterial membrane lipids, *Annu. Rev. Biochem.*, 47, 163, 1978.
82. **Daniels, M. J.,** Lipid synthesis in relation to the cell cycle of *Bacillus megaterium* KM and *Escherichia coli, Biochem. J.*, 115, 697, 1969.
83. **Lubochinsky, B. and Burger, M. M.,** Cyclic variations of ^{32}P incorporation into phospholipids during synchronous growth of *Bacillus licheniformis*, paper presented at 6th Meet. Fed. Eur. Biochem. Soc., Madrid, 1969, 345.
84. **Sargent, M. G.,** Membrane synthesis in synchronous cultures of *Bacillus subtilis* 168, *J. Bacteriol.*, 116, 397, 1973.
85. **Galdiero, F.,** The growth and partition of cell membranes during synchronized division cycle of *Caulobacter crescentus, Arch. Mikrobiol.*, 94, 125, 1973.
86. **Churchward, G. G. and Holland, I. B.,** Envelope synthesis during the cell cycle in *Escherichia coli* B/r, *J. Mol. Biol.*, 105, 245, 1976.
87. **Fraley, R. T., Lueking, D. R., and Kaplan, S.,** Intracytoplasmic membrane synthesis in synchronous cell populations of *Rhodopseudomonas sphaeroides*. Polypeptide insertion into growing membrane, *J. Biol. Chem.*, 253, 458, 1978.
88. **Fraley, R. T., Lueking, D. R., and Kaplan, S.,** The relationship of intracytoplasmic membrane assembly to the cell division cycle in *Rhodopseudomonas sphaeroides*, personal communication, 1978.
89. **Lueking, D. R., Fraley, R. T., and Kaplan, S.,** Intracytoplasmic membrane synthesis in synchronous cell populations of *Rhodopseudomonas sphaeroides*. Fate of "old" and "new" membrane, *J. Biol. Chem.*, 253, 451, 1978.
90. **Boyd, A. and Holland, I. B.,** personal communication, 1978.
91. **Lee, N. and Inouye, M.,** Outer membrane proteins of *Escherichia coli:* biosynthesis and assembly, *FEBS Lett.*, 39, 167, 1974.
92. **Boyd, A. and Holland, I. B.,** Protein d, an iron-transport protein induced by filtration of cultures of *Escherichia coli, FEBS Lett.*, 76, 21, 1977.
93. **Pierucci, O.,** unpublished data cited in Reference 87.
94. **Sargent, M. G.,** Control of membrane protein synthesis in *Bacillus subtilis, Biochim. Biophys. Acta*, 406, 564, 1975.
95. **Fraley, R. T., Yen, G. S. L., Lueking, D. R., and Kaplan, S.,** The physical state of the intracytoplasmic membrane of *Rhodopseudomonas sphaeroides* and its relationship to the cell division cycle, personal communication, 1978.
96. **Fraley, R. T., Jameson, D. M., and Kaplan, S.,** The use of the fluorescent probe α-parinaric acid to determine the physical state of the intracytoplasmic membranes of the photosynthetic bacterium *Rhodopseudomonas sphaeroides, Biochim. Biophys. Acta*, 511, 52, 1978.
97. **Hartmann, W., Tan, I., Hüttermann, A. and Kühlwein, H.,** Studies on the cell cycle of *Myxobacter* AL-1. II. Activities of seven enzymes during the cell cycle, *Arch. Microbiol.*, 114, 13, 1977.
98. **Wraight, C. A., Lueking, D. R., Fraley, R. T., and Kaplan, S.,** Synthesis of photopigments and electron-transport components in synchronous phototrophic cultures of *Rhodopseudomonas sphaeroides, J. Biol. Chem.*, 253, 465, 1978.
99. **Ohki, M. and Mitsui, H.,** Defective membrane synthesis in an *E. coli* mutant, *Nature (London)*, 252, 64, 1974.
100. **Ohki, M.,** personal communication, 1978.
101. **Haddock, B. A. and Jones, C. W.,** Bacterial respiration, *Bacteriol. Rev.*, 41, 47, 1977.
102. **Shipp, W. S.,** Cytochromes of *Escherichia coli, Archiv. Biochem. Biophys.*, 150, 459, 1972.
103. **Ferretti, J. J. and Gray, E. D.,** Enzyme and nucleic acid formation during synchronous growth of *Rhodopseudomonas sphaeroides, J. Bacteriol.*, 95, 1400, 1968.
104. **Daneo-Moore, L. and Shockman, G. D.,** The bacterial cell surface in growth and division, in *The Synthesis, Assembly and Turnover of Cell Surface Components*, Poste, G. and Nicolson, G. L., Eds., Elsevier/North-Holland Biomedical Press, Amsterdam, 1977, chap. 9.

105. **Jacob, F., Ryter, A., and Cuzin, F.,** On the association between DNA and membrane in bacteria, *Proc. R. Soc. London Ser. B.,* 164, 267, 1966.
106. **Ryter, A.,** Flagella distribution and a study of the growth of the cytoplasmic membrane in *Bacillus subtilis, Ann. Inst. Pasteur Paris,* 121, 271, 1971.
107. **Morrison, D. C. and Morowitz, H. J.,** Studies on membrane synthesis in *Bacillus megaterium* KM, *J. Mol. Biol.,* 49, 441, 1970.
108. **Autissier, F. and Kepes, A.,** Segregation of membrane markers during cell division in *Escherichia coli.* II. Segregation of Lac-permease and Mel-permease studied with a penicillin technique, *Biochim. Biophys. Acta,* 249, 611, 1971.
109. **Kepes, A. and Autissier, F.,** Topology of membrane growth in bacteria, *Biochim. Biophys. Acta,* 265, 443, 1972.
110. **Kepes, A. and Autissier, F.,** Membrane growth and cell division in *E. coli,* in *Mechanism and Regulation of DNA Replication,* Kolber, A. R. and Kohiyana, M., Eds., Plenum Press, New York, 1974, 383.
111. **Poole, R. K., Scott, R. I., and Britnell, C. H.,** Cytochrome segregation during cell division in *Escherichia coli, Soc. Gen. Microbiol. Q.,* 6, 22, 1978.
112. **Haddock, B. A. and Schairer, H. U.,** Electron transport chains of *Escherichia coli*: reconstitution of respiration in a δ-amino laevulinic acid requiring mutant, *Eur. J. Biochem.,* 35, 34, 1973.
113. **Green, E. W. and Schaechter, M.,** The mode of segregation of the bacterial cell membrane, *Proc. Natl. Acad. Sci. U.S.A.* 69, 2312, 1972.
114. **Tsukagoshi, N., Fielding, P., and Fox, C. F.,** Membrane assembly in *Escherichia coli.* I. Segregation of preformed and newly formed membrane into daughter cells, *Biochem. Biophys. Res. Commun.,* 44, 497, 1971.
115. **Wilson, G. and Fox, C. F.,** Membrane assembly in *Escherichia coli,* II. Segregation of preformed and newly formed membrane proteins into cells and minicells, *Biochem. Biophys. Res. Commun.,* 44, 503, 1971.
116. **Mindich, L. and Dales, S.,** Membrane synthesis in *Bacillus subtilis.* III. The morphological localization of the sites of membrane synthesis, *J. Cell Biol.,* 55, 32, 1972.

Chapter 4

THE RESPIRATORY SYSTEM OF *ESCHERICHIA COLI*

P. D. Bragg

TABLE OF CONTENTS

I. INTRODUCTION

Escherichia coli is a facultative organism capable of growth on a wide variety of substrates. Under anaerobic conditions, fermentative growth is supported by substrates such as glucose, galactose, and maltose. ATP is formed by substrate-level phosphorylation in the glycolytic pathway and by phosphoroclastic cleavage of pyruvate to acetate.[1] If a terminal electron acceptor such as oxygen, nitrate, or fumarate is present, growth will occur on a wider range of substrates.[2,3] Thus, glycerol, acetate, and intermediates of the tricarboxylic acid cycle can be used. In the presence of these terminal electron acceptors, formation of ATP coupled to electron transfer through the respiratory chain can occur. The nature of the respiratory chain present in the cell varies with the substrate and the terminal electron acceptor.[2-4] Thus, anaerobic growth in the presence of glycerol and fumarate results in the formation of the glycerol-fumarate oxidoreductase pathway. Similarly, the formate-nitrate reductase pathway is formed when formate and nitrate are present in the growth medium. Both of these pathways involve quinones and *b*-type cytochromes, as does the respiratory chain which uses oxygen as the terminal electron acceptor ("aerobic respiratory chain"). Although the NADH oxidase and formate-nitrate reductase pathways are separate systems,[2-4] it is not clear if the NADH-nitrate reductase pathway involves components of both systems. The extent of overlap and interaction between the different respiratory systems remains to be determined. In this chapter, the aerobic respiratory pathway of *E. coli* and its energy transducing function will be discussed. For the present purposes, it will be considered to be a system distinct from the other respiratory pathways. These are discussed elsewhere in these books.

II. THE AEROBIC RESPIRATORY CHAIN

A. Oxidase Activities

Intact cells or whole cell extracts of *E. coli* can oxidize a variety of different substrates (glucose, galactose, fructose, glycerol, succinate, malate, formate, acetate, pyruvate, glutamate, α-ketoglutarate, citrate, and isocitrate) depending on the conditions of growth. By contrast, only a limited number of substrates are oxidized by membrane preparations (NADH, succinate, D- and L-lactate, L-glycerol-3-phosphate, L-malate, formate, α-hydroxybutyrate, dihydro-orotate, and pyruvate).[4-7] The capacity of the membrane preparations to oxidize these substrates, and the occurrence of the bulk of the cytochromes in the membrane fraction, indicates that the aerobic respiratory pathway is membrane-bound. Separation of the inner (cytoplasmic) from the outer membrane confirms that the respiratory system is located in the cytoplasmic membrane. This holds also for the associated energy transducing systems such as the ATPase and transhydrogenase.[4,6,7]

The activity of the oxidases depends on the growth conditions, since a number of the dehydrogenases are inducible. The activity of NADH dehydrogenases varies little with the growth substrate.[8] D-lactate dehydrogenase behaves similarly, except that it is 1.5-fold higher in cells grown aerobically on DL-lactate compared with its level in cells grown on glucose medium. By contrast, L-lactate dehydrogenase is generally present at low levels in cells until induced by growth on DL-lactate. Similarly, L-glycerol-3-phosphate dehydrogenase associated with the aerobic respiratory pathway is induced by aerobic growth on glycerol.[9] Succinate dehydrogenase is inducible as well as being controlled by catabolite repression.[10,11] Pyruvate oxidase, a pyruvate dehydrogenase distinct from the enzymes of the pyruvate dehydrogenase complex and the phosphoroclastic system, is induced by the accumulation of pyruvate in the medium.[12] This enzyme is a flavoprotein which converts pyruvate to acetate and CO_2 with the transfer of reducing equivalents into the aerobic respiratory chain.

There may be other factors which determine respiratory chain activity besides induction and repression of the dehydrogenases. As will be discussed later, the level and nature of the cytochromes present in the aerobic respiratory chain is affected by the oxygen content of the growth medium and by the availability of intracellular cAMP.[13,14]

B. Composition of the Respiratory Chain

Reduced *minus* oxidized difference spectra of intact cells of *E. coli* show that several cytochromes are present. In the visible region, the α-band absorption peaks measured at room temperature are at 630, 590, and 560 nm. These have been attributed to cytochromes *d*, a_1, and b_1.[15] A shoulder on the cytochrome b_1 peak at 562 to 565 nm is due to a soluble cytochrome probably not associated with the aerobic respiratory pathway.[16] Also seen in the difference spectrum are absorption minima at about 465 and 650 nm due to nonheme-iron flavoprotein and cytochrome *d*. As will be discussed later, the cytochrome b_1 peak can be resolved into several components. Besides the cytochromes detected in the reduced *minus* oxidized difference spectrum, another cytochrome, cytochrome *o*, can be demonstrated by its reactivity with carbon monoxide. Absorption peaks at 416, 538, and 567 nm, with minima at 430 and 555 nm, in reduced plus carbon monoxide *minus* reduced difference spectra are characteristic of this cytochrome.[17] Cytochrome *d* also combines with carbon monoxide to show absorption peaks in the difference spectrum at about 440, 537, and 640 nm. Absorption minima are found at 443 and 620 nm.[15,19]

The presence of nonheme-iron-sulfur protein and flavorprotein, detected spectroscopically, can also be measured by direct analysis.[18,20-22] Both FMN and FAD have been found in the respiratory-chain-linked dehydrogenases. Ubiquinone-8, menaquinone-8 and/or demethylmenaquinone-8 are also found associated with the respiratory chain.[23-25] At present, there is no direct evidence for the presence of copper-containing proteins in the respiratory pathway.[26] Functional copper centers are found in mitochondrial cytochrome oxidase and, so, might be expected to occur in bacterial respiratory chains. Some typical analyses of the respiratory chain components of membranes are presented in Table 1.

C. Properties of the Respiratory Chain Dehydrogenases

Several respiratory-chain-linked dehydrogenases have been purified from the membranes of *E. coli*. The dehydrogenases for NADH, D- and L-lactate, L-glycerol-3-phosphate, and pyruvate (pyruvate oxidase) have been brought to a satisfactory state of purity (Table 2).[8,27-31] Succinate and dihydro-orotate dehydrogenases have been partially purified only.[32,33]

There have been several attempts to purify the NADH dehydrogenase from *E. coli* membrane fragments. Earlier work[34-36] showed that at least three types of NADH dehydrogenase activity could be solubilized from membrane preparations: (1) an FAD-containing enzyme having NADH oxidase and NADH-cytochrome c reductase activities. The enzyme is specific for NADH (K_m, 59 to 71 μM); (2) NAD(P)H diaphorase of molecular weight 35,000 to 38,000, stimulated by both FAD and FMN, and with a K_m for NADH of 8 to 14 μM; and (3) NAD(P)H diaphorase of molecular weight 64,000, stimulated by both FAD and FMN, and with a K_m for NADH of 250 to 300 μM. Similar enzymes to last two have been found in the cytoplasmic fraction of the cell.[37] Thus, it is likely that they are either loosely bound to the membranes or are cytoplasmic enzymes trapped in the sealed vesicles of the membrane preparation.

The multiplicity of NADH-utilizing enzymes in the membranes has made it difficult to determine which enzyme is the real respiratory-chain-linked enzyme. A likely possibility is that purified by Dancey et al.[27,28,39] The close relationship of this enzyme to the respiratory chain is suggested by the following evidence.

TABLE 1

Composition of the Respiratory Chain in Membrane Vesicles[a]

	Flavin					Cytochrome				
Strain	Total	Acid-soluble	DMK[b]	UQ[b]	Fe[b]	b_1	o	d	a_1	Ref.
NRC 482[c]	0.11	0.053		4.0	3.4	0.36	0.087			20
NRC 482[c]						0.36	0.066	0.076	0.038	19
NRC 482[d]						0.96	0.02	0.64	0.25	19
C-1(S)[e]	0.53				4.3	0.25	0.10			22
AN387[c]	0.24		0.37	2.26		0.27	0.025	0.071		50

[a] ng-atom or nmol/mg protein
[b] DMK, demethylmenaquinone; UQ, ubiquinone; Fe, nonheme iron.
[c] The cells were grown to the late exponential phase.
[d] The cells were grown to the stationary phase.
[e] The cells were grown in continuous culture with limiting glycerol.

1. The enzyme is specific for NADH.
2. The K_m (30 μM) for NADH is similar to that of the respiratory chain (50 μM).
3. Both the solubilized and membrane-bound NADH dehydrogenase activities are inhibited by 5′-AMP with the same K_i value (500 μM).
4. Antibody against the solubilized enzyme inhibits the membrane-bound NADH oxidase and dehydrogenase activities.

Dancey et al. did not report if their enzyme could reduce oxygen or cytochrome *c*. However, the absolute specificity for NADH, K_m value, FAD as a noncovalently bound prosthetic group, and the ability to reduce the same artificial electron acceptors, suggest that the NADH-cytochrome *c* reductase (solubilized NADH oxidase) described above may be identical to the enzyme of Dancey et al.

The preparation of Dancey et al. has a monomer molecular weight of 38,000. This value is somewhat lower than that of 65,000 to 70,000 determinated for the NADH dehydrogenase of beef-heart mitochondria. The purified dehydrogenase was inhibited by AMP (K_i, 0.6 mM), ADP (K_i, 1 mM), and ATP (K_i, 8.5 mM), but the most effective inhibitor was NAD^+ (K_i, 0.02 mM). This suggests that the enzyme may be regulated by the $NADH/NAD^+$ ratio in the cell.

Three other dehydrogenases which have been obtained in a homogeneous form are those oxidizing L-glycerol-3-phosphate[30] and D-[28,29] and L-lactate.[8] The first two contain noncovalently bound FAD as a prosthetic group, whereas FMN is the prosthetic group of L-lactate dehydrogenase. The two lactate dehydrogenases convert lactate to pyruvate with transfer of reducing equivalents into the respiratory pathway. They differ, therefore, from the cytoplasmic L-lactate dehydrogenase which catalyzes the reduction of pyruvate using NADH as a coenzyme. Nothing is known about the regulation of the lactate dehydrogenases. However, purified L-glycerol-3-phosphate dehydrogenase is inhibited at high (10 mM) and stimulated at low (0.5 mM) levels of ATP. It is possible that ATP, as a product of oxidative phosphorylation, may regulate the rate at which L-glycerol-3-phosphate is oxidized by the respiratory chain.[30]

An interesting property of both the D-lactate and L-glycerol-3-phosphate dehydrogenases is their ability to interact with right-side-out membrane vesicles of mutants lacking the dehydrogenase to reconstitute D-lactate and L-glycerol-3-phosphate oxidase activities.[30,40,41] The oxidase activity so generated was capable of supporting the transport of Rb^+, lactose, and amino acids. The added dehydrogenase interacted with the outer surface of the vesicles and so had a different orientation to that in vivo.[42,43]

TABLE 2

Properties of the Purified Respiratory Chain Dehydrogenases of *E. coli*

Enzyme	K_m (m*M*)	Specific activity[a]	Molecular weight (monomer)	Prosthetic group	Ref.
NADH dehydrogenase	0.03	0.65	38,000	FAD	27, 38
D-Lactate dehydrogenase	0.6—0.9	75—82	71,000—74,000	FAD	28, 29
L-Lactate dehydrogenase	0.12	31	43,000	FMN	3
L-Glycerol-3-phosphate dehydrogenase	0.8	37.7	58,000	FAD	30, 99
Pyruvate oxidase	10	200	66,000	FAD	31
Succinate dehydrogenase	0.15		61,000		93
Dihydro-orotate dehydrogenase		3.63	67,000		33

[a] μmol/min/mg protein.

Pyruvate oxidase, the first of the dehydrogenases to be highly purified, contains 1 mol FAD per polypeptide chain.[31] It exists as a tetramer in the purified form. Besides FAD, the enzyme requires thiamine pyrophosphate and a divalent cation for activity. This enzyme differs from the other dehydrogenases previously discussed in being readily released from the membrane by sonic oscillation. This suggests that it is a peripheral protein of the membrane.[44] By contrast, the other dehydrogenases appear to be intrinsic membrane proteins since chaotropic agents or detergents are required to solubilize them. As expected from their membrane location, all of the isolated dehydrogenases are stimulated by phospholipids.[39,44-47] In some cases at least, this is due to dissociation of the aggregated enzyme into monomers on insertion into the phospholipid micelle.

D. Ubiquinone

Both ubiquinone and menaquinone are found in membranes of *E. coli*.[23-25] Approximately 85% of the total ubiquinone found in wild-type *E. coli* is ubiquinone-8. Ubiquinones-5, 6, and 7 constitute 1, 2, and 10% of the total amount of ubiquinone present. Ubiquinones-1, 2, 3, and 4 together account for less than 0.1% of the total ubiquinone. The lower isoprenologues presumably represent stages in the biosynthesis of the major ubiquinone.[24] A similar situation occurs with the menaquinones. The two major menaquinones are menaquinone-8 and 2-demethylmenaquinone-8. However, smaller amounts of the hexa-, octa-, and nonaprenyl derivatives of menaquinone, and the heptaprenyl derivative of 2-demethylmenaquinone, are present.[25]

The total amount of quinone is generally in great excess over the other respiratory chain components with the exception of nonheme iron (Table 1). The relative amounts of ubiquinone and menaquinone present in cells depends on the degree of aeration of the growth medium. Polglase et al.[48] found that 20-fold more ubiquinone than menaquinone was formed when the culture was vigorously aerated. Menaquinone predominated in anaerobically grown cells. Higher levels of menaquinone are also formed in cells in which respiratory function is impaired. Thus, cells grown in the presence of cyanide-,[49] ubiquinone-[50] or heme-deficient mutants,[51] or cells growing on limiting sulfate and showing alterations in the respiratory chain at site 1,[22] all show higher than normal levels of menaquinone. These results suggest that menaquinone can substitute for ubiquinone under some circumstances. The extent to which this can occur has been investigated by Wallace and Young[50,52] using mutants unable to form ubiquinone (*ubi*$^-$ *men*$^+$), menaquinone/demethylmenaquinone (*ubi*$^+$ *men*$^-$), or both quinones (*ubi*$^-$ *men*$^-$). NADH, succinate, D-lactate, and α-glycerophosphate oxidase activities were greatly impaired in the *ubi*$^-$ *men*$^-$ strain compared with the *ubi*$^+$ *men*$^+$ strain. These oxidases were unaffected in the *ubi*$^+$ *men*$^-$ strain, suggesting that ubiquinone was the major quinone of the aerobic respiratory pathway.[50,52,53] However, in the *ubi⁻ men⁺* mutant, substantial D-lactate and α-glycerophosphate oxidase activities were retained, indicating that menaquinone could substitute for ubiquinone in these pathways. However, it could not replace ubiquinone in the NADH and succinate oxidase chains. Menaquinone is primarily involved in anaerobic electron transport pathways such as the reduction of fumarate by L-glycerol-3-phosphate,[54] NADH,[54] and dihydroorotate.[55]

E. Nonheme Iron

Nonheme iron is a major component of the respiratory pathway, being at a 10-fold higher level than the other respiratory chain components with the exception of ubiquinone (Table 1). However, the level of nonheme iron is not constant.[18,21,22] The major factor determining the level of nonheme iron in the membranes is the level available to the growing cells. Rainnie and Bragg[56] examined the effect of allowing growing cells to deplete the iron content of the medium. When growth was limited by iron depletion, the amounts of nonheme iron and heme iron in the membrane were 69 and 9.2 ng atoms/g cells, respectively. Within 15 min of adding 12 μM ferric citrate to the medium, the level of nonheme iron had risen to 272 ng atoms/g cells, whereas the level of heme iron remained at about 9.9 ng atoms/g. This ratio of nonheme iron to heme iron of about 20 to 30:1 has been found in the membrane of cells grown under iron-sufficient conditions.[18,21]

There has been only one report of the concentration of acid-labile sulfide in membranes of *E. coli.*[22] Bacteria grown in continuous culture in the presence of excess sulfate (15 mM) contained 4.28 and 1.43 nmol/mg membrane protein of nonheme iron and acid-labile sulfide, respectively. When cell growth was limited by sulfate in the growth medium (50 μM), the levels of nonheme iron and acid-labile sulfide were 2.42 and 0.63 nmol/mg protein, respectively. Thus, a decrease in the amount of acid-labile sulfide caused by sulfate limitation was accompanied by a decrease in the level of nonheme iron, as would be expected from the occurrence together of these elements in iron-sulfur proteins.

The presence of iron-sulfur proteins can also be detected as a signal at $g = 1.94$ in electron paramagnetic resonance spectra of membrane preparations which have been reduced by dithionite or substrates.[22,57,58] Spectra obtained at 12°K also show a predominant peak at $g = 1.94$.[22] Further signals can be observed, but these have not been related to the numerous iron-sulfur centers found in the NADH and succinate dehydrogenases of mitochondria.[59] In agreement with the results of chemical analysis, the $g = 1.94$ signal is lower in sulfate-limited than in normal cells.[22]

F. Cytochromes

Reduced *minus* oxidized difference spectra of membranes from aerobically grown cells of *E. coli* (measured at room temperature) show absorption bands attributable to cytochromes b_1, a_1, d, and o. Cytochrome o cannot be distinguished from cytochrome b_1 in these spectra since their absorption bands overlap.

1. Cytochromes b and c

Cytochrome b_1 shows an α-absorption peak at 559 to 560 nm in difference spectra measured at room temperature. At 77°K, the α-band absorption peaks are narrower and show a shift of 2 to 3 nm towards lower wavelengths. Under these conditions, the cytochrome b_1 peak is partially resolved to show absorption maxima at 548 to 552, 556, 558, and 562 nm.[22,51,60-62] In fourth-order finite-difference spectra, the 548 to 552 nm peak resolves into two components (*c*-type cytochromes) with absorption maxima at 548 to 549 nm and 552 to 553 nm.[63] The remaining three peaks have been attributed to *b*-type cytochromes.

The presence of several *b*-type cytochromes has been confirmed by redox titration.[60,64] Cytochrome b_{556} and b_{558} showed midpoint oxidation-reduction potentials of +15 to +34 mV (n = 1) and +165 to +205 mV (n = 1), respectively.[60] The exact values depended on the growth conditions. The relative contribution of the two cytochromes to the cytochrome b_1 peak was estimated to be 73 and 27% in membranes from exponential phase cells, and 60 and 40% in membranes from stationary phase cells. The presence of 14 m*M* KCN did not affect the midpoint potentials, suggesting that cytochrome o, which reacts with KCN, cannot be either of these two cytochromes. The redox potentials of the *c*-type cytochromes and cytochrome b_{562} could not be estimated due to the small absorption peaks of these components. However, cytochrome b_{562} was fully reduced by ascorbate (E'_0, +80 mV) and so must have a redox potential of at least +100 mV. In contrast to these results, Hendler et al.[64] found that the cytochrome b_1 peak was composed of three components with midpoint oxidation-reduction potentials of −50 mV, +110, and +220 mV. The cytochromes responsible for these potentials were not identified.

Only one cytochrome, cytochrome b_1, has been solubilized and purified from the membranes of *E. coli*.[65,66] Its prosthetic group is noncovalently bound protoporphyrin IX. The purified cytochrome exists as an aggregate (mol wt, 500,000) of the monomer (mol wt, 66,000). The midpoint oxidation-reduction potential is −340 mV.[66] This value is considerably lower than the value of −10 to −20 mV found with crude preparations of the cytochrome,and may be due to the removal of a potential-modifying protein during purification of the cytochrome. A potential-modifying protein able to increase the midpoint potential from −340 mV to −120 mV was isolated, but not further characterized.[66]

It is not easy to relate the purified cytochrome b_1 to one of the spectroscopically detectable cytochromes. The α-band of the purified cytochrome is at 557.5 nm at room temperature. This peak should shift to 554.5 to 555.5 nm at 77°K. Although no band is observed at this wavelength, the expected absorption maximum is close to that of cytochrome b_{556}. The difference could be accounted for by the modification of the cytochrome which is observed when it is removed from its environment in the membrane. It is not clear if there is only one heme-b-containing polypeptide in the membrane since only 34% of the cytochrome b_1 was solubilized in Deeb and Hager's procedure.[66] The remaining nonsolubilized cytochrome could represent the other *b*-cytochromes detected spectroscopically. However, there is no information on this point.

2. *Cytochrome Oxidases*

Cytochromes a_1 (peak at 594 nm), *d* (peak at 628 nm; trough at 650 nm), and *o* (characteristic spectrum with carbon monoxide) have been detected in the membranes of *E. coli.*

Although cytochrome a_1 is the sole cytochrome oxidase of *Acetobacter pasteurianum*,[67] it is unlikely to have a significant role as a terminal oxidase in *E. coli.* Photodissociation spectra of the carbon monoxide complexes of the cytochrome oxidases in *E. coli* indicated that cytochrome *o* was the major oxidase in exponential phase cells, whereas both cytochromes *o* and *d* were active in the stationary phase.[67] There was no indication that cytochrome a_1 functioned as an oxidase. This has been confirmed by stopped-flow kinetic measurements of the rate of reoxidation of the reduced cytochromes.[61] Cytochromes *o* and *d* were reoxidized with a half-time of less than 3.3 msec compared with 25 msec for cytochrome a_1. Thus, the function of cytochrome a_1 in *E. coli* membranes is presently unknown. It has a midpoint oxidation-reduction potential of +147 mV (n = 1).[60]

The evidence from photodissociation spectra on the relative roles of cytochrome *o* and *d* as oxidases in exponential and stationary phase cells has been supported by the results of other studies.[51,68] It is likely that there is a relatively low level of dissolved oxygen in the medium during late exponential and stationary phases of growth when the cell density is high. Haddock and Schairer[51] have suggested that the respiratory chain formed under conditions of high aeration contains, predominantly, cytochromes b_{556}, b_{562}, and *o*, whereas under conditions of low aeration an additional cytochrome segment consisting of cytochromes b_{558} and *d* is incorporated into the respiratory system. The K_m values for oxygen of cytochrome *o* (K_m, 0.2 μM) and cytochrome *d* (K_m, 0.024 μM) are consistent with their roles as the primary oxidases under conditions of high and low aeration, respectively.[68]

Little is known about the biochemical properties of cytochrome *o* since it has not been purified from *E. coli.* It is probably a *b*-type cytochrome[69] which can react with carbon monoxide in the reduced state and with cyanide.[15,17,70,71] Cytochrome *o* is identified by its characteristic spectrum with carbon monoxide.[17] In reduced *minus* oxidized difference spectra, its absorption peaks are masked by those of the other *b* cytochromes. In carbon monoxide difference spectra recorded at 77°K, absorption peaks at 557 and 430 nm are eliminated on reaction with carbon monoxide.[19] If cytochrome *o* does have an α-absorption peak at 557 nm, then it is not detectable as a distinct peak in reduced *minus* oxidized difference spectra at 77°K where it must be hidden by the peaks of cytochromes b_{556} and b_{558}. Of the absorbance at 556 nm, 20 to 40% may be due to cytochrome *o*.[61] The midpoint oxidation-reduction potential of cytochrome *o* in *E. coli* membranes has not been determined, but it should be at least +100 mV since the cytochrome is reduced fully by ascorbate (E'_o, +80 mV).[72] A purified cytochrome *o* from *Vitreoscilla* had a midpoint oxidation-reduction potential of +100 mV.[69]

The second cytochrome oxidase of the aerobic respiratory pathway is cytochrome *d*. Although it has not been solubilized and purified from the membranes of any organism, its prosthetic group has been identified as an iron-chlorin.[73] Cytochrome *d* is detected in reduced *minus* oxidized difference spectra as an absorption peak at 628 nm with a trough at 648 nm.[15,19,70] The Soret band is at 442 nm. The trough at 648 nm is usually considered to be due to the oxidized form of cytochrome d as it is observed in membrane preparations in the absence of substrate.[19,70] In the presence of substrate, it is replaced by the peak of reduced cytochrome *d* at 628 nm. The conversion of oxidized to reduced cytochrome *d* proceeds through an intermediate, cytochrome *d**, which does not have a detectable absorption band in the 600 to 700 nm region of the spectrum.[70]

Cytochrome *d* behaves in a complex manner in experiments to determine its redox potential.[60] The midpoint potential of cytochrome *d* measured from absorption changes in the 628 nm peak is +260 mV (n = 1). However, the absorption band at 648 nm does not follow normal redox behaviour. Increasing the redox potential of the system from 0 to 450 mV does not result in the appearance of the band at 648 nm due to the oxidized cytochrome, unless oxygen is present. The significance of this is not clear. One possible interpretation of these results is that cytochrome *d** is a partially reduced species of cytochrome *d*. The measured redox potential of +260 mV (n = 1) would then apply to the interconversion of fully reduced cytochrome *d* and cytochrome *d**.

Cytochrome *d** is present at very low levels in the aerobic steady state. Increasing levels are found as the temperature is decreased.[74] At 1°C, 28% of the cytochrome *d* is present in this form with ascorbate as reductant. This amount is increased to 57% at −38°C. Obviously, there is a temperature-sensitive step in the oxidation-reduction cycle of cytochrome *d*. Cytochrome *d**, not the fully oxidized or fully reduced forms, reacts with cyanide.[70] The rate of formation of cyanocytochrome *d* is directly proportional to the rate of electron flux through cytochrome *d*.[70,71] From these results, it can be concluded that cytochrome *d** is an intermediate in the normal redox cycle of cytochrome *d* in the respiratory chain. Since reduction of one molecule of oxygen requires the transfer of four electrons, the existence of several intermediate states of reduction of the cytochrome oxidase would be expected. The presence of four redox carriers (two molecules of heme and two atoms of copper) in the catalytic unit of the mitochondrial cytochrome oxidase is consistent with the need to transfer more than one electron at a time to the oxygen molecule.[75] There is no evidence that more than one heme molecule is present in cytochrome *d.* Although substantial amounts of copper have been detected in *E. coli* membranes, it does not give an electron paramagnetic resonance signal.[26] It may be similar to the "invisible" copper of the mitochondrial oxidase which, although undectable by spectroscopic techniques, appears to have a role in oxygen reduction.[75] Further studies on the mechanism of reduction of oxygen by cytochrome *d* must await the solubilization and purification of this cytochrome.

G. Reversal of Electron Flow in the Aerobic Respiratory Chain

Energy-linked reduction of NAD^+ by succinate involves the transfer of reducing equivalents via succinate and NADH dehydrogenases to NAD^+ by an energy-dependent process in which there is reversal of the normal direction of electron transport at site 1. In mitochondria, energy can be supplied either by hydrolysis of ATP or by generation of an energized state by electron transfer through sites 2 or 3.

The ATP-dependent reaction has been demonstrated in *E. coli* membrane preparations.[76,77] The activity is low, being about 8 to 11 nmol NAD^+ reduced/min/mg protein. Sweetman and Griffiths[76] showed that ATP, and to a lesser extent other nucleoside triphosphates, could act as an energy source for the reaction. The ability of the nucleoside triphosphates to act as energy donors was related to their ability to act as substrates of the membrane-bound ATPase system. Hydrolysis of 1 to 2 molecules of ATP was required for the reduction of 1 molecule of NAD^+. Energy-dependent reduction of NAD^+ energized by substrate oxidation through the respiratory chain has not yet been demonstrated in *E. coli* or other bacteria.

NAD^+ reduction is inhibited by uncoupling agents and by inhibitors of succinate dehydrogenase. The pathway of transfer of reducing equivalents involves ubiquinone, but not cytochrome, as shown by the use of mutants lacking one or other of these components.[77] Besides succinate, DL-lactate and glycerol-3-phosphate can supply reducing equivalents for the reduction of NAD^+. These pathways are also dependent on the presence of ubiquinone being absent in membranes from a ubiquinone-deficient mutant unless the membranes are supplemented by exogenous ubiquinone-1.[77] It is probable that ubiquinone is the locus at which reducing equivalents from several flavin-linked dehydrogenases can enter the respiratory chain.

III. ENERGY COUPLING TO THE AEROBIC RESPIRATORY CHAIN

In *E. coli*, a number of energy-dependent processes such as ATP formation (oxidative phosphorylation), active transport, transhydrogenation of $NADP^+$ by NADH, reversal of electron flow in the respiratory chain, and flagella movement can be driven by the energy derived by substrate oxidation through the aerobic respiratory chain.[4,19,78] Although outside the scope of this chapter, some of these processes can also be coupled to the anaerobic pathways of electron transfer which result in the reduction of fumarate and nitrate.[2,3] The reactions occurring in the formation of ATP by oxidative phosphorylation are reversible. Thus, ATP hydrolysis can also supply energy to drive these energy-dependent processes.

Energy is supplied to the energy-dependent processes as an "energized state" of the membrane.[4,19,78-81] The nature of the energized state is still a matter of some dispute.[82]

However, many recent findings with *E. coli* are consistent with the "Chemiosmotic Hypothesis" of Mitchell.[83] According to this hypothesis, the components of the respiratory pathway are organized in the membrane such that electron transfer through the respiratory chain is coupled to the transfer of protons across the membrane. Thus, the energized state is the transmembrane electrochemical gradient of protons ("protonmotive force") ($\Delta\bar{\mu}_{H}+$). This is composed of electrical and chemical parameters related, in electrical units (usually millivolts), by the equation:

$$\Delta\bar{\mu}_{H^+} = \Delta\psi - Z \cdot \Delta pH$$

where $\Delta\psi$ is the electrical potential across the membrane, ΔpH is the pH difference between the internal and external compartments, and Z is a factor to convert pH to electrical units. At 25°C, Z has the numerical value of 59 mV.

A. Sites of Proton Translocation

As originally demonstrated by Reeves,[84] the addition of a pulse of oxygen to an anaerobic suspension of cells results in extrusion of protons from the cell in the presence of a permeant ion such as thiocyanate. The stoichiometry of this process can be measured. Oxidation of endogenous substrates yielded H^+ to O ratios of about 4 with cells grown under a variety of conditions (substrate-, oxygen-, sulfate-, or ammonium ion-limitation).[85-87] Oxidation of L-malate or formate gave H^+ to O ratios of 3.1 to 3.6 and 3.63, respectively, whereas ratios of 1.9 to 2.42 were obtained with succinate, D-lactate, and glycerol.[22,85,88] On the basis of these results, two energy-conserving sites appear to be present in the aerobic respiratory chain of *E. coli.* One of the sites would be in the NADH dehydrogenase region of the chain and equivalent to site 1 in mitochondria. The second site would be associated with the respiratory chain between oxygen and the site, probably ubiquinone, where reducing equivalents from succinate, D-lactate, and L-glycerol-3-phosphate are introduced. A site of proton translocation would also be present in the formate dehydrogenase region in those cells having formate oxidase activity.[85] The presence of the proton translocation site between NADH and ubiquinone has been confirmed by adding pulses of ubiquinone-1 to anaerobic cell suspensions. H^+ to $2e^-$ ratios of 1 to 2 were obtained.[89] The presence of these two coupling sites associated with the aerobic respiratory pathway has also been indicated by molar growth-yield experiments,[6,87,90] P to O ratios,[91] reversal of electron transfer through NADH dehydrogenase,[76,77] and coupling to energy-dependent transhydrogenation.[92,93]

Mitchell predicted that $2H^+$ would be translocated across the membrane during the passage of $2e^-$ through each site of energy conservation in the respiratory chain. However, Lehninger and his co-workers have recently shown that in mitochondria 3 to 4 H^+ may be translocated per site.[94] Compensatory movements of phosphate across the membrane had led previous workers to underestimate the H^+ to O ratio. Although recent results make it unlikely that the movement of phosphate is responsible for the low H^+ to O ratios in *E. coli*,[95] the movement of other ions has not been eliminated. It is of interest that H^+ to O ratios of greater than two per site have been observed in *E. coli.* Meyer and Jones[96] obtained a H^+ to O ratio of six for the oxidation of endogenous substrates, and oxidation of succinate, glycerol, and D-lactate has given H^+ to O ratios of up to 2.46.[22,85,88] Thus, there could be 3 H^+ translocated per site in the respiratory chain of *E. coli.* Recently, Meijer et al.[97] have found that 3 to 4 H^+ per site are translocated in *Paracoccus denitrificans*.

B. Magnitude of the Protonmotive Force

There have been several attempts to measure the magnitude of the protonmotive force. The results of some of these investigations are summarized in Table 3. The

TABLE 3

Protonmotive Force in Respiring *E. coli* Cells, Spheroplasts, and Right-Side-Out Vesicles

Preparation	Substrate	Buffer pH	Buffer K⁺ (m*M*)	Buffer Valinomycin[a]	ΔpH	Δψ(mV)	$\Delta\bar{\mu}_{H^+}$(mV)	Ref.
tris-EDTA-treated cells	Succinate	7	0.01	+	0.13	122	129	102
		7	1	+	0.62	84	120	
		7	150	+	0.89	16.2	68	
		6	150	+	2.10	12.8	137	
		8	150	+	−0.10	13.8	8	
Spheroplasts	Endogenous	6.45—6.75	0.1	+	1.65 ± 0.2	132 ± 10	230 ± 15	100
Vesicles	Ascorbate (+ phenazine methosulfate)	5.5	50	−	1.95[b]	74	189	101
	D-Lactate	5.5	50	−	1.7[b]	70	172	
	Succinate	5.5	50	−	0	64	64	

[a] +, −: indicates presence or absence of valinomycin.
[b] Calculated from Reference 101.

maximum protonmotive force recorded is 230 mV.[100] According to thermodynamic calculations, a force of 210 mV is required to obtain an ATP/ADP ratio of 1 with 10 m*M* phosphate.[83] Thus, the observed protonmotive force should support ATP synthesis. Ascorbate and succinate are the most efficient energy donors in right-side-out membrane vesicles.[101] The protonmotive force generated by oxidation of these substrates would arise solely from the second site of energy coupling in the respiratory chain.

The relative contributions of the membrane potential (Δψ) and pH difference (ΔpH) to the protonmotive force depends on the pH and the concentration of K⁺ in the buffer in which the preparation is suspended.[102] In *tris*-EDTA treated cells in the presence 0.01 m*M* K⁺, Δψ makes the major contribution to the protonmotive force. This contribution is decreased at higher concentrations of K⁺ if valinomycin is present, since this ion can then move into the cells. This results in an increase in ΔpH concomitant with the decline in Δψ. The contribution of ΔpH to the protonmotive force is highest at pH 5.5 to 6.0.[101,102] In both cells and right-side-out vesicles, the maximum ΔpH is about 2 pH units. At increasing pH values, the contribution of ΔpH to the protonmotive force declines to become zero at about pH 7.5 to 8.0. Thus, at pH values greater than pH 7.5 to 8.0, the protonmotive force is entirely due to Δψ. In vesicles, Δψ changed little between pH 5.5 and 8.5, being approximately 75 mV over this pH range.[101] Uncoupling agents decrease both ΔpH and Δψ, as would be expected from the ability of these compounds to conduct protons across the membrane. Valinomycin and nigericin reduce the contribution of Δψ and ΔpH, respectively, to the protonmotive force.[101] This is consistent with their known modes of action. Valinomycin provides a pathway for K⁺ to cross the membrane in response to Δψ, whereas nigericin causes electroneutral exchange of K⁺ for protons.

The generation of an uncoupler-sensitive protonmotive force has also been observed with everted vesicles.[103-105] Using the permeant fluorescent dye 9-aminoacridine, Singh and Bragg[104,105] found that at pH 7 oxidation of NADH, succinate, D-lactate, or glycerol-3-phosphate, or the hydrolysis of ATP, generated a ΔpH of 3.3 to 3.7 pH units. These experiments were carried out in 300 m*M* KCl where Δψ was zero. A protonmotive force of 195 to 218 mV can be calculated from these results. In the absence of KCl, a $\Delta\psi$ of up to 150 mV was generated by substrate oxidation.[123]

C. Coupling of the Protonmotive Force to the Formation of ATP

Maloney et al.[106] showed that the application of a membrane potential could drive ATP formation in *E. coli.* The membrane potential was generated by placing cells with a normal complement of K^+ into a K^+-free buffer. The addition of valinomycin resulted in the diffusion of K^+ from the cells and the generation of a potential, interior negative, across the cell membrane. In further experiments,[107] it was shown that imposition of a membrane potential increased the concentration of ATP in starved cells from 0.1 to 1.6 m*M*. ATP could also be synthesized by ΔpH in the absence of a membrane potential by diluting cells at pH 8 into a medium at pH 3 which contained 100 m*M* KCl in the presence of valinomycin to prevent the formation of a membrane potential. Intracellular ATP levels increased from 0.1 m*M* to 2.1 m*M*. ATP synthesis can also be driven by a combination of $\Delta\psi$ and ΔpH. Somewhat similar results to these have been obtained by Grinius et al.[108] using intact cells, and by Tsuchiya and Rosen[109,110] with right-side-out membrane vesicles. Wilson et al.[107] found that there was little synthesis of ATP until the protonmotive force attained a value of 200 mV. This compares favorably with the value of 210 mV calculated by Mitchell[83] required to maintain an ATP/ADP ratio of 1 in the presence of 10 m*M* phosphate.

The mechanism by which the protonmotive force is coupled to the formation of ATP is beyond the scope of this review. However, there is clear evidence that the membrane-bound Ca^{2+}-Mg^{2+}-activated ATPase system is involved in this process.[4,19] Thus, mutants in which the ATPase is inactive (*unc* A mutants), or active but functionally uncoupled from energization of the membrane (*unc* B mutants), cannot form ATP by oxidative phosphorylation and cannot hydrolyze ATP to form an energized state of the membrane. Moreover, ATP cannot be formed by the imposition of an artificial protonmotive force in *unc* A mutants.[107]

IV. ORGANIZATION OF THE AEROBIC RESPIRATORY CHAIN AND THE GENERATION OF THE ENERGIZED STATE

A number of schemes have been proposed for the arrangement of the components in the respiratory chain of *E. coli*,[20-23,51,58,112-114] and have been discussed previously by the author.[19,21] Two recent schemes are shown in Figure 1.[4,62] These will serve as the basis to discuss evidence for the respiratory chain sequence.

In the proposal of Haddock and Jones,[4] the respiratory chain is branched at the level of ubiquinone, which serves both to collect reducing equivalents from the various dehydrogenases and to pass them to oxygen by either of two cytochrome chains. The main chain involves cytochromes b_{556} and *o*, and probably, b_{562}. It may be the only chain present in cells grown with high aeration. The other chain of cytochromes b_{558} and *d* is formed under conditions of inadequate aeration, growth in the presence of cyanide, or during sulfate limitation.

The respiratory chain sequence of Downie and Cox[62] differs from that of Haddock and Jones in several features. Thus, ubiquinone, as ubisemiquinone complexed with nonheme iron, is suggested to occur at two sites, one on either side of cytochromes b_{562}-b_{556} in the respiratory chain sequence. Downie and Cox postulated that the cytochrome b_{558}-*d* chain originates from the main cytochrome b_{562}-b_{556}-*o* pathway at the second ubisemiquinone-nonheme-iron site. In contrast to the respiratory chain of Haddock and Jones, electrons passing from substrate to cytochromes b_{558}-*d* must traverse cytochromes b_{562} and b_{556}.

The scheme of Downie and Cox was derived from studies of a mutant unable to

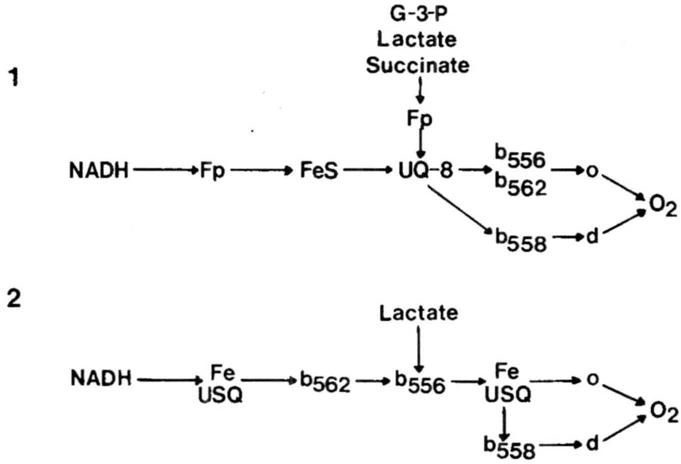

FIGURE 1. Schemes for the sequence of the aerobic respiratory chain components of *E. coli* as proposed by Haddock and Jones (1)[4] and Downie and Cox (2).[62] Fp, flavoprotein; FeS, iron-sulfur protein; UQ-8, ubiquinone-8; Fe-USQ, complex of nonheme iron and ubisemiquinone; and b_{556}, b_{558}, b_{562}, *o*, *d*, cytochromes b_{556}, etc.

form ubiquinone. Aerobic steady-state levels of reduction of cytochromes b_{556} and b_{558} were higher (50 to 60% reduced) in membranes from ubiquinone-deficient cells compared with those from the normal strain (5 to 10% reduced). When ubiquinone-1 was added to the mutant membranes, the level of cytochrome reduction returned to normal. These results suggest that ubiquinone is located between these b cytochromes and oxygen. Ubiquinone also appears to be located prior to cytochromes b_{556} and b_{562} since they were reduced more slowly in ubiquinone-deficient than in normal membranes. Downie and Cox stated that cytochrome b_{558}, like cytochrome *d*, was not appreciably reduced in the aerobic steady state in ubiquinone-deficient and normal membranes and must, therefore, be closer than ubiquinone to oxygen. In studies in the author's laboratory,[124] cytochrome b_{558}, as well as cytochromes b_{556} and b_{562}, were reduced in the aerobic steady state to levels commensurate with their respective midpoint oxidation-reduction potentials.[60] Thus, it seems more likely that Downie and Cox[62] underestimated the extent of reduction of cytochrome b_{558} since it was masked by the presence of cytochromes b_{556} and b_{562}. These cytochromes have been completely resolved only in the spectroscopic studies of Shipp.[63] Shipp found that cytochrome b_{558} was always present in cells grown under a variety of conditions. Thus, the apparent absence of this cytochrome in membranes from cells grown at high rates of aeration may be due to inadequate resolution of the cytochrome *b* peak even at 77°K. In support of this view, Pudek and Bragg,[60] using redox titration methods, found that cytochrome b_{558} was present in membranes from such cells.

The arrangement of the cytochromes in the scheme of Haddock and Jones[4] is to some extent dependent on the apparently simultaneous induction of cytochromes b_{558} and *d* during growth under a variety of different conditions. However, as recently found,[111] the coordinate synthesis of cytochrome c_{552} and NADH-nitrite reductase activity could not be taken as evidence that cytochrome c_{552} was a component of the NADH-nitrite reductase pathway. Thus, the coordinate synthesis of cytochromes b_{558} and d may not mean that these cytochromes form a separate branch of the respiratory chain, synthesized, and integrated into the main respiratory pathway. A second line of evidence in support of the scheme of Haddock and Jones comes from stopped-flow experiment on the rate of reoxidation of the reduced cytochromes following a pulse of oxygen.[61] Cytochrome d was reoxidized with a half-time of less than 3.3 msec. The oxidation of the *b* cytochromes showed two phases with half-times of less than 3.3 msec and about 25 msec, respectively. The two pools of cytochrome *b* were of about equal size. The fast oxidizing pool is kinetically competent to function as an oxidase and, perhaps, represents cytochrome *o*.[61] In these experiments, the spectral characteristics of the *b* cytochromes reacting in the two phases were not determined. The possibility that the two phases do not represent individual cytochrome species, but pools of several *b* cytochromes, cannot be eliminated. Thus, we have observed[12] that the reduction of the b cytochromes by NADH, succinate, and D-lactate is biphasic. Both cytochromes b_{556} and b_{558} are reduced in both the slow and the fast phases of reduction.

Downie and Cox[62] have suggested that cytochrome b_{562} is more strongly reduced by NADH than by D-lactate, and they conclude that cytochrome b_{562} precedes cytochrome b_{556} in the respiratory chain sequence. If this conclusion is correct, the sequence is different from the order of the redox potentials of these cytochromes. Cytochrome b_{556} has a midpoint oxidation-reduction potential of +15 to +34 mV, whereas the redox potential of cytochrome b_{562} must be at least +100 mV since it is completely reduced by ascorbate (E'_o, +80 mV).[60]

It can be concluded from the above discussion that the sequence of the cytochromes has not been established unambiguously. The position of ubiquinone and nonheme iron in the respiratory chain is also unsure. The suggestion by Downie and Cox[62] that there are two sites for ubiquinone in the respiratory chain of *E. coli* is supported by some earlier studies.[20,72,112,115] However, it is not clear how two pools of ubiquinone, presumably having different redox potentials, could exist in the membrane. A plausible explanation of the *apparent* existence of ubiquinone at two sites comes from a suggestion of Mitchell[6,116,117] on the flow of reducing equivalents in the site 2 region of the respiratory chain. This proposal, the protonmotive Q cycle, adapted to the respiratory chain of *E. coli*, is shown in Figure 2. Electrons from the dehydrogenases together with electrons from a *b*-type cytochrome (possibly cytochrome b_{562}) are transferred to ubiquinone which, with the uptake of protons at the cytoplasmic side of the membrane, becomes reduced to ubiquinol. Ubiquinol diffuses to the outer face of the membrane where protons are discharged into the medium, while electrons are disproportionated between the cytochrome oxidase (cytochromes *o* or *d*) and a *b* cytochrome (possibly cytochrome b_{556}). Electrons from the *b* cytochrome are transferred via the other *b* cytochrome to reduce ubiquinone on the inner face of the membrane. The pathway of electrons to oxygen will be transmembranous if the reaction site of oxygen is on the cytoplasmic surface of the membrane. Cytochrome b_{558} can be integrated into this pathway either as part of a cytochrome b_{558}-*d* pathway or as a parallel route to allow the return of electrons from ubiquinol to ubiquinone. In this scheme, ubiquinone is needed both for the reduction and oxidation of the *b* cytochromes. Thus, it can account for the apparent location of ubiquinone at two sites in the respiratory chain. Perhaps a simpler explanation of the requirement of ubiquinone for the oxidation and reduction of the *b* cytochromes is that ubiquinone forms a complex with the cytochrome, or a protein interacting with it, to affect the redox properties of the cytochrome. Thus, lack of ubiquinone could influence both its reduction and oxidation. Because the amount of ubiquinone greatly exceeds that of the dehydrogenases and cytochromes, and kinetically acts as a single pool, the possibility that it might be compartmentalized has not been seriously considered. However, the recent results of Yu et al.[118] suggest that protein-bound, not free, ubiquinone is the true carrier form of ubiquinone in the respiratory chain. The results of Wallace and Young[50,52] on the apparent specificity of the *E. coli* dehydrogenases for ubiquinone, menaquinone, and their derivatives, and the results of Baillie et al.[20] on the different requirements for ubiquinone to reconstitute the NADH and succinate oxidase pathways in a quinone-

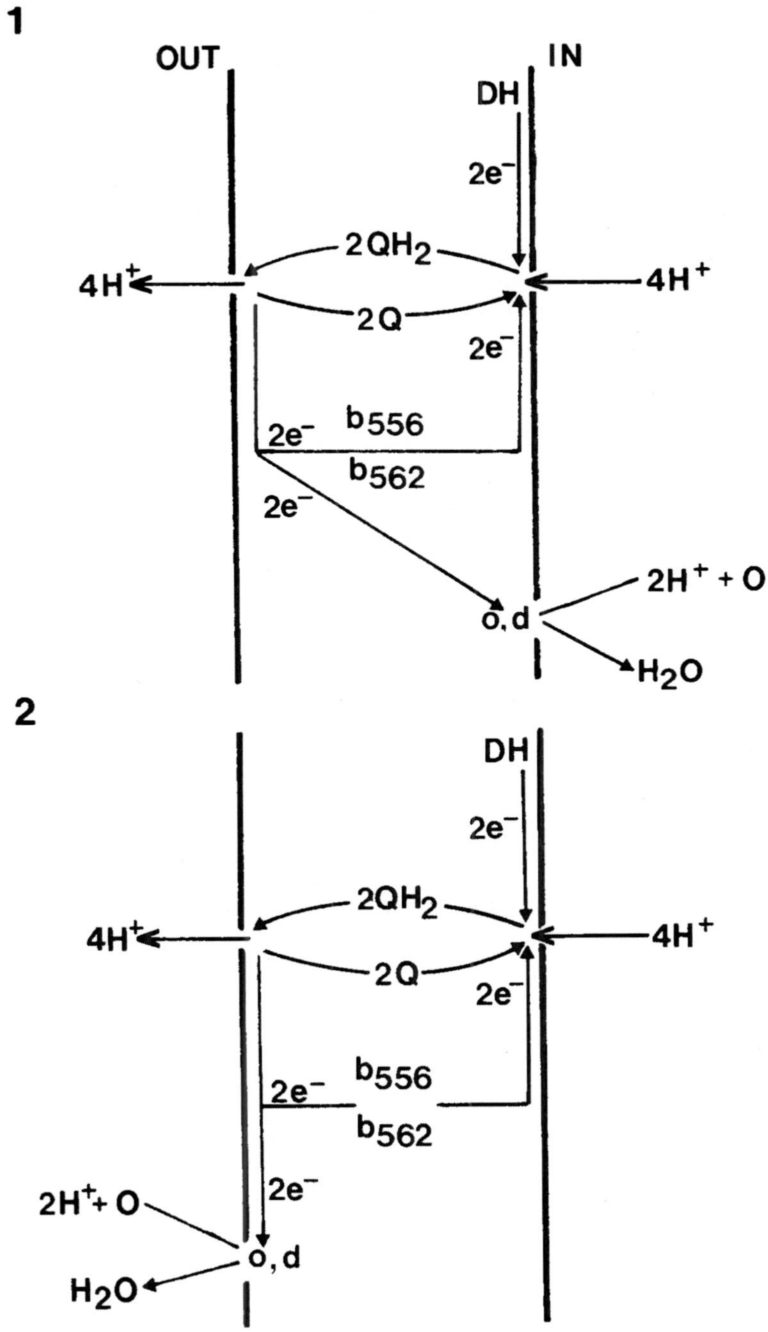

FIGURE 2. Protonmotive Q cycle in the cell membrane of *E. coli* linked to the reduction of oxygen at the cytoplasmic (1) or the external (2) face of the membrane. DH, dehydrogenase; Q, ubiquinone; and QH_2, ubiquinol.

depleted respiratory chain complex from *E. coli,* are more readily understood if ubiquinone has distinct binding sites on the proteins of the respiratory chain.

Cox et al.[112] have speculated on the nature of the reduced form of ubiquinone in the respiratory chain. Hamilton et al.[57] found that membranes of normal, but not ubiquinone-deficient, cells gave an electron paramagnetic resonance signal at g =

2.003 which was attributed to ubisemiquinone. The signal could account for only 2% of the ubiquinone present in the membranes. However, indirect evidence from ubiquinone extraction experiments[112] suggested that the ubiquinone might be present entirely as the semiquinone in the absence of substrate. The low resonance signal was attributed to interaction of the ubisemiquinone with nonheme iron. The evidence supporting this hypothesis is very indirect, and more convincing results are required before it can be accepted. Unfortunately, little is known about the position of nonheme iron in the respiratory chain of *E. coli.*[19,21,22,58,72,119] It is likely that nonheme iron is associated with the dehydrogenases, but direct evidence for this has not been presented.[58] The loss of site 1, associated with diminished levels of iron-sulfur proteins, suggests that nonheme iron is present in the NADH dehydrogenase region of the respiratory chain.[22] The presence of a nonheme iron species in the cytochrome b_1 region of the respiratory chain has been indicated by experiments with chelating agents.[72,119] However, there is no evidence at present that it is associated with ubiquinone.

As discussed in a previous section, there are two regions (NADH dehydrogenase, cytochromes) in the respiratory chain closely associated with the translocation of protons coupled to electron transport. Mitchell[83] has suggested that an arrangement of alternating hydrogen and electron carriers across the membrane could result in proton translocation. This idea has been applied to the respiratory chain of *E. coli.*[4,22] One possible arrangement is shown in Figure 3. The first proton translocating loop is associated with NADH dehydrogenase. Hydrogen atoms are carried from NADH to the external face of the membrane by the FAD prosthetic group of the dehydrogenase. Each hydrogen atom is disproportionated to a proton, which is discharged into the external medium, and an electron, which is returned to the cytoplasmic face of the membrane by the iron-sulfur protein of the NADH dehydrogenase. Lack of the iron-sulfur center in sulfate- and iron-limited cells would result in the loss of proton translocation in this region. Under these conditions, hydrogen atoms would presumably bypass the loop and be transferred directly to ubiquinone.[22] The second loop would be in the ubiquinone-cytochrome region of the respiratory chain. The mechanism by which the protonmotive cycle could result in proton translocation has been discussed already. A simpler mechanism is shown in Figure 3. Here, ubiquinone carries hydrogen atoms to the external face of the membrane, protons are discharged into the medium, and electrons are returned via the cytochromes to reduce oxygen at the inner face of the membrane. It is not known if oxygen is reduced on the cytoplasmic or external face of the membrane. Reduction of oxygen at the cytoplasmic face of the membrane is mandatory if a H^+ to O ratio of 4 is to be obtained with the scheme shown in Figure 3. If the protonmotive cycle is present, reduction of oxygen at the external face of the membrane should yield a H^+ to O ratio of 4. A H^+ to O ratio of 6 would be obtained if oxygen was reduced at the cytoplasmic surface (Figure 2). The last alternative would provide an average H^+/site ratio of 3. This is more in line with recent values obtained with mitochondria.[94]

There is little direct evidence for the existence of proton-translocating loops in the respiratory chain of *E. coli.* The spatial arrangement in the membrane of the components of the respiratory chain of *E. coli* has not been investigated, although it is clear that the active sites of the NADH, L-glycerol-3-phosphate, D- and L-lactate dehydrogenases are on the cytoplasmic face of the membrane.[42,43,120] The existence of proton pumps, perhaps involving the redox-dependent protonation and deprotonation of a carrier molecule, are possible.[121,122] As presently envisioned, proton-translocating loops cannot account for stoichiometries of greater than 2 H^+/site. Higher stoichiometries could be accommodated by a proton-pumping system. More precise measurements of the stoichiometry of proton translocation are required. The components of the respiratory chain, especially the cytochromes, need to be purified; and their spatial

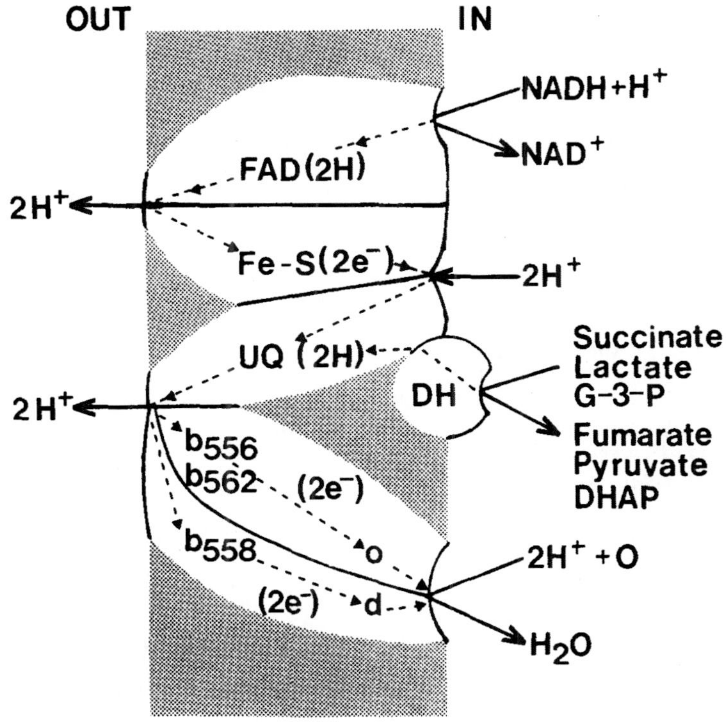

FIGURE 3. Scheme for the organization of the components of the respiratory chain in the cell membrane of *E. coli.* The pathway for the reducing equivalents is shown by the broken line. UQ, ubiquinone; DH, dehydrogenases for the oxidation of succinate, D- and L-lactate, and L-glycerol-3-phosphate; G-3-P, L-glycerol-3-phosphate; and DHAP, dihydroxyacetone phosphate.

location and interactions in the membrane need to be defined. Then, perhaps, the investigation of the pathway of electrons and protons in the membrane can be tackled in a meaningful way.

ACKNOWLEDGMENT

The author acknowledges the generous financial support of the Medical Research Council of Canada.

REFERENCES

1. **Sanwal, B. D.**, Allosteric controls of amphibolic pathways in bacteria, *Bacteriol. Rev.*, 34, 20, 1970.
2. **Konings, W. N. and Boonstra, J.**, Anaerobic electron transfer and active transport in bacteria, *Curr. Top. Membr. Transp.*, 9, 177, 1977.
3. **Kröger, A.**, Phosphorylative electron transport with fumarate and nitrate as terminal hydrogen acceptors, in *Microbial Energetics*, Haddock, B. A. and Hamilton, W. A., Eds., Cambridge University Press, Cambridge, 1977, 61.
4. **Haddock, B. A. and Jones, C. W.**, Bacterial respiration, *Bacteriol. Rev.*, 41, 47, 1977.
5. **Hendler, R. W.**, Respiration and energy transduction in *Escherichia coli*, in *The Enzymes of Biological Membranes*, Vol. 3, Martonosi, A., Ed., Plenum Press, New York, 1976, 75.

6. **Jones, C. W.,** Aerobic respiratory systems in bacteria, in *Microbial Energetics,* Haddock, B. A. and Hamilton, W. A., Eds., Cambridge University Press, Cambridge, 1977, 24.
7. **Hendler, R. W., Burgess, A. H., and Scharff, R.,** Respiration and protein synthesis in *Escherichia coli* membrane-envelope fragments. I. Oxidative activities with soluble substrates, *J. Cell Biol.,* 42, 715, 1969.
8. **Futai, M. and Kimura, H.,** Inducible membrane-bound L-lactate dehydrogenase from *Escherichia coli, J. Biol. Chem.,* 252, 5820, 1977.
9. **Lin, E. C. C.,** Glycerol dissimilation and its regulation in bacteria, *Annu. Rev. Microbiol.,* 30, 535, 1976.
10. **Ruiz-Herrera, J. and Garcia, L. G.,** Regulation of succinate dehydrogenase in *Escherichia coli, J. Gen. Microbiol.,* 72, 29, 1972.
11. **Takahashi, Y.,** Effect of glucose and cyclic adenosine 3′,5′-monophosphate on the synthesis of succinate dehydrogenase and isocitrate lyase in *Escherichia coli, J. Biochem.,* 78, 1097, 1975.
12. **Gounaris, A. D. and Hager, L. P.,** A resolution of the *Escherichia coli* pyruvate dehydrogenase complex, *J. Biol. Chem.,* 236, 1013, 1961.
13. **Dills, S. S. and Dobrogosz, W. J.,** Cyclic adenosine 3′,5′-monophosphate regulation of membrane energetics in *Escherichia coli, J. Bacteriol.,* 131, 854, 1977.
14. **Broman, R. L., Dobrogosz, and White, D. C.,** Stimulation of cytochrome synthesis in *Escherichia coli* by cyclic AMP, *Arch. Biochem. Biophys.,* 162, 595, 1974.
15. **Lemberg, R. and Barrett, J.,** *Cytochromes,* Academic Press, New York, 1973.
16. **Fujita, T. and Sato, R.,** Soluble cytochromes in *Escherichia coli, Biochim. Biophys. Acta,* 77, 690, 1963.
17. **Revsin, B. and Brodie, A. F.,** Carbon monoxide-binding pigments of *Mycobacterium phlei* and *Escherichia coli, J. Biol. Chem.,* 244, 3101, 1969.
18. **Kim, I. C. and Bragg, P. D.,** Properties of nonheme iron in a cell envelope fraction from *Escherichia coli, J. Bacteriol.,* 107, 664, 1971.
19. **Bragg, P. D.,** Electron transport and energy transducing systems of *Escherichia coli,* in *Membrane Proteins in Energy Transduction,* Capaldi, R. A., Ed., Marcel Dekker, New York, in press.
20. **Baillie, R. D., Hou, C., and Bragg, P. D.,** The preparation and properties of a solubilized respiratory complex from *Escherichia coli, Biochim. Biophys. Acta,* 234, 46, 1971.
21. **Bragg, P. D.,** Nonheme iron in respiratory chains, in *Microbial Iron Metabolism, A Comprehensive Treatise,* Nielands, J. B., Ed., Academic Press, New York, 1974, 303.
22. **Poole, R. K. and Haddock, B. A.** Effects of sulfate-limited growth in continuous culture on the electron-transport chain and energy conservation in *Escherichia coli K12, Biochem. J.,* 152, 537, 1975.
23. **Kashket, E. R. and Brodie, A. F.,** Oxidative phosphorylation in fractionated bacterial systems. X. Different roles for the natural quinones of *Escherichia coli* W in oxidative metabolism, *J. Biol. Chem.,* 238, 2564, 1963.
24. **Daves, G. D., Muraca, R. F., Whittick, J. S., Friis, P., and Folkers, K.,** Discovery of ubiquinones-1, -2, -3, and -4 and the nature of biosynthetic isoprenylation, *Biochemistry,* 6, 2861, 1967.
25. **Campbell, I. M. and Bentley, R.,** Inhomogeneity of vitamin K_2 in *Escherichia coli, Biochemistry,* 8, 4651, 1969.
26. **Lund, T. and Raynor, J. B.,** Electron spin resonance of bacterial respiratory membranes, *J. Bioenerg.,* 7, 161, 1975.
27. **Dancey, G. F., Levine, A. E., and Shapiro, B. M.,** The NADH dehydrogenase of the respiratory chain of *Escherichia coli.* I. Properties of the membrane-bound enzyme, its solubilization, and purification to near homogeneity, *J. Biol. Chem.,* 251, 5911, 1976.
28. **Kohn, L. D. and Kaback, H. R.,** Mechanisms of active transport in isolated bacterial vesicles. XV. Purification and properties of the membrane-bound D-lactate dehydrogenase from *Escherichia coli, J. Biol. Chem.,* 248, 7012, 1973.
29. **Futai, M.,** Membrane D-lactate dehydrogenase from *Escherichia coli,* Purification and properties, *Biochemistry,* 12, 2468, 1973.
30. **Schryvers, A., Lohmeier, E., and Weiner, J. H.,** Chemical and functional properties of the native and reconstituted forms of the membrane-bound, aerobic glycerol-3-phosphate dehydrogenase of *Escherichia coli, J. Biol. Chem.,* 253, 783, 1978.
31. **Williams, F. R. and Hager, L. P.,** Crystalline flavin pyruvate oxidase from *Escherichia coli.* I. Isolation and properties of the flavoprotein, *Arch. Biochem. Biophys.,* 116, 168, 1966.
32. **Kim, I. C. and Bragg, P. D.,** Some properties of the succinate dehydrogenase of *Escherichia coli, Can. J. Biochem.,* 49, 1098, 1971.
33. **Karibian, D. and Couchoud, P.,** Dihydroorotate oxidase of *Escherichia coli*: purification properties, and relation to the cytoplasmic membrane, *Biochim. Biophys. Acta,* 364, 218, 1974.

34. **Brodie, A. F.**, DPNH cytochrome c reductase (bacterial), in *Methods in Enzymology*, Vol. 2, Colowick, S. P. and Kaplan, N. O., Eds., Academic Press, New York, 1955, 693.
35. **Bragg, P. D. and Hou, C.**, Reduced nicotinamide adenine dinucleotide oxidation in *Escherichia coli* particles. I. Properties and cleavage of the electron transport chain, *Arch. Biochem. Biophys.*, 119, 194, 1967.
36. **Bragg, P. D. and Hou, C.**, Reduced nicotinamide adenine dinucleotide oxidation in *Escherichia coli* particles. II. NADH dehydrogenases, *Arch. Biochem. Biophys.*, 119, 202, 1967.
37. **Bragg, P. D. and Hou, C.**, Reduced nicotinamide-adenine dinucleotide oxidation in *Escherichia coli* particles. III. Cellular location of menadione reductase and ATPase activities, *Can. J. Biochem.*, 45, 1107, 1967.
38. **Dancey, G. F. and Shapiro, B. M.**, The NADH dehydrogenase of the respiratory chain of *Escherichia coli*. II. Kinetics of the purified enzyme and the effects of antibodies elicited against it on membrane-bound and free enzyme, *J. Biol. Chem.*, 251, 5921, 1976.
39. **Dancey, G. F. and Shapiro, B. M.**, Specific phospholipid requirement for activity of the purified respiratory chain dehydrogenase of *Escherichia coli*, *Biochim. Biophys. Acta*, 487, 368, 1977
40. **Reeves, J. P., Hong, J. S., and Kaback, H. R.**, Reconstitution of D-lactate-dependent transport in membrane vesicles from a D-lactate dehydrogenase mutant of *Escherichia coli*, *Proc. Natl. Acad. Sci. U.S.A.*, 70, 1917, 1973.
41. **Futai, M.**, Reconstitution of transport dependent on D-lactate or glycerol-3-phosphate in membrane vesicles of *Escherichia coli* deficient in the corresponding dehydrogenases, *Biochemistry*, 13, 2327, 1974.
42. **Weiner, J. H.**, The localization of glycerol-3-phosphate dehydrogenase in *Escherichia coli*, *J. Membr. Biol.*, 15, 1, 1974.
43. **Futai, M. and Tanaka, Y.**, Localization of D-lactate dehydrogenase in membrane vesicles prepared using a French press or ethylenediamine tetraacetate-lysozyme from *Escherichia coli*, *J. Bacteriol.*, 124, 470, 1975.
44. **Russell, P., Schrock, H. L., and Gennis, R. B.**, Lipid activation and protease activation of pyruvate oxidase. Evidence suggesting a common site of interaction on the protein, *J. Biol. Chem.*, 252, 7883, 1977.
45. **Tanaka, Y., Anraku, Y., and Futai, M.**, *Escherichia coli* membrane D-lactate dehydrogenase. Isolation of the enzyme in aggregated form and its activation by Triton X-100 and phospholipids, *J. Biochem.*, 80, 821, 1976.
46. **Kimura, H. and Futai, M.**, Effect of phospholipids on L-lactate dehydrogenase from membranes of *Escherichia coli*. Activation and stabilization of the enzyme with phospholipids, *J. Biol. Chem.*, 253, 1095, 1978.
47. **Blake, R., Hager, L. P., and Gennis, R. B.**, Activation of pyruvate oxidase by monomeric and micellar amphiphiles, *J. Biol. Chem.*, 253, 1963, 1978.
48. **Polglase, W. J., Pun, W. T., and Withaar, J.**, Lipoquinones of *Escherichia coli*, *Biochim Biophys. Acta*, 118, 425, 1966.
49. **Ashcroft, J. R. and Haddock, B. A.**, Synthesis of alternate membrane-bound redox carriers during aerobic growth of *Escherichia coli* in the presence of potassium cyanide, *Biochem. J.*, 148, 349, 1975.
50. **Wallace, B. J. and Young, I. G.**, Role of quinones in electron transport to oxygen and nitrate in *Escherichia coli*. Studies with a *ubi* A^- *men* A^- double mutant, *Biochim. Biophys. Acta*, 461, 84, 1977.
51. **Haddock, B. A. and Schairer, H. U.**, Electron-transport chains of *Escherichia coli*. Reconstitution of respiration in a 5-aminolevulinic acid-requiring mutant, *Eur. J. Biochem.*, 35, 34, 1973.
52. **Wallace, B. J. and Young, I. G.**, Aerobic respiration in mutants of *Escherichia coli* accumulating quinone analogues of ubiquinone, *Biochim. Biophys. Acta*, 461, 75, 1977.
53. **Cox, G. B., Snoswell, A. M., and Gibson, F.**, The use of a ubiquinone-deficient mutant in the study of malate oxidation in *Escherichia coli*, *Biochim. Biophys. Acta*, 153, 1, 1968.
54. **Singh, A. P. and Bragg, P. D.**, Reduced nicotinamide adenine dinucleotide dependent reduction of fumarate coupled to membrane energization in a cytochrome deficient mutant of *Escherichia coli* K12, *Biochim. Biophys. Acta*, 396, 229, 1975.
55. **Newton, N. A., Cox, G. B., and Gibson, F.**, The function of menaquinone (vitamin K_2) in *Escherichia coli*, *Biochim. Biophys. Acta*, 244, 155, 1971.
56. **Rainnie, D. J. and Bragg, P. D.**, The effect of iron deficiency on respiration and energy-coupling in *Escherichia coli*, *J. Gen. Microbiol.*, 77, 339, 1973.
57. **Hamilton, J. A., Cox, G. B., Looney, F. D., and Gibson, F.**, Ubisemiquinone in membranes from *Escherichia coli*, *Biochem. J.*, 116, 319, 1970.
58. **Hendler, R. W.**, Respiration and protein synthesis in *Escherichia coli* membrane-envelope fragments. V. On the reduction of nonheme iron and the cytochromes by nicotinamide adenine dinucleotide and succinate, *J. Cell Biol.*, 51, 664, 1971.

59. **Albracht, S. P. J. and Subramanian, J.,** The number of Fe atoms in the sulfur centers of the respiratory chain, *Biochim. Biophys. Acta,* 462, 36, 1977.
60. **Pudek, M. R. and Bragg, P. D.,** Redox potentials of the cytochromes in the respiratory chain of aerobically grown *Escherichia coli, Arch. Biochem. Biophys.,* 174, 546, 1976.
61. **Haddock, B. A., Downie, J. A., and Garland, P. B.,** Kinetic characterization of the membrane-bound cytochromes of *Escherichia coli* grown under a variety of conditions by using a stopped-flow dual wavelength spectrophotometer, *Biochem. J.,* 154, 285, 1976.
62. **Downie, J. A. and Cox, G. B.,** Sequence of b cytochromes relative to ubiquinone in the electron transport chain of *Escherichia coli, J. Bacteriol.,* 133, 477, 1978.
63. **Shipp, W. S.,** Cytochromes of *Escherichia coli, Arch. Biochem. Biophys.,* 150, 459, 1972.
64. **Hendler, R. W., Towne, D. W., and Shrager, R. I.,** Redox properties of b-type cytochromes in *Escherichia coli* and rat liver mitochondria and techniques for their analysis, *Biochim. Biophys. Acta,* 376, 42, 1975.
65. **Fujita, T., Itagaki, E., and Sato, R.,** Purification and properties of cytochrome b_1 from *Escherichia coli, J. Biochem.,* 53, 282, 1963.
66. **Deeb, S. S. and Hager, L. P.,** Crystalline cytochrome b_1 from *Escherichia coli, J. Biol. Chem.,* 239, 1025, 1964.
67. **Castor, L. N. and Chance, B.,** Photochemical determination of the oxidases of bacteria, *J. Biol. Chem.,* 234, 1587, 1959.
68. **Rice, C. W. and Hempfling, W. P.,** Oxygen-limited continuous culture and respiratory energy conservation in *Escherichia coli, J. Bacteriol.,* 134, 115, 1978.
69. **Webster, D. A. and Hackett, D. P.,** The purification and properties of cytochrome o from *Vitreoscilla, J. Biol. Chem.* ,241, 3308, 1966.
70. **Pudek, M. R. and Bragg, P. D.,** Inhibition by cyanide of the respiratory chain oxidases of *Escherichia coli, Arch. Biochem. Biophys.,* 164, 682, 1974.
71. **Pudek, M. R. and Bragg, P. D.,** Reaction of cyanide with cytochrome d in respiratory particles from exponential phase *Escherichia coli, FEBS Lett.,* 50, 111, 1975.
72. **Bragg, P. D.,** Reduction of nonheme iron in the respiratory chain of *Escherichia coli, Can. J. Biochem.,* 48, 777, 1970.
73. **Barrett, J.,** The prosthetic group of cytochrome a_2, *Biochem. J.,* 64, 627, 1956.
74. **Pudek, M. R. and Bragg, P. D.,** Trapping of an intermediate in the oxidation-reduction cycle of cytochrome d in *Escherichia coli, FEBS Lett.,* 62, 330, 1976.
75. **Erecinska, M. and Wilson, D. F.,** Cytochrome c oxidase: a synopsis, *Arch. Biochem. Biophys.,* 188, 1, 1978.
76. **Sweetman, A. J. and Griffiths, D. E.** Studies on energy-linked reactions. Energy-linked reduction of oxidized nicotinamide-adenine dinucleotide by succinate in *Escherichia coli, Biochem. J.,* 121, 117, 1971.
77. **Poole, R. K. and Haddock, B. A.,** Energy-linked reduction of nicotinamide-adenine dinucleotide in membranes derived from normal and various respiratory-deficient mutant strains of *Escherichia coli* K12, *Biochem. J.,* 144, 77, 1974.
78. **Simoni, R. D. and Postma, P. W.,** The energetics of bacterial active transport, *Annu. Rev. Biochem.,* 44, 523, 1975.
79. **Harold, F. M.,** Conservation and transformation of energy by bacterial membranes, *Bacteriol. Rev.,* 36, 172, 1972.
80. **Harold, F. M.,** Membranes and energy transduction in bacteria, *Curr. Top. Bioenerg.* 84, 1977.
81. **Garland, P. B.,** Energy transduction and transmission in microbial systems, in *Microbial Energetics,* Haddock, B. A. and Hamilton, W. A., Eds., Cambridge University Press, Cambridge, 1977, 1.
82. **Boyer, P. D., Chance, B., Ernster, L., Mitchell, P., Racker, E., and Slater, E. C.,** Oxidative phosphorylation and photophosphorylation, *Annu. Rev. Biochem.,* 46, 955, 1977.
83. **Mitchell, P.,** Chemiosmotic coupling in oxidative and photosynthetic phosphorylation, *Biol. Rev.,* 41, 445, 1966.
84. **Reeves, J. P.,** Transient pH changes during D-lactate oxidation by membrane vesicles, *Biochem. Biophys. Res. Commun.,* 45, 931, 1971.
85. **Garland, P. B., Downie, J. A., and Haddock, B. A.,** Proton translocation and the respiratory nitrate reductase of *Escherichia coli, Biochem. J.,* 152, 547, 1975.
86. **Jones, C. W., Brice, J. M., Downs, A. J., and Drozd, J. W.,** Bacterial respiration-linked proton translocation and its relationship to respiratory-chain composition, *Eur. J. Biochem.,* 52, 265, 1975.
87. **Farmer, I. S. and Jones, C. W.,** The energetics of *Escherichia coli* during aerobic growth in continuous culture, *Eur. J. Biochem.,* 67, 115, 1976.
88. **Lawford, H. G. and Haddock, B. A.,** Respiration-driven proton translocation in *Escherichia coli, Biochem. J.,* 136, 217, 1973.

89. **Haddock, B. A., Downie, J. A., and Lawford, H. G.,** The function of ubiquinone respiration studied in cytochrome-deficient mutants of *Escherichia coli, Proc. Soc. Gen. Microbiol.,* 1, 50, 1974.
90. **Jones, C. W., Brice, J. M., and Edwards, C.,** The effect of respiratory chain composition on the growth efficiencies of aerobic bacteria, *Arch. Microbiol.,* 115, 85, 1977.
91. **Hertzberg, E. L. and Hinkle, P. C.** Oxidative phosphorylation and proton translocation in membrane vesicles prepared from *Escherichia coli, Biochem. Biophys. Res. Commun.,* 58, 178, 1974.
92. **Bragg, P. D. and Hou, C.,** Energization of energy-dependent transhydrogenase of *Escherichia coli* at a second site of energy conservation, *Arch. Biochem. Biophys.,* 163, 614, 1974.
93. **Singh, A. P. and Bragg, P. D.,** Ascorbate-phenazine methosulfate-dependent energization in respiratory chain mutants of *Escherichia coli, Biochem. Biophys. Res. Commun.,* 72, 195, 1976.
94. **Brand, M. D. and Lehninger, A. L.,** H^+/ATP ratio during ATP hydrolysis by mitochondria: modification of the chemiosmotic theory, *Proc. Natl. Acad. Sci. U.S.A.,* 74, 1955, 1977.
95. **Cox, J. C. and Haddock, B. A.,** Phosphate transport and the stoicheiometry of respiratory driven proton translocation in *Escherichia coli, Biochem. Biophys. Res. Commun.,* 82, 46, 1978.
96. **Meyer, D. J. and Jones, C. W.,** Oxidative phosphorylation in bacteria which contain different cytochrome oxidases, *Eur. J. Biochem.,* 36, 144, 1973.
97. **Meijer, E. M., van Verseveld, H. W., van der Beek, E. G., and Stouthamer, A. H.,** Energy conservation during aerobic growth in *Paracoccus denitrificans, Arch. Microbiol.,* 112, 25, 1977.
98. **Cowell, J. L., Raffeld, M., and Friedberg, I.,** Isolation, partial purification and properties of succinic dehydrogenase from *Escherichia coli, Bacteriol. Proc.,* 157, 1973.
99. **Weiner, J. H. and Heppel, L. A.,** Purification of the membrane-bound and pyridine nucleotide-independent L-glycerol 3-phosphate dehydrogenase from *Escherichia coli, Biochem. Biophys. Res. Commun.,* 47, 1360, 1972.
100. **Collins, S. H. and Hamilton, W. A.,** Magnitude of the protonmotive force in respiring *Staphylococcus aureus* and *Escherichia coli, J. Bacteriol.,* 126, 1224, 1976.
101. **Ramos, S., Schuldiner, S., and Kaback, H. R.,** The electrochemical gradient of protons and its relationship to active transport in *Escherichia coli* membrane vesicles, *Proc. Natl. Acad. Sci U.S.A.,* 73, 1892, 1976.
102. **Padan, E., Zilberstein, D., and Rottenberg, H.,** The proton electrochemical gradient in *Escherichia coli* cells, *Eur. J. Biochem.,* 63, 533, 1976.
103. **West, I. C. and Mitchell, P.,** The proton-translocating ATPase of *Escherichia coli, FEBS Lett.,* 40, 1, 1974.
104. **Singh, A. P. and Bragg, P. D.,** Effect of inhibitors on the substrate-dependent quenching of 9-aminoacridine fluorescence in inside-out vesicles of *Escherichia coli, Eur. J. Biochem.,* 67, 177, 1976.
105. **Singh, A. P. and Bragg, P. D.,** ATP-dependent proton translocation and quenching of 9-aminoacridine fluorescence in inside-out membrane vesicles of a cytochrome-deficient mutant of *Escherichia coli, Biochim. Biophys. Acta,* 464, 562, 1977.
106. **Maloney, P. C., Kashket, E. R., and Wilson, T. H.,** A protonmotive force drives ATP synthesis in bacteria, *Proc. Natl. Acad. Sci. U.S.A.,* 71, 3896, 1974.
107. **Wilson, D. M., Alderete, J. F., Maloney, P. C., and Wilson, T. H.,** Protonmotive force as the source of energy for adenosine 5′-triphosphate synthesis in *Escherichia coli, J. Bacteriol.,* 126, 327, 1976.
108. **Grinius, L., Slusnyte, R., and Griniuviene, B.,** ATP synthesis driven by protonmotive force imposed across *Escherichia coli* cell membranes, *FEBS Lett.,* 57, 290, 1975.
109. **Tsuchiya, T., and Rosen, B. P.,** Adenosine 5′-triphosphate synthesis energized by an artificially imposed membrane potential in membrane vesicles of *Escherichia coli, J. Bacteriol.,* 127, 154, 1976.
110. **Tsuchiya, T.,** Adenosine-5′-triphosphate synthesis driven by a protonmotive force in membrane vesicles of *Escherichia coli, J. Bacteriol.,* 129, 763, 1977.
111. **Newman, B. M. and Cole, J. A.,** The chromosomal location and pleiotropic effects of mutations of the *nir* A^+ gene of *Escherichia coli:* the essential role of *nir* A^+ in nitrite reduction and in other anaerobic redox reactions, *J. Gen. Microbiol.,* 106, 1, 1978.
112. **Cox, G. B., Newton, N. A., Gibson, F., Snoswell, A. M., and Hamilton, J. A.,** The function of ubiquinone in *Escherichia coli, Biochem. J.,* 117, 551, 1970.
113. **Hendler, R. W. and Nanninga, N.,** Respiration and protein synthesis in *Escherichia coli* membrane-envelope fragments. III. Electron microscopy and analysis of the cytochromes, *J. Cell Biol.,* 46, 114, 1970.
114. **Birdsell, D. C. and Cota-Robles, E. H.,** Electron transport particles released upon lysis of spheroplasts of *Escherichia coli* B by Brij 58, *Biochim. Biophys. Acta,* 216, 250, 1970.
115. **Jones, R. G. W.,** Ubiquinone deficiency in an auxotroph of *Escherichia coli* requiring 4-hydroxybenzoic acid, *Biochem. J.,* 103, 714, 1967.
116. **Mitchell, P.,** The protonmotive Q cycle: a general formulation, *FEBS Lett.,* 59, 137, 1975.

117. **Garland, P. B., Clegg, R. A., Boxer, D., Downie, J. A., and Haddock, B. A.,** Proton translocating nitrate reductase of *Escherichia coli*, in *Electron transfer chains and oxidative phosphorylation*, Quagliariello, E., Ed., North-Holland, Amsterdam, 1975, 351.
118. **Yu, C. A., Yu, L., and King, T. E.,** The existence of an ubiquinone binding protein in the reconstitutively active cytochrome b-c_1 complex, *Biochem. Biophys. Res. Commun.*, 78, 259, 1977.
119. **Crane, R. T., Sun, I. L., and Crane, F. L.,** Lipophilic chelator inhibition of electron transport in *Escherichia coli*, *J. Bacteriol.*, 122, 686, 1975.
120. **Futai, M.,** Orientation of membrane-vesicles of *Escherichia coli* prepared by different procedures, *J. Membrane Biol.*, 15, 15, 1974.
121. **Skulachev, V. P.,** Energy coupling in biological membranes: current state and perspectives, in *Energy Transducing Mechanisms*, Racker, E., Ed., University Park Press, Baltimore, 1975, 31.
122. **Papa, S.,** Proton translocation reactions in the respiratory chains, *Biochim. Biophys. Acta*, 456, 39, 1976.
123. **Singh, A. P. and Bragg, P. D.,** unpublished results.
124. **Pudek, M. R. and Bragg, P. D.,** unpublished results.

Chapter 5

OXYGEN REACTIVE HEMOPROTEIN COMPONENTS IN BACTERIAL RESPIRATORY SYSTEMS

Peter Jurtshuk, Jr. and Tsan-yen Yang

TABLE OF CONTENTS

I. INTRODUCTION

"It is remarkable how little is known about oxidases in bacteria. For example, one of the most active oxidases known — that in *Azotobacter* — has never been isolated in a state sufficiently pure and active to permit characterization of its functional moiety. The list of oxidase systems in which the active terminal enzyme can be asserted to be a heme protein, and for which an oxidase function has been shown or suggested, is very short. Thus, there may be mentioned (i) the *a*-type cytochromes, such as occur in *Staphylococcus albus, Acetobacter* species, *Bacillus subtilis* etc., (ii) the "cytochrome *o*" pigments found in a wide variety of organisms, (iii), the soluble cytochrome *d* (formerly known as 'a_2'), detected in *E. coli* and *Acetobacter peroxidans,* (iv) the soluble diheme protein of *Pseudomonas aeruginosa,* classified as cytochrome *cd,* and (v) the soluble RHP-type proteins of the purple photosynthetic bacteria."

Martin D. Kamen, 1965.[46]

As indicated by the above statement, after almost 2 decades of study, it has been found that bacterial electron transport systems are quite complex and diverse with respect to the types of electron carriers found within their membranes.[25,26,30,34,43-45,51,67,68,79] This diversity extends particularly to the oxygen-reactive hemoproteins, i.e., the cytochrome oxidases and the mixed function oxidases which are found in bacteria as multiple species. Cytochrome oxidases function as terminal oxidases which are membrane-bound enzyme complexes that utilize molecular oxygen as a terminal electron acceptor for bioenergetic processes like oxidative phorphorylation.[26,34,43] The bacterial terminal oxidases are cytochromes o, aa_3, a_1, and d (formerly called a_2). Mixed-function oxidases also are multienzyme complexes, which possess oxygen-reactive hemoproteins that will use an electron transport carrier system to energize or activate molecular oxygen so that while one atom of oxygen serves as an electron acceptor, the other oxygen atom will "oxygenate" an organic molecule like an aliphatic or aromatic hydrocarbon. An example of a hemoprotein that functions as a mixed function oxidase in bacteria is cytochrome P-450.[41] All of the above cytochrome components react with carbon monoxide, and such reactions cause characteristic spectral shifts that allow for "tentative" identification of the hemoprotein component. There is presumptive evidence that other b or c′-type cytochromes may function as oxidases, or at least can react with carbon monoxide, but the physiological role of these hemoproteins is still not understood today. With regard to terminal oxidases, it is *rare* that a single cytochrome oxidase serves as the sole oxygen-reactive component in a bacterial species, as cytochrome aa_3 serves for beef-heart mitochrondria. When such examples do arise, they are noteworthy, as is the case of the "soluble" cytochrome o which appears to serve as the sole terminal oxidase for an aerobic organism, *Vitreoscilla* spp.[73-76]. In this instance, even the c-type cytochromes are missing in this microorganism.

The diversity of bacterial cytochrome oxidases is also reflected in the combinations in which they appear within the cell. In most cases, they occur in combinations of three (and on occasion, two) hemoprotein components. In many organisms, they occur as cytochromes a_1:d:o, or aa_3:d:o, and sometimes, as aa_3:o. In some cases, the presence of a cytochrome oxidase in a bacterial cell will depend upon: (1) the chemical nature of the substrate used for the growth of the organism, i.e., whether it allows for an oxidative- or fermentative-type metabolism, (2) the age of the bacterium in the growth culture cycle, and (3) whether growth occurs aerobically, anaerobically, or under microaerophilic conditions. For example, cytochrome *d* (plus a_1) synthesis is usually more pronounced in bacterial cells grown under microaerophilic conditions rather than under highly aerobic environments.[30,43] This diversity of cytochrome oxidases found in bacteria probably reflects evolutionary divergence that occurred after the appearance of oxygen in the atmosphere. Prior to the appearance of oxygen, procaryotic cells were "probably" highly evolved species with respect to their electron transport systems, which would have been required for anaerobic photosynthesis and anaerobic respiration. As oxygen appeared, such microorganisms, through selection and adaptation, evolved with *different capabilities* for utilizing molecular oxygen, or both biological oxidations and oxygenation reactions. Thus, one could explain the great diversity in these oxygen-reactive hemoprotein components found today in procaryots.

There is now evidence in the literature which suggests that cytochrome *o* is perhaps the most important terminal oxidase found in bacterial respiratory systems. Cytochrome o is found in bacteria with greater frequency than cytochrome $a + a_3$.[34,43] Other preliminary reports indicate that cytochrome *o* may be found in some eucaryotic cells like yeast,[54] as well as in *Tetrahymena*.[60] However, the presence of cytochrome *o* in many microorganisms can be effectively questioned[51] since the *rigorous* criterion of demonstrating this specific hemoprotein by the photochemical action spectrum technique[12,16] is seldom used.

II. HISTORICAL AND CURRENT ASPECTS

Cytochrome *o* is a carbon-monoxide-binding hemoprotein that is found primarily in bacterial electron transport particles.[7,13,43,51,68,79] Like the *b*-type cytochromes, cytochrome *o* contains protoheme as its prosthetic group,[70] but unlike other *b*-type cytochromes, cytochrome *o* does react with carbon monoxide and, most probably, with cyanide.[5,73,81] Cytochrome *o* is also autooxidizable and reacts with molecular oxygen, while cytochrome *b* in general, does not behave in this manner.[51] Bacterial cytochrome *o* is usually a membrane-bound hemoprotein.[43] Nevertheless, in some microorganisms cytochrome *o* has been reported to be a soluble species.[63,73]

Chance and associates were the first to recognize cytochrome *o* as a unique and distinct type of hemoprotein in *Micrococcus pyogenes* var. *albus* (*Staphylococcus albus*) by carbon monoxide difference spectra studies.[13,14,67] The peak and trough of the carbon monoxide absorption bands were reported at 416 and 432 nm, respectively. The α- and β-bands of the CO-cytochrome *o* lie roughly at 578 and 540 nm, and these have a much less pronounced absorption band than that noted for the Soret peak. It was first reported by Chance that the Soret absorption band of the CO-compound of the newly discovered cytochrome was found at a wavelength characteristic of protoheme, instead of the green-type hemin enzymes such as the *a*-type cytochromes.[14] Chance was able to further analyze the multiple carbon monoxide reacting hemoproteins of *Proteus vulgaris* and *Aerobacter aerogenes*, in (stationary phase) cells, and found these contained not only cytochrome *o*, but also cytochrome *d* and/or a_1.[14] Using the action-spectrum technique[12,16] and measuring the monochromatic light relief of carbon monoxide inhibition of oxygen uptake, it was confirmed that the new carbon monoxide-reacting hemoprotein was a respiratory enzyme.[12,13] In 1959, Castor and Chance named this hemoprotein cytochrome *o*, and proposed that this hemoprotein served as an oxidase in the electron transport chain of bacteria.[13] There is now ample evidence in the literature to verify this early finding. In fact, it seems that cytochrome *o* might be the most abundantly found cytochrome oxidase that reacts with carbon monoxide and which can be detected readily in most bacterial systems. However, even today, there are only a few instances[13,51] of action-spectrum analysis being used to verify the presence of cytochrome *o* in bacteria.

Cytochrome *o*, in some species of bacteria, is the only carbon monoxide-reacting hemoprotein that can be detected, suggesting that cytochrome *o* might serve as the sole oxidase in these organisms.[13,43,45,51] In earlier studies, cytochrome *o* was found to be the only detectable oxidase in *Staphlococcus albus* and in *Acetobacter suboxydans*, and later, in *Vitreoscilla* species as determined by the criteria of action spectra. Cytochrome *o* was also the only oxidase detected in dark, aerobically grown cells of *Rhodospirillum rubrum*[28,51] and in log-phase cells of *Escherichia coli*, and in *Proteus vulgaris*, and *Aerobacter aerogenes*.[13]

Cytochrome *o* is generally detected and recognized by its characteristic CO-cytochrome *o* difference spectra.[14,43,51] This type of spectral analysis is also used to measure the cytochrome *o* concentration in either the whole cells or subcellular preparations.[15,18,70] A survey of the literature shows that CO-reduced difference spectra of cytochrome *o* derived from various bacterial species have α- absorption peaks at 578 to 565 nm, β- peaks at 539 to 535 nm, and the Soret peak at 419 nm. The troughs appear at 560, 547, and most prominently, in the 430 to 432 nm region.[51] The position of the α- peak in reduced *minus* oxidized difference spectra of cytochrome *o*, which often is used to detect the presence of *b*-type cytochrome, varies from 565 (in *S. albus*) to 556 nm (in *S. aureus*). This, along with some biochemical and other spectral characteristics, suggests that there may be several classes of cytochrome *o* present in bacterial electron transport systems.[51,70]

III. CYTOCHROME *o* AND OTHER TERMINAL OXIDASES IN *AZOTOBACTER VINELANDII*

In recent years, major attempts have been made toward understanding the role of branched electron transport systems and their concomitant terminal oxidases in the respiration of bacteria.[7,25,26,30,34,43-45,51,68,79,82] This has been particularly true for *Azotobacter vinelandii*.[1,2,23,24,31-40,42-44,47-49,55,59,69,80-82] Although a tentative branched scheme of the *A. vinelandii* electron transport chain was first proposed in 1967 by Jones and Redfearn,[33] the exact sequence and nature of the terminal branched chain remains uncertain even today.[43,82] The branching-point site(s) through which the electrons are sequentially transferred, and their relationship to substrate oxidations and to the multiple oxidase pathways, have been subjected to several revisions. Even today, the current schemes proposed are grossly oversimplified, and there is evidence which suggests that the actual sequences of the electron transport carriers in *A. vinelandii* may be more complicated than it was originally thought. Most of the information used in constructing pathway schemes were from the studies on: (1) substrate-oxidation and reduction kinetics, (2) inhibition patterns, and (3) photochemical action spectra. From such studies, which utilize either whole cells or the membrane particles isolated from the electron transport chain, some ideas can be obtained of the possible sequence of electron transfer carriers in association with their specific substrate oxidations and their terminal oxidase function.

A. vinelandii, a free-living, nitrogen-fixing, obligate aerobe, uses molecular oxygen as terminal electron acceptor for repiration. It also possesses the highest cellular respiratory rate of any known organism.[52] Because of these unique features, a great deal of work has been done in attempting to elucidate the nature of its complex electron transport chain in which the oxidases play a integral part. In addition to coenzyme Q-8, nonheme iron proteins, and phospholipid, multiple *c*-, and *b*-, and *a*-type cytochromes have also been implicated as functional components in the electron transport system of *A. vinelandii*.[31-33,35-40,42-44,69,81,82] Numerous membrane-associated oxidoreductases have been found in *A. vinelandii* which dehydrogenate L-malate,[31,38] D-lactate,[31-36] succinate,[31,39] L-glutamate,[42] as well as NADH[31,59] and NADPH.[2,23] In addition to these membrane-bound flavoprotein-type oxidoreductases, there are at least three cytochrome oxidases found in *A. vinelandii*[13,33,44,47,48] that can react with molecular oxygen. They are cytochromes a_1,*d*, and *o*. Cytochrome oxidase activity has been demonstrated using assays which have employed artificial electron donors like ascorbate-TMPD[35,39,44] and ascorbate-DCIP[33,44] as well as "natural" donors like reduced *Azotobacter* cytochrome c_4 + c_5[33,37] and reduced mammalian cytochrome *c*.[37] All the above cytochromes and related electron carrier components are membrane-bound and integrated functionally. The isolation and purification of each of these individual components for reconstitution-type studies has yet to be accomplished for the *A. vinelandii* respiratory chain system. To date, only the *c*-type cytochromes have been isolated and purified from the membrane fractions of *A. vinelandii.*[69] The "harsh" solubilization procedures (butanol extraction) used for purifying cytochromes c_4 and c_5 undoubtedly affect their capability for serving as electron donors for measuring cytochrome oxidase activity. Hence, the functional role of these *c*-type cytochromes are still essentially unknown. Consequently, only indirect biochemical and spectral information has been obtained from such purification studies.[69]

Most recently, the terminal oxidase cytochrome *o* has been isolated and purified from the electron transport chain of *A. vinelandii.*[80,81] These studies represent the first successful isolation and purification of a "membrane-bound" cytochrome oxidase from *A. vinelandii* that was free of any other terminal oxidase hemoprotein component. Because of the difficulties encountered in assaying terminal oxidases in bacteria, as well as isolating membrane-bound cytochromes, very little work has been done on isolating and purifying bacterial cytochrome oxidases.[43] Consequently, very little is known about any terminal oxidase components in bacteria. This is in contrast to the eucaryotic and mammalian cytochrome aa_3 which has been extensively purified and studied. This could be done mainly because of the ease with which the mammalian cytochrome oxidase could be assayed and, thereby, established as the major oxidase present in the inner membranes of mammalian and eucaryotic mitochondria.

IV. ACTION SPECTRA STUDIES

In *Azotobacter vinelandii*, cytochromes a_1 and *o* were first detected by action spectra in 1959 and reported to be the functional oxidases by Castor and Chance.[13] Although the initial action spectra analyses did not reveal the presence of cytochrome *d*, this hemoprotein could not be excluded as a significant oxidase of this organism. L. Smith[67] had demonstrated spectrally that cytochrome *d* was present in *A. vinelandii* membrane fractions.[13] Eventually, Jones and Redfearn,[33] and, more recently, Erikson and Diehl[24] showed by action spectra studies, that cytochrome *d* is unquestionably present and functions as an oxidase in *A. vinelandii.* As noted earlier,[13] photochemical action spectra cannot discriminate at a *given* wavelength between the "relief" of inhibition of a single or multiple species of a carbon monoxide-reacting hemoprotein(s), and as such, the data obtained may reflect the presence of one or *more* than one hemoprotein component. Also, there is no simple relationship between the degree or extent of oxidase activity inhibition and the peak height of the carbon monoxide-hemoprotein absorbance noted for the "relief" bands of the photochemical action spectrum.[13] This could account for the difference between the earlier studies of Castor and Chance[13] and the later studies of Jones and Redfearn.[33] They are not comparable with regard to the action spectra obtained. This might be due to (1) different strains of *A. vinelandii* being used by these two groups of investigators, as was shown to be the case in studies with *Acetobacter suboxydans,*[51] and/or (2) the difference in growth (or aeration) conditions, which would then reflect the difference in the stage of the growth curve at which the cells were harvested. Castor and Chance showed earlier that stationary and log phase cells of *E. coli* and *Aerobacter aerogenes* contained different terminal cytochrome oxidases.[13]

Most recently, Erikson and Diehl,[24] using action spectra studies, reinvestigated the respiratory electron transport chain of *Azotobacter vinelandii.* These investigators confirmed that cytochromes a_1, *d* and *o* were present and functioned as terminal oxidases for physiological substrates such as NADH and L-malate. However, in studies using the TMPD-ascorbate electron donor system (which did not allow for coupled phosphorylation) the photochemical action spectra indicated the involvement of cytochromes a_1 and, possibly, and *o*-type hemoprotein as oxidases. The electrons generated by the ascorbate-DCIP system did give coupled phosphorylation, but used *neither* cytochromes a_1 nor *d* as terminal oxidases. All three terminal oxidases (a_1, *d*, and *o*) were involved in NADH-dependent respiration, and further confirmation was obtained that the TMPD-ascorbate electron donor system did not use cytochrome *d* as an electron acceptor. The studies of Erikson and Diehl[24] also indicate the possible presence of an unidentified *o*-type cytochrome, different from the one involved in the NADH oxidation pathway, that possibly functions as an active oxidase in *A. vinelandii.*

V. INHIBITOR AND SPECTRAL STUDIES

Jones and Redfearn,[33] examining the *A. vinelandii* electron transport particle, found that at *low* concentrations of cyanide (50 μM) the ascorbate-DCIP oxidase was *maximally* inhibited, whereas NADH oxidase activity was not significantly affected at this cyanide concentration. These results, together with other inhibitor data, showed that cyanide and azide could increase or alter the aerobic steady-state reduction of cytochrome c_4 + c_5, but not of b_1. This suggested the presence of a branched cytochrome pathway, each branch linked functionally to a *separate* oxidase system. This idea was supported by the findings of Jurtshuk et al. who examined the cyanide sensitivity of all the substrate-linked oxidases in the electron transport particle of *A. vinelandii.*[43] These latter investigators showed that *all* the oxidases were cyanide sensitive. However, the variation and degree of the cyanide sensitivity exhibited was very great. Of all the electron-transport-dependent oxidases examined, the one most sensitive to cyanide was the succinate oxidase, which showed 50% inhibition at 6×10^{-6} *M*. The least cyanide-sensitive activity was the D-lactate oxidase which required almost 1 m*M* cyanide to attain the same degree of inhibition. The membrane-bound *Azotobacter vinelandii* succinate oxidase was 150 times more sensitive to cyanide than the membrane-bound D-lactate oxidase, which suggested that there were probably *at least* two (or more) different oxidases functioning at the terminal end of the electron transport chain.

Jurtshuk et al.[39] in earlier studies noted that, spectrally, a substantial reduction of cytochromes c_4 + c_5 occurred by the sole addition of ascorbate to the *A. vinelandii* electron transport particle. Cytochromes a_1 and *d* were not reduced by the sole addition of ascorbate, but these two oxidases were subsequently reduced when TMPD was added to the "ascorbate reduced" electron transport particle.[39] Electrons from ascorbate were capable of reducing the *c*-type cytochromes (c_4 + c_5), but these electrons were not transferred to cytochromes a_1 and *d* until TMPD was added. This suggested that *A. vinelandii* *c*-type cytochromes are similar in function to the mammalian cytochrome *c* in that they can be directly reduced by ascorbate, but ferrocytochrome *c* oxidation requires TMPD in addition to a terminal oxidase (like cytochromes a + a_3) as demonstrated for the mammalian oxidase[64] or possibly another terminal oxidase like cytochrome *o*) in *Azotobacter vinelandii*,[44,55] It is also possible cytochromes a_1 and *d* may also serve in this same capacity, as does the cytochrome *o*, in reoxidizing the bacterial-type ferrocytochrome *c*. The branched chain, cytochrome-dependent electron transport system in *A. vinelandii*, was first postulated by Jones and Redfearn.[33] In its original form it was illustrated as:

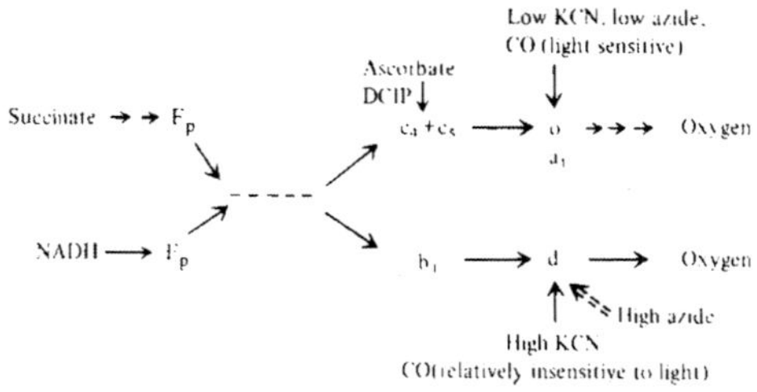

The aerobic steady-state reduction values indicated that ascorbate-DCIP donated electrons at the level of cytochromes c_4 + c_5. The relatively high aerobic steady-state reduction of the cytochrome c_4 + c_5 that occurred after the addition of succinate and NADH, and the observation that an increase in *this* reduction occurred upon the addition of *low* concentrations of KCN or azide, supported the concept of a $c_4 + c_5 \rightarrow o/a_1 \rightarrow O_2$ terminal oxidases pathway. Jurtshuk and associates[44,55] were able to isolate and purify by detergent fractionation an *A. vinelandii* cytochrome oxidase which catalytically oxidized reduced TMPD. They found that this terminal oxidase contained *c*-type cytochrome(s) (c_4 + c_5) integrated with cytochrome *o*. More recent studies from this laboratory[80] have provided evidence which indicates that the purified cytochrome oxidase of *A. vinelandii*, is a "solubilized" enzyme complex consisting of cytochrome c_4 + *o*. No cytochrome c_5 could be detected by cold-temperature spectroscopy in highly purified preparations. This suggests that cytochromes c_4 and c_5 may not be closely associated components in the electron transfer pathway of *A. vinelandii*. This is supported by the finding that purified cytochromes c_4 and c_5 from *A. vinelandii* yielded significantly different P/O ratios when used as electron mediators between ascorbate and the respiratory electron transport chain.[1] These two cytochromes may actually operate in parallel pathways rather than in series in a single pathway, as previously postulated. In the original formulation of the branched-chain cytochrome respiratory system of *A. vinelandii*, the splitting of the chain was *postulated* to occur at the quinone level to yield two terminal oxidase pathways: the major pathway ($b_1 \rightarrow d$) and a minor pathway ($c_4 + c_5 \rightarrow a_1/o$).

The evidence for cytochrome *d* serving as the functional oxidase for the major NADH oxidizing pathway was obtained from action spectra studies using NADH as an electron donor with both particles and *A. vinelandii* whole cells. The pertinent findings were (1) carbon monoxide inhibition of NADH oxidation was relatively *insensitive* to white light, (2) only *high intensity* red light (but not blue light) could substantially "relieve" this carbon monoxide inhibition, and (3) low concentrations of cyanide could *not abolish* this red light relief of carbon monoxide inhibition. The NADH-oxidizing electron transport pathway was found to be relatively insensitive to both cyanide and azide inhibition. Subsequently, Kauffman and van Gelder, studying the binding of cyanide to cytochrome *d*, reported a decrease in the absorption intensity of this cytochrome after cyanide addition.[48] This decrease corresponded with a gradual *increase in the inhibition* of NADH oxidation. This confirmed that cytochrome *d* was involved in the oxidation of NADH,[52] supporting the previous conclusions of Jones and Redfearn.[33]

The existence of a major cytochrome $b_1 \rightarrow d$ pathway was partially supported by other data, namely the partial purification of an NADH oxidase in the earlier studies of Repaske and Josten.[59] The concentration of cytochromes b_1 and *d*, in this partially purified NADH oxidase preparation, was higher than the concentration of cytochromes c_4 + c_5. However, cytochrome a_1 was also found in this partially purified oxidase preparation, suggesting that cytochrome a_1 as well as *d* might be responsible for carrying out NADH oxidation. In this early study, cytochrome c_4 + c_5 was also readily detected in the NADH oxidase preparation. It was demonstrated that *all* of the cytochromes associated with the NADH oxidation underwent rapid oxidation and reduction, and that the membranous cytochrome c_4 + c_5 was enzymatically *more* active than exogenously purified cytochrome c_4 + c_5 isolated from this *Azotobacter* strain. These findings suggested that cytochrome c_4 + c_5 was involved in NADH oxidation. Since the concentration of cytochrome a_1 found was small, as judged from the spectra shown (although not quantitatively estimated), no comment was made concerning its role in NADH oxidation. However, it was subsequently interpreted[33] that, since the concentrations of cytochromes a_1 along with c_4 + c_5 had actually decreased during the purification of this NADH oxidase preparation, only a "casual relationship" existed between these two cytochrome components relative to NADH oxidation.

The exclusion of cytochrome b_1 from the *minor* electron transport pathway was deduced from the apparent inability of 2-*n*-alkyl-4-hydroxy-quinoline-*N*-oxide (HQNO) to inhibit oxidation completely. However, Jurtshuk et al.[36,38,39,42,43] eventually showed complete inhibition by HQNO of most of the flavoprotein-dependent substrate oxidoreductases that are directly associated with the terminal oxidases of the respiratory chain of *A. vinelandii.* The pathways shown below represent the *A. vinelandii* electron transport chain based on all current information:

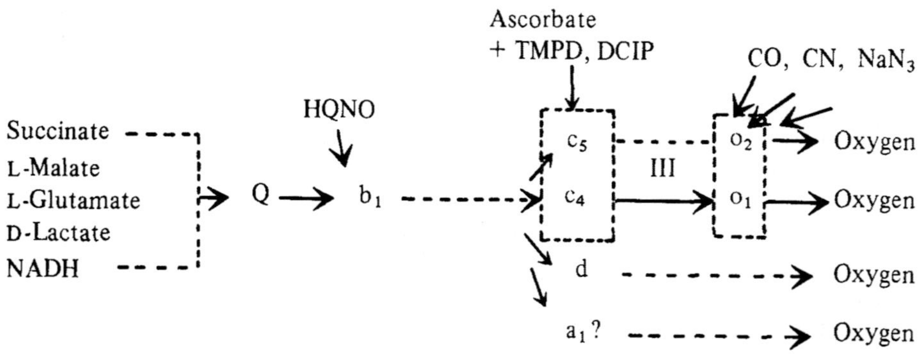

Unlike the past illustrated schemes, the above pathway is modified so that *most* branches of the chain pass through the *c*-type cytochromes which carry the major share of electrons from NADH and succinate. In considering respiration in *A. vinelandii,* (1) all oxidases are readily inhibited to some extent by cyanide, and most are probably inhibited by carbon monoxide, (2) the ascorbate-DCIP and ascorbate-TMPD assays measure the *c*-type dependent oxidase rate, (3) the cytochrome c_4 + *o* complex, isolated using ascorbate-TMPD assay, is the major electron transprt pathway that is highly reactive with carbon monoxide, cyanide, and oxygen, and lastly (3) TMPD oxidase activity in the electron transport particle quantitatively reflects the actual high respiration rate commonly associated with whole cells of *A. vinelandii.* It is therefore concluded that cytochrome $c_4 \rightarrow$ oxygen probably represents the major electron transport pathway in *A. vinelandii,* and it appears to be responsible for coupled site III phosphorylation.[1]

Yates and Jones,[82] in a more recent, revised electron transport scheme, also show separate cytochrome c_4 and c_5 pathways which channel into different terminal oxidases, cytochromes a_1 and *o*, respectively. The represented $c_5 \rightarrow o$ and $c_4 \rightarrow a_1$ pathways are *not* accurate, and there is no evidence to support existence of the $c_4 \rightarrow a_1$ pathway. Recent studies in our laboratory on the purified cytochrome oxidase suggest that the major terminal oxidase pathway is the cytochrome $c_4 \rightarrow o$ pathway for the respiratory chain of *A. vinelandii.*[44,80,81]

There also has been some doubt cast over the role of cytochrome *o* serving as a terminal oxidase in *A. vinelandii*[77] because of the low redox potential recorded in measurements employing whole cells. However, certain data are currently available which might alter this interpretation. Recent studies[80,81] show that the α peak of cytochrome *o* absorbs maximally at 557 nm rather than at 560 nm, which is the wavelength used in determining the redox potential of this hemoprotein. The absorbance measurements at 560 nm would predominantly reflect absorbance changes attributed to cytochrome b_1, and not cytochrome *o*. The redox potential of purified *A. vinelandii* cytochrome o has yet to be determined, and it might well exist in *A. vinelandii* as a functional low-potential oxidase. Unlike the case of mammalian mitochondria, the redox potentials for the individual electron carriers in bacterial systems have not been precisely determined. Therefore, the various cytochromes found cannot be assigned in an "accurate" sequence scheme of electron carrier.

This point is exemplified in earlier studies of Jones and Redfearn.[33] They examined various substrate-mediated cytochrome reductions in the electron transport particle of *A. vinelandii* and found that cytochromes a_1 and *o* could not be subjected to such reduction analysis. In addition, the heterogeneous nature of some of the electron carriers, as revealed for *b*-type cytochrome(s), can also complicate the positioning and sequencing of electron carriers. This point might also be true for cytochrome *o*, as there is some preliminary evidence to suggest that multiple species exist. Consequently, the functional roles and the significance of each of the electron carriers in the overall scheme remains in doubt. Unlike the mitochondrial electron transport system which has been physically and functionally fractionated into four segments and then reconstituted together to demonstrate the overall sequence of electron carriers,[27] no such major reconstitution studies have as yet been successfully undertaken for any microbial electron transport system.[26,43,51,79] Only recently has it been possible to isolate functional enzyme complexes that have been "solubilized" from the *Azotobacter* electron transport particle by detergent fractionation procedures.[44,55,80,81] From the limited knowledge already accumulated, the earlier suppositions[25,43,51,67,68,79] which indicated the high degree of complexity of bacterial electron transport systems have proven true when compared to those of mitochondria.

VI. MEMBRANE SOLUBILIZATION STUDIES

Membrane-bound cytochrome-containing enzyme complexes can now be isolated from membrane fragments in a "solubilized state." These solubilized components can then be further purified by conventional biochemical techniques. The isolation and purification of a membrane segment (or of a complex) of the electron transfer chain, such as the flavoprotein oxidoreductase or a terminal cytochrome oxidase, is a mandatory requirement for reconstitution-type studies. Such simplified units can be then examined for the appropriate enzyme activity. The attempt can be made to reconstitute the entire respiratory chain pathway, as exemplified by the elegant studies of Hatefi with the mitochondrial system.[27] This approach was most effectively used by David E. Green's Laboratory at the Enzyme Institute, University of Wisconsin, in elucidating the function of membrane-bound enzyme complexes, I, II, III, and IV in beef-heart mitochondria.

Repaske and Josten[59] first attempted to solubilize an NADH-oxidizing particle (by sonication and differential centrifugation) in *Azotobacter vinelandii.* Later, Jones and Redfearn[32] successfully fractionated the electron transport particle of *A. vinelandii,* using sodium deoxycholate in the presence of high concentrations of KCl followed by a sucrose density-gradient centrifugation procedure. The "classical" red-green split commonly observed in mammalian mitochondrial fractionation studies was obtained. The color difference observed between these two types of solubilized electron transport particles reflected their cytochrome contents, as determined by spectral analyses. The red-solubilized particle in the supernatant fraction contained greater concentrations of cytochromes c_4 + c_5, b_1, and *o*. The concentration levels of cytochromes a_1 and *d* were relatively low in this fraction. The other fraction, the green-residue particle, was enriched for cytochromes a_1 and *d*.[32] Analyses on the distribution of cytochrome a_1 in the red supernatant and green-residue fractions suggested that the a_1 hemoprotein component was more closely linked to cytochrome *d* than to the *c*-type cytochrome. The *a*- and *d*-type cytochromes, which were found in higher concentrations in the green-residue pellet, exhibited *no* cytochrome oxidase activity. However, the red supernatant fraction, which contained high concentration levels of flavoprotein, coenzyme Q, non-heme iron, and cytochromes c_4 + c_5, *b*, and *o*, exhibited high specific activities for the oxidation of succinate, ascorbate-DCIP, and ascorbate-TMPD. This finding strongly suggested that terminal cytochrome oxidase activity in *A. vinelandii* required the presence of *c*-type cytochromes, which in turn could be linked directly to cytochrome *o*.

Subsequently, Jurtshuk and associates[44,55] were able to isolate a membrane-bound cytochrome oxidase from *A. vinelandii* using a detergent-salt fractionation procedure. This enzyme complex was purified 20-fold and exhibited very high specific activities for TMPD oxidation. The V_m for TMPD oxidation ranged from 60 to 78 μ atoms oxygen consumed per minute per milligram protein. This oxidase activity was also inhibited by carbon monoxide, cyanide, and hydroxylamine. Spectral analyses on the purified oxidase at 77°K permitted identification of two hemoprotein components, cytochromes c_4 and *o*. All active oxidase preparations exhibited the split-peak-trough absorbance (416 nm to 432 nm) of the carbon monoxide-dithionite-reduced difference spectra characteristic for cytochrome *o*. Both carbon monoxide and cyanide caused *marked* spectral changes when added to this purified terminal oxidase similar to that previously noted for the highly purified cytochrome *o* isolated from this same organism.[80,81] Cytochrome c_4 in the highly purified oxidase preparation was reduced by ascorbate, but not the cytochrome *o*. Carbon monoxide also caused *marked* spectral changes in the ascorbate-TMPD-reduced cytochrome oxidase, suggesting that this inhibitor reacts with reduced cytochrome *o* hemoprotein. This enzyme represents the first active cytochrome oxidase preparation that has been purified in a bacterial system using a TMPD-oxidase assay system, and it appears to be an integrated cytochrome c_4 + *o* enzyme complex.

Yang and Jurtshuk,[80,81] by an almost identical solubilization procedure, were able to isolate a membrane-bound cytochrome *o* from the *A. vinelandii* electron transport particle. By further use of conventional chromatographic procedures, the cytochrome *o* was purified further as a detergent-containing (Triton X-100®) hemoprotein complex, which contained 1.2 nmol of heme per milligram of protein. Cold-temperature spectral analyses revealed the absence of any *other* cytochrome component in the purified preparation. Electrophoretic gel analyses revealed that only one type of hemoprotein was present. The purified cytochrome o reacted with both carbon monoxide and cyanide readily. Only in the reduced form did it combine with carbon monoxide, whereas the oxidized form reacted with cyanide.[80,81] An "oxygenated" form of the cytochrome o was demonstrated to be spectrally distinguishable from both the oxidized and the reduced forms.[81] The prosthetic group for cytochrome *o* was found to be protoheme.[80] Amino acid analyses revealed the purified cytochrome *o* to be an acid protein which contained high concentrations of hydrophobic residues, indicating a possible structural relationship with membrane phospholipids. Lipid analyses[21,40] on the purified cytochrome *o* preparation revealed the presence of a high concentration of phospholipid. The amount of phospholipid present was 40.5% by weight. Radioautographic analyses, using two-dimensional thin-layer chromatography, revealed the presence of phosphatidylethanolamine and phosphatidylglycerol.

It was possible to isolate and purify the *A. vinelandii* cytochrome o because it represented one of two cytochrome components (c_4 + o) required for high TMPD-oxidase activity. The cytochrome *o* is in a solubilized form that enables one to separate it from the cytochrome c_4 by chromatography. Like other terminal oxidases, cytochrome *o* in *A. vinelandii* is tightly bound to the membrane. Together with cytochrome c, it appears to function as an enzyme complex which carries out carbon monoxide- and cyanide-sensitive oxidation of TMPD, DCIP, and phenazine-methosulfate (PMS) oxidase reactions that are associated with the electron transport particle of this organism.[44] Of these two hemoproteins, only cytochrome *o* reacts readily with carbon monoxide and cyanide, producing recognizable spectral shifts.[80,81] The *A. vinelandii* cytochrome *o* is also highly autooxidizable, and its ferrous form reacts with oxygen to form an oxygenated species.[81] In many respects, the biochemical properties the *A. vinelandii* of cytochrome o resemble that of the classical cytochrome $a + a_3$ oxidase found in mammalian mitochrondria. The highly purified cytochrome *o* preparation contained two of the three phospholipids found in the electron transport particle of *A. vinelandii*.[40,43] It is possible that these phospholipids may be essential for oxidase activity, as is the case for the mammalian cytochrome oxidase $a + a_3$. Cytochrome *o* also aggregates when the purified preparation is delipidated, another property that is associated with the mitochondrial cytochrome oxidase. The phospholipid (and detergent) may well play a role essential for maintaining the native or "solubilized" state of the cytochrome *o*. Amino acid analyses on cytochrome *o* and mitochondrial cytochrome aa_3 oxidase show that both contain concentrations of amino acids rich in hydrophobic residues.[11] This same finding was true for the amino acid composition of two forms of cytochrome P_{450}, but even more striking was the similarity in amino acid content found for cytochrome *o* and those reported for cytochrome P_{450}.[19,20] Table 1 shows this comparison. Because cytochrome *o* contains a CO-reacting protoheme as a prosthetic group (as does cytochrome P_{450}), and since the amino acid content of cytochrome *o* appears to be very similar to that of cytochrome P_{450}, there is the suggestion that cytochrome *o* might function both as a terminal oxidase as well as an oxygenating hemoprotein.

VII. MULTIPLICITY OF CYTOCHROME *o* COMPONENTS

As referred to previously, Jones and Redfearn[33] carried out substrate-specific reductions and examined spectrally the changes in the individual cytochrome components in the electron transport particle of *A. vinelandii*. Although the reduction of cytochromes $c_4 + c_5$, b_1, and *d* could be followed with accuracy, cytochromes a_1 and *o* could not be examined in this manner. Therefore, no reduction data were reported for these two cytochrome oxidases. Since the absorption spectrum of cytochrome b_1 obviously overlapped with that of cytochrome *o*, the consequent complication in the reduction kinetics observed for cytochrome b_1 was not surprising. More than one *b*-type cytochrome component was implicated as being present in *A. vinelandii* based on the above reduction analyses. This finding is compatible with the observations reported by other workers. Shipp,[65] studying the spectral characteristics of bacterial cytochrome α-bands that could be resolved by fourth-order finite difference analyses in the low-temperature (77°K) absorption spectra, found that in *A. vinelandii* at least two *b*-type cytochromes were present. At least one of these cytochrome *b* components was undoubtly cytochrome *o*.

Erikson and Diehl[24] showed that the photochemical action spectra obtained with NADH and phosphorylating-type particles from *A. vinelandii* was very similar in the visible region to that obtained for whole cells which oxidized endogenous substrate(s). All three terminal oxidases, i.e., cytochromes a_1, *d*, and *o*, were involved in the NADH-dependent respiration. Cytochrome *o* was identified with characteristic light relief maxima at 570, 537, and 417 nm. This was similar to the results reported much earlier by Castor and Chance.[13] However, when ascorbate-TMPD was used as the reducing substrate, the action spectrum obtained showed a light maximum at 560 nm, indicating yet another identified *o*-type cytochrome which was obviously different from the cytochrome o_{570} found when NADH was used as the reducing substrate. This finding suggests the presence of two cytochrome *o* species. There is also other evidence suggesting the presence of multiple cytochrome *o* in different bacterial electron transport systems, as indicated by studies on *Vitreoscilla* spp. and *Acetobacter suboxydans*.[51,73] The preliminary studies of Yang and Jurtshuk[84] revealed that in fractionating the electron transport particle of *A. vinelandii* there seemed to be two multiple cytochrome o components. One was tightly bound to the cytochrome oxidase c_4 + *o* enzyme complex of *A. vinelandii*,[81] the other solubilized in the (0 to 27%) ammonium sulfate fraction.[44] Spectral analysis revealed that these two cytochrome *o* preparations reacted differently with cyanide, as judged from the distinct spectral products formed, from CN-cytochrome *o* complexes.[84] The question to be asked is what, if any, functional difference does exist between these two distinct types of cyanide-reacting cytochromes *o*? Are both cytochromes *o* serving as terminal oxidases, as indicated by the electron transport pathway presented previously? If this were the case, could it be that both cytochromes *o* (designated as cytochrome o_1 and o_2) serve in different electron transport pathways? Is it possible that of the components of these two cytochromes *o* is not a terminal oxidase, but an oxygenating-type cytochrome like P_{450}? When the amino acid composition of our purified cytochrome *o* was compared to the amino acid content of the cytochrome P_{450} (cam) found in the camphor-oxidizing *Ps. putida* and the cytochrome P_{450} (lm) of liver microsomes, the similarity noted was *quite* striking (see Table 1). The fact that both cytochromes *o* and P_{450} have carbon-monoxide-reacting protoheme as prosthetic groups and almost a similar amino acid content leads one to conclude that there may be a functional similarity between these two hemoproteins. If so, then one cytochrome *o* might well serve as the terminal oxidase and the other as an oxygenase. Thus, a *mixed function oxidase reaction* would be among the biological oxidations reactions that cytochrome *o* may carry out.

There is other evidence in the literature which suggests that cytochrome o may function as a mixed-function oxidase. For example, the cytochrome *o* content in *Pseudomonas oleovorans* cells grown in *n*-hexane as the sole carbon source was found to be 4.5 times higher than in cells grown on glucose. Since the concentration of cytochrome *o* was so markedly high in hexane-grown cells, the possibility exists that cytochrome o served to oxygenate *n*-hexane in order that the microorganism could utilize this hydrocarbon as a growth substrate.[56] Even though the ω-hydroxylation of *n*-hexane is not inhibited by carbon monoxide,[57] this observation might not be a significant deterrent to the above possibility. There is evidence which indicates that some species of cytochrome *o* may have low affinities for carbon monoxide, and it is possible that the presence of such a low-affinity carbon monoxide-reacting hemoprotein would account for the uninhibited hydroxylation that would be required for bacterial growth.

VIII. CYTOCHROME *o* IN *VITREOSCILLA* SP.

To date, the most extensively studied cytochrome *o* has been that isolated from *Vitreoscilla*, an obligately aerobic gliding bacterium now classified as *Beggiatoaceae*.[10] Webster and Hackett first examined the respiratory chain of these so-called colorless "algae" and found that at room temperature the difference spectrum of the whole cells showed no evidence of the presence of *a*- or *c*-type cytochromes.[74] A carbon monoxide-reacting component having absorption peaks at 570 to 535 nm and at 416 nm was found to be present in large concentrations. The carbon monoxide-difference spectra revealed the presence of hemoprotein with α- and β-absorption peaks at 570 and 535 nm and a Soret peak at 416 nm, which is characteristic for cytochrome *o*. The

TABLE 1

Amino Acid Composition of the Highly Purified Cytochrome *o* of *Azotobacter vinelandii* and Two Cytochrome P_{450} Components

Amino acid	Cytochrome *o*[a]	P_{450}[b] cam	P_{450}[b] LM
Asp	22	27	21
Thr	19	19	23
Ser	17	21	26
Glu	26	42	24
Pro	13	27	24
$CysO_3H$	ND[c]	6	6
Gly	31	26	30
Ala	32	34	23
Val	20	24	27
Met	4	9	8
Ile	20	24	19
Leu	43	40	46
Tyr	10	9	11
Phe	24	17	28
His	6	12	11
Lys	12	13	19
Agr	15	24	29
Trp	ND	1	1

[a] HCl hydrolysis carried out for 24 hr.[84]

[b] P_{450} values were taken from References 19 and 20 and represent those of the camphor oxidizing *Pseudomonas putida* and the membrane-bound P_{450} of liver microsomes.

[c] ND, not determined.

From Dus, K., Litchfield, W. J., Miguel, A. G., van der Hoeven, T. A., Haugen, D. A., Dean, W. L., and Coon, M. J., Immunochemical and comparison of cytochrome P_{450cam} of *Pseudomonas putida* and P_{450LM} of phenobarbitol-induced rabbit liver microsomes, in Cytochromes P_{450} and b_5: Structure, Function and Interaction, Cooper, D. Y., Rosenthal, O., Snyder, R., and Witmer, C., Eds., Plenum Press, New York, 1975, 47. With permission.

action spectrum for light reversal of carbon monoxide inhibition of *Vitreoscilla* respiration showed absorption maxima at 568 and 534 nm with a monochromatic light source, but 416 nm was found to be the most effective wavelength for reversal of the carbon monoxide inhibition. The carbon monoxide difference spectrum and action spectrum of this cytochrome *o* supported the evidence that it *o* functioned as a respiratory oxidase in *Vitreoscilla* species. This hemoprotein was subsequently purified and characterized extensively by Webster and his associates.[73-76]

Two types of cytochrome *o* preparations were obtained originally from the early work of Webster and Hackett.[73] The purification steps used to isolate this hemoprotein involved freezing and thawing of *Vitreoscilla* cells in the presence of 1% sodium deoxycholate, followed by protamine sulfate and ammonium sulfate fractionation. These were followed by Sephadex® and TEAE-cellulose chromatography. Two highly purified cytochrome *o* fractions were obtained which were designated as Fractions I and II. Fraction I had two α-bands and two Soret bands, when examined spectrally under liquid nitrogen temperature, which indicated that two heme groups might have been present in this fraction. Using a gel filtration technique, a molecular weight of 27,000 was found for Fraction I, and a molecular weight of 22,000 was reported for Fraction II. Fraction I had carbon monoxide difference spectra maxima at 570,534, and 419 nm, with a trough at 436 nm. Fraction II had absorption maxima at 566,532, and 418 nm, with a trough at 436. The cytochrome *o* in Fraction I had a redox potential of 1 V and was slowly autooxidizable. In Fraction II, it had a redox potential of −0.09 V and was rapidly autooxidizable. Because of its extreme autooxidizability, Fraction II was considered to be a soluble form of the membrane-bound cytochrome *o*. Even though the redox potential of this fraction was too low to be considered an oxidase, Webster and Hackett argued that its true redox potential could be much higher in the membrane milieu.[73,74] This argument could also be equally applied to explain the low autooxidizability of Fraction I, which had a much higher redox potential. In earlier studies, no NADH oxidation (for which high activities were reported in both whole cells and membrane particles of *Vitreoscilla*) could be carried out by either of these two cytochrome *o* preparations. In 1974, Webster and Liu, while reinvestigating the purification of cytochrome *o* from another *Vitreoscilla* species (Murray strain 389), reported on its successful reduction by NADH.[75] An NADH-reductase enzyme was copurified with the cytochrome *o* preparation. Using NADH as the reductant, the difference spectra obtained for this cytochrome *o* preparation showed absorption maxima at 577,544, and 420 nm. This was believed to be an oxygenated form of reduced cytochrome *o*. Dithionite-reduced difference spectra of this cytochrome *o* had absorption maxima at 560 and 435 nm. The carbon monoxide difference spectra of this preparation resembled Fraction I in the earlier original studies. However, the dithionite-reduced difference spectra showed absorption characteristics common to both Fractions I and II (see Table 2). These closely related, but not identical, spectral properties of the cytochrome *o* purified from *Vitreoscilla* sp. seem to indicate that more than one type of cytochrome *o* may actually be present in this organism. Oxygen uptake exhibited by the various purified preparations from *Vitreoscilla* sp. (Murray strain 389) were found to be associated with NADH-cytochrome *o* reductase activity. It was concluded from the oxygen uptake experiments that cytochrome *o* and the NADH reductase(s) constituted an electron transfer pathway for the oxidation of NADH by molecular oxygen. However, the attempts to separate cytochrome *o* from NADH-cytochrome *o* reductase were unsuccessful. Therefore, the nature of the reaction mechanism could not be resolved. The oxygenated form of this *soluble* cytochrome *o* was identified as the predominant steady-state species during turnover of the cytochrome in vivo. This was based on the observation of the spectra of actively respiring cells which were compared directly to the purified, oxygenated cytochrome *o* preparation. Evidence was also presented which indicated that the cyclic changes of the oxidized, reduced, and oxygenated cytochrome *o* were all involved in the terminal respiration of *Vitreoscilla*. The oxygenated form was undoubtly an active intermediate species of this cytochrome *o*.[75,76] The results, presented by Webster and Liu, represent the first evidence for a functional oxygenated compound.[75] A comparison of the spectral characteristics reported for the two cytochrome *o* preparations from *Vitreoscilla* sp. and the one purified from *Azotobacter vinelandii* are shown in Table 2.

IX. CYTOCHROME *o* IN *STAPHYLOCOCCUS AUREUS*

Taber and Morrison, in 1964, first identified protoheme as the prosthetic group of cytochrome *o* by examining the pyridine hemochromogen derivatives of partially purified electron transport particles of *Staphylococcus aureus*.[70] It was shown that the particulate respiratory enzyme system in this organism contained an *a*-type cytochrome and two types of cytochrome *b*. The latter two *b*-type cytochromes were resolved by low-temperature spectroscpy. They were shown to possess α- absorption peaks at 555 and 557 nm. Ascorbate-DCIP additions did not reduce cytochrome b_{557}, suggesting that it had a lower redox potential than did the other *b*-type cytochrome, b-$_{555}$. Cytochrome b-$_{557}$ was also sensitive to 2-heptyl-4-hydroxyquinoline-*N*-oxide (HQNO) inhi-

TABLE 2

The Spectral Absorbance Characteristics of Three Purified Bacterial Cytochrome *o* Preparations

		Vitreoscilla sp.[73,76]			*Azotobacter vinelandii*[80,81]		
Absolute		α	β	Soret	α	β	Soret
Oxidized I	—	—	—	398	—	—	—
Reduced I	—	599	—	428	—	—	—
Oxidized II	—	—	—	399	—	—	412
Reduced II	—	553	—	423	557	525	426
CO:reduced	—	—	—	—	557	525	420
Difference							
CO:reduced	—	570	534	(peak) 418	—	—	—
I				(trough) 436			
Reduced II	—	560	—	428	557	526	429
CO:reduced	—	566	532	(peak) 418	572	538	(peak) 416
II				(trough) 436			(trough) 431
Cold-Temp							
(−190°C)							
I	580 (shoulder)	565,551	530,518 (shoulder)	430,433	—	—	—
II	580 (shoulder)	566,555	530 (shoulder)	430	555	527	—

bition, but it did not react with carbon monoxide. It was thus concluded that cytochrome b_{557} was an intermediate electron carrier like a cytochrome *b* or b_1 type, rather than a terminal oxidase. However, cytochrome b_{555} had a higher redox potential than b_{557}. It was not reduced by succinate in a HQNO-inhibited system, but was reduced by ascorbate-DCIP. When reduced by succinate or ascorbate-DCIP and then exposed to carbon monoxide, the cytochrome b_{555} preparation showed a spectrum characteristic of the cytochrome *o*-CO complex. From subsequent carbon monoxide action spectra studies, it was concluded that cytochrome b_{555} was an *o*-type hemoprotein that served as a major oxidase in *S. aureus.*[70] Although cyanide (5×10^{-2} *M*) inhibited this cytochrome-dependent respiratory system, the formation of a specific hemoprotein-CN complex could not be detected spectrally in this organism.

X. CYTOCHROME *o* IN PHOTOSYNTHETIC BACTERIA

Rhodospirillum rubrum, when grown aerobically in the dark, like *S. aureus,* possessed two types of cytochrome *b* in membrane fractions.[71] One *b*-type cytochrome, at liquid nitrogen temperature, had α- and β-absorption maxima at 564 and 540 nm, respectively. The carbon monoxide CO-difference and CO-action spectra revealed this *b*-type cytochrome to be cytochrome *o,* and it was confirmed to be the sole oxidase in *R. rubrum.* Both ascorbate-DCIP and succinate served equally well as reducing agents for this cytochrome *o.* However, when dithionite was used as the reductant, the amount of CO-compound formed was 25% greater than that formed by the ascorbate-DCIP or succinate-reduced enzyme complex.[71] The second *b*-type cytochrome present in the *R. rubrum* membrane fragments was associated with succinate-dehydrogenase activity. Cytochrome *a*-types and *cc′*-type were *not* detected in the aerobically dark-grown cells.

In the anaerobic, light-grown cells of *Rhodopseudomonas spheroides,* cytochrome *o* was the only oxidase present in the electron transport particles.[62] However, when these cells were grown aerobically in the dark, unlike the cells of *R. rubrum,* an *a*-type cytochrome was reported to be the major oxidase. Cytochrome *o* was found to be a minor component.[62] Cytochrome *o* formation could be induced with low oxygen tension and light energy. Cytochrome *a* synthesis under these conditions was repressed.

Horio and Taylor conducted both photochemical and action-spectrum studies on *R. rubrum* and concluded that cytochrome *o* was indeed the oxidase of dark-grown cells.[23] Although they failed to measure an oxidase action spectrum in light-grown cells because of their photosynthetic activity, they reported that cytochrome *o* also functioned as an oxidase in these cells.[28] Their evidence for believing that cytochrome *cc′* was an extracted form of cytochrome *o* was later to be challenged on the grounds that the prosthetic groups of the *cc′*-type and *o*-type cytochromes were different. Chance et al. also presented evidence which excluded cytochrome *cc′* from being an oxidase in photosynthetic bacteria.[51]

XI. CYTOCHROME *o* IN *BACILLUS SPP.*

The electron transport particle of *Bacillus megaterium* strain *KM* has been shown by Broberg and Smith to contain two carbon-monoxide-binding pigments, cytochromes a_3 and *o.*[9] Both hemoproteins were membrane-bound, were induced during endogenous metabolism in whole cells, and could be reduced by NADH in isolated membrane fractions. These two cytochrome components could be separated by treatment with lipase. The cytochrome a_3 component remained bound to the particle while the cytochrome *o* was released into the supernatant fraction. In comparing these two hemoproteins, cytochrome *o* was found to have lower affinity for carbon monoxide than cytochrome a_3. The lipase treatment did not seem to alter the carbon monoxide-binding capacity of cytochrome *o*, nor was any unusual spectral alteration caused by this solubilization procedure.[9] The differential solubility, upon lipase treatment, indicated a difference in the binding (or association) of cytochromes a_3 and *o* to the membrane. The almost identical conclusion was drawn from the studies of Revsin and Brodie,[61] who examined the carbon monoxide-binding pigments in *Mycobacterium phlei.* In spite of the difference in their linkage to the membrane, it was reported that the two CO-binding cytochromes, a_3 and *o*, must be positioned on the membrane so that they could react with the other electron carriers. Both are potentially capable of serving as terminal oxidases in *B. megaterium.* Although cytochrome *o* was shown to be selectively solubilized from the membrane, further purification of this hemoprotein was not attempted. The lipase treatment used "to remove" the cytochrome *o* component did not impair the respiratory activity of the remaining electron transport particle. If cytochrome *o* could be completely removed by lipase solubilization and the remaining cytochrome a_3 component was capable of catalyzing respiration maximally, this would imply that cytochrome a_3 serves as the major terminal oxidase in *B. megaterium.* The question of whether or not cytochrome *o* could function as a terminal oxidase in the absence of active cytochrome a_3 might be answered tentatively by the data presented on the carbon monoxide inhibition of respiration. A large proportion of the respiration was inhibited by a low CO/O_2 ratio. Since cytochrome a_3 showed a higher affinity for carbon monoxide spectrally, the inhibition observed must have been due to the cytochrome a_3-CO complex formed. Thus, cytochrome a_3 must have been responsible for a large proportion of the cellular respiration. The small, yet relatively CO-resistant respiration, could be attributed to the cytochrome *o* component. It was concluded that in the *B. megaterium* electron transport chain, although significant quantities of both cytochromes a_3 and *o* were present, cytochrome *o* could only catalyze *at most* about one fourth of the cell's total respiration.[9]

Bacillus subtilus also synthesized these same two carbon monoxide-binding hemoproteins, cytochromes a_3 and *o*.[72] The differential solubilization studies obtained by lipase treatment gave essentially the same results described previously by Broberg and Smith.[9] Of interest is the fact that the cytochrome *o* concentration was found to be *highest* in dormant spores of *B. subtilus* strain PC 1219. Germinating spores, young vegetative cells, and vegetative cells had only half as much cytochrome *o* as did the dormant spores.[72] Dormant spores of *B. subtilus* strain JB 69 contained cytochrome a_3 as the sole carbon monoxide-binding respiratory pigment. Cytochrome *o* could not be detected. Thus, cytochrome a_3 presumably serves as the only oxidase functioning in dormant spores of strain JB 69. Since cytochrome *o* was not essential for oxidation of NADH in both *B. megaterium* and *B. subtilus,* the physiological significance of this cytochrome is still uncertain.[7,72]

XII. CYTOCHROME *o* IN *MYCOBACTERIUM PHLEI*

Brodie and co-workers, studying the respiratory system of *Mycobacterium phlei,* found cytochromes aa_3 and *o* were both present in the electron transport chain at the third phosphorylation site.[61] Triton X-100® was used to solubilize a cytochrome oxidase aa_3 + *o* enzyme complex which could be further digested with trypsin and partially purified by ammonium fractionation. Although a complete separation of both these terminal oxidase components was never achieved, as described for *B. megaterium,* a solubilized cytochrome *o* fraction was obtained from *M. phlei* by a similar lipase treatment. This cytochrome *o* preparation contained only small amounts of cytochromes *b, c,* and aa_3. The CO-cytochrome *o* peaks absorbed at 416, 538, and 567 nm. As part of the electron transport particle, the 416 nm carbon monoxide-Soret peak of cytochrome *o* could be observed after reduction with succinate, NADH, and malate-FAD. Such substrate reductions utilized the entire respiratory chain. The reduction obtained by ascorbate-TMPD indicated that a segment of the respiratory chain located between cytochrome *c* and molecular oxygen, and the formation of the CO-cytochrome *o* peak(s), were time dependent. Thus, it appears that in *M. phlei* the cytochrome *o* hemoprotein component does not react immediately with carbon monoxide. Increasing the length of time that carbon monoxide was bubbled through the solution did not affect the final absorption maxima of the CO-cytochrome *o* complex.[61] Again, no further attempt was made to purify the solubilized membrane-bound cytochrome *o* in this organism.

XIII. CYTOCHROME *o* IN *RHIZOBIUM JAPONICUM*

The relationship of cytochrome *o* to cytochrome a_3 in free living *Rhizobium japonicum* has been examined by Appleby.[4] At the oxygen tension of air, the respiration of succinate by air-grown *Rhizobium* was substantially inhibited by carbon monoxide, and a photochemical action spectrum showed that both cytochromes aa_3 and *o* functioned as terminal oxidases. However, in the nitrogen-fixing bacteriods of *R. japonicum*, neither of the cytochrome oxidase components were detected. The oxidases of free living cells were reported to be very sensitive to cyanide, complete inhibition occurring at a concentration level of 10 μM.[4-6] No major spectral changes that could be attributed to the cyanide-cytochrome *o* complex were observed as in the case of *S. aureus*. Due to the concentration level of cyanide used, some of the inhibition observed may not have been entirely attributed to the formation of the CN-heme complex, suggesting that perhaps another reaction site was involved. At liquid nitrogen temperatures, the spectrum of the succinate-reduced free-living cells showed the presence of two cytochrome *b*-type components, one with α peak at 556 nm and the other at 559.5 nm. When this difference spectrum was rerun in the presence of cyanide, the absorbance of the 556 nm peak remained undiminished, while that of the 559.5 nm peak partially collapsed. It was proposed that cytochrome b_{556} represented a cyanide-in-sensitive component which also was present in bacteriods. The other cyanide-sensitive cytochrome, $b_{559.5}$, was found only in free-living cells. It was presumed to be cytochrome *o*.[4,5] Inhibitor studies on the electron carrier components of nitrogen-fixing bacteriods in *R. japonicum* suggested the presence of another metalloprotein oxidase that was not a cytochrome. Also, several new CO-reactive hemoproteins were detected in bacteriods. Two of them were tentatively characterized as cytochromes c_{552} and c_{554}. No statement concerning their function was made.

The spectral characterization of cytochrome P_{450} in bacteriods has been well documented.[6] The inhibition observed by carbon monoxide and *N*-phenyl-imidazol seemed to suggest that the bacteroid's cytochrome P_{450}, rather cytochrome c_{552} or c_{554}, was the terminal oxidase which functioned during efficient respiration, yielding a characteristic high ATP production and nitrogenase activity. Both of these *c*-type cytochromes were soluble and capable of combining with carbon monoxide. Unlike the cytochrome *o* components of *R. rubrum* or *S. aureus* (which remained oxidized in the presence of HQNO and were, thereby, clearly distinguishable from other cytochromes *b* in low-temperature difference spectra), all the cytochrome components in whole cells (and particles) of *R. japonicum* were completely reducible by succinate respiration, even after equilibration with 8 μM HQNO. This observation, and the fact that *Rhizobium* cytochrome *o* reduction was inhibited by cyanide, suggests that cytochrome *o* may have different properties in different microorganisms and thus be associated with different types of electron transfer systems.

XIV. CYTOCHROME *o* IN *MICROCOCCUS DENITRIFICANS*

Scholes et al. reported on the similarity of the respiratory chain of mammalian mitochondria to that of *Micrococcus (Paracoccus) denitrificans.*[63] The carbon monoxide spectra of (1) anaerobically grown cells reduced with substrate, or (2) dithionite-reduced membranes from either aerobically or anaerobically grown cells, showed CO-compounds characteristic of both cytochromes a_3 and *o*. The absorption peak at 415 nm showed an absorbance increase after treating the membrane fraction with either Triton X-100® or deoxycholate, a result similar to that reported earlier by Porra and Lascelles.[58] The carbon monoxide spectrum of cytochrome *o* was *not* observed in the membrane fraction of 12-hr-old aerobically grown cells when redued anaerobically with substrate. Thus, this hemoprotein was *not* considered to be functional in aerobically grown cells. Comment based on unpublished observations indicated that cytochrome *o* had been detected in the soluble fraction of anaerobically grown cells and that deoxycholate increased the absorption of the carbon monoxide compound of cytochrome *o*. It was suggested that cytochrome *o* may be an altered form of some other hemoprotein-type pigment.[63] No evidence was presented for a functional role for cytochrome *o* in the electron transport system of *M. denitrificans.*

XV. CYTOCHROME *o* IN *ACETOBACTER SUBOXYDANS*

A cytochrome b_{558} was isolated and identified as the effective cytochrome *o* in cells of *Acetobacter suboxydans* strain IAM 1828 grown on lactate. It was also reported to be crystallized.[29] Oxidation of lactate was stimulated when this partially purified cytochrome b_{558} was added to *A. suboxydans* membrane fragments which contained cytochrome *a* and another *o*-type cytochrome. No lactate oxidase activity occurred without added cytochrome b_{558}. There was no report that the crystallized hemoprotein b_{558} possessed any cytochrome oxidase activity. Another *A. suboxydans* strain (ATCC 621) that was examined by Daniel also was unique in that it contained cytochrome *o* in the absence of *a*-type cytochromes.[18] The low-temperature carbon monoxide difference spectrum demonstrated that two carbon monoxide-binding hemoproteins were present in this organism when cells were grown on glycerol as the carbon source. The carbon monoxide inhibition of oxidase activity, and the relief of this inhibition with light, showed that at least one of these CO-reactive cytochromes was a terminal oxidase.[18] The Lineweaver-Burk plot of the oxygen affinity for NADH oxidase activity in the presence and absence of carbon monoxide indicated the presence of two terminal oxidases. Daniel suggested that both CO-binding *b*-type cytochromes served as terminal oxidases in this strain of *A. suboxydans.* The two CO-binding cytochromes were designated as cytochrome o_{565} and cytochrome o_{558}. Evidence was then presented to show that these two cytochromes *o* had different affinities toward carbon monoxide. Cytochrome o_{565} had a high carbon monoxide affinity and cytochrome o_{558} had the low carbon monoxide affinity.[18]

XVI. CYTOCHROME a_1 IN BACTERIAL ELECTRON TRANSPORT SYSTEMS

Cytochrome a_1 has been isolated and partially purified from *Nitrosomonas europaea.*[22] This partially purified cytochrome a_1 preparation possessed weak cytochrome *c* oxidase activity which could not account for the oxygen uptake exhibited by the whole cell respiration. The cytochrome *c* oxidase activity was inhibited by cyanide (10^{-5} *M*). Carbon monoxide inhibition could not be demonstrated. Cytochrome a_1 was also noted in "semianaerobic" *Hemophilus parainfluenzae* cells.[78] There, the concentration of cytochrome a_1 was found to be greater in cells grown anaerobically in the presence of nitrate.[66,77] In such cells, cytochrome a_1 served as a nitrate reductase. Cytochrome a_1 has been found in membrane-particle preparations in *Nitrobacter,*[3] *Achromobacter,*[53] *Acetobacter,*[13] *P. vulgaris,*[13] *E. coli,*[7] *Ferrobacillus ferrooxidans,*[8] *Holobacterium cutirubrum,*[50] *H. halobium,*[17] and *A. vinelandii.*[13]

The early studies of Jones and Redfearn[32] and Jurtshuk et al.[39] revealed that the concentration of cytochrome a_1 in *A. vinelandii* was low in cells harvested at log or late-log phases of growth. Kauffman and van Gelder[47-49] also reported that in some of their cell preparations of *A. vinelandii* cytochrome a_1 could not be detected. These latter investigators examined NADH and ascorbate-DCIP oxidations. In some cellular preparations which lacked cytochrome a_1, these activities were found to be as high as in those preparations which contained a *substantial* amount of cytochrome a_1.[49] This observation suggests that cytochrome a_1 plays only a minor role as an integral functional component in the *A. vinelandii* respiratory chain.Other data give additional, yet conflicting, information on the role played by cytochrome a_1 in *A. vinelandii.* For example, Erickson and Diehl[24] showed the presence of cytochrome a_1 and possibly another *o* type cytochrome with action spectrum studies using ascorbate-TMPD as the reductant. This cytochrome *o* was different from the cytochrome *o* found when NADH was used as the reductant. This latter finding suggests that cytochrome a_1 does functionally participate in the ascorbate-TMPD oxidation pathway as was originally proposed by Jones and Redfearn.[33]

In examining respiratory activity in the *A. vinelandii* electron transport system, Kauffman and van Gelder observed spectrally that cyanide had *no effect* on cytochrome a_1. They concluded that cytochrome a_1 was unable to function as a cyanide-sensitive oxidase.[49] This is in direct contrast to the earlier findings reported by Jones and Redfearn. Jurtshuk et al. further demonstrated that a purified cyanide-sensitive cytochrome oxidase, which carried out active ascorbate-TMPD oxidation, consisted of a cytochrome c_4 + *o* complex.[44,55] No cytochrome a_1 was found to be associated with this cytochrome oxidase preparation. This suggested that cytochrome a_1, while insensitive to cyanide inhibition, functioned, perhaps, in another pathway. Yang and Jurtshuk[84] found that when ascorbate-TMPD was used as a reductant for monitoring the spectral changes in the electron transport particle, the majr reduction occurs at 551 nm. No absorption changes are noted in the red region (580 to 600 nm), further indicating the lack of participation of *a*-type cytochromes in reduction studies employing ascorbate-TMPD.[39]

While there is sufficient evidence for linking cytochromes c_4 and *o* in the pathway of the electron transport chain of *A. vinelandii,* there is no evidence to indicate the functional association of cytochromes c_5 and a_1 in any *Azotobacter* electron transfer reactions, to date. Only limited work has been done in attempting to establish the significance of cytochrome a_1 in serving as a functional bacterial oxidase. There is also conflicting evidence as to the actual electron transferring pathway in which cytochrome a_1 serves as a terminal electron acceptor. Its true role remains uncertain.

REFERENCES

1. **Ackrell, B. A. C. and Jones, C. W.,** The respiratory system of *Azotobacter vinelandii.* I. Properties of phosphorylating respiratory membranes, *Eur. J. Biochem.,* 20, 22, 1971.
2. **Ackrell, B. A. C., Erickson, S. K., and Jones, C. W.,** The respiratory chain NADPH dehydrogenase of *Azotobacter vinelandii, Eur. J. Biochem.,* 26, 387, 1972.

3. **Aleem, M. I. H. and Nason, A.**, Nitrite oxidase, a particulate cytochrome electron transport system from Nitrobacter, *Biochem. Biophys. Res. Commun.*, 1, 323, 1959.
4. **Appleby, C. A.**, Electron transport systems of *Rhizobium japonicum*. I. Haemoprotein p-450, other CO-reactive pigments, cytochromes and oxidases in bacteriods from N_2-fixing root nodules. *Biochim. Biophys. Acta*, 172, 71, 1969.
5. **Appleby, C. A.**, Electron transport systems of *Rhizobium japonicum*. II. *Rhizobium* hemoglobin cytochromes, and oxidases in free-living (cultured) cells, *Biochim. Biophys. Acta*, 172, 88, 1969.
6. **Appleby, C. A.**, Function of p-450 and other cytochromes in *Rhizobium* respiration, in *Proc. 11th FEBS Meeting*, Degn, H., Lloyd, D., and Hill, G. C., Eds., Pergamon Press, Oxford, 1978, 11.
7. **Bartsch, R. G.**, Bacterial cytochromes, *Annu. Rev. Microbiol.*, 22, 181, 1968.
8. **Blaylock, B. A. and Nason, A.**, Electron transport studies on the chemoautograph *Ferrobacillus ferrooxidans*, *Fed. Proc. Fed. Am. Soc. Exp. Biol.*, 20, 46, 1961.
9. **Broberg, P. L. and Smith, L.**, The cytochrome system of *Bacillus megaterium* KM. The presence and some properties of two CO-binding cytochromes, *Biochim. Biophys. Acta*, 131, 479, 1967.
10. **Buchanan, R. E. and Gibbons, N. E.**, Eds., *Bergey's Manual of Determinative Bacteriology*, 8th ed., William & Wilkins, Baltimore, 1974.
11. **Capaldi, R. A. and Briggs, M.**, The Structure of Cytochrome Oxidase, in *Enzymes of Biological Membranes*, Martonosi, A., Ed., Plenum Press, New York, 1976, 87.
12. **Castor, L. N. and Chance, B.**, Photochemical action spectra of carbon monoxide-inhibited respiration, *J. Biol. Chem.*, 217, 453, 1955.
13. **Castor, L. N. and Chance, B.**, Photochemical determinations of the oxidases of bacteria, *J. Biol. Chem.*, 234, 1587, 1959.
14. **Chance, B.**, The carbon monoxide compounds of the cytochrome oxidases. I. Difference spectra, *J. Biol. Chem.*, 202, 383, 1953.
15. **Chance, B.**, in *Hematin Enzymes*, Falk, J. E., Lemberg, R., and Morton, R. K., Eds., Pergamon Press, New York, 1961, 433.
16. **Chance, B., Smith, L., and Castor, L. N.**, New methods for the study of the carbon monoxide compounds of respiratory enzymes, *Biochim. Biophys. Acta*, 12, 289, 1953.
17. **Cheah, K. S.**, Properties of electron transport particles from *Halobacterium cutirubrum*. The respiratory chain system, *Biochim. Biophys. Acta*, 180, 320, 1969.
18. **Daniel, R. M.**, The electron transport system of *Acetobacter suboxydans* with particular reference to cytochrome *o*, *Biochim. Biophys. Acta*, 216, 328, 1970.
19. **Dus, K.**, On the structure and function of cytochromes p-450, in *Enzymes of Biological Membranes*, Martonosi, A., Ed., Plenum Press, New York, 1976, 230.
20. **Dus, K., Litchfield, W. J., Miguel, A. G., van der Hoeven, T. A., Haugen, D. A., Dean, W. L., and Coon, M. J.**, Immunochemical and comparison of cytochrome p-450_{cam} of *Pseudomonas putida* and p-450_{LM} of phenobarbitol-induced rabbit liver microsomes, in *Cytochromes p-450 and b_5:Structure, Function and Interaction*, Cooper, D. Y., Rosenthal, O., Snyder, R., and Witmer, C., Eds., Plenum Press, New York, 1975, 47.
21. **Eichberg, J., Shein, H. M., Schwartz, M., and Hauser, G.**, Stimulation of $^{32}P_i$ incorporation into phosphatidylinositol and phosphatidylglycerol by catecholamines and β-adrenergic receptor blocking agents in rat pineal organ cultures, *J. Biol. Chem.*, 248, 3615, 1973.
22. **Erickson, S. K., Hooper, A. B., and Terry, K. R.**, Solubilization and purification of cytochrome a_1 from *Nitrosomonas*, *Biochim. Biophys. Acta*, 283, 155, 1972.
23. **Erickson, S. K., Ackrell, B. A. C., and Jones, C. W.**, The respiratory chain dehydrogenases of *Azotobacter vinelandii*, *Biochem. J.*, 127, 73, 1972.
24. **Erickson, S. K. and Diehl, H.**, The terminal oxidases of *Azotobacter vinelandii*, *Biochem. Biophys. Res. Commun.*, 50, 321, 1973.
25. **Gel'man, N. S., Lukoyanova, M. A., and Ostrovskii, D. N.**, *Respiration and Phosphorylation in Bacteria*, Plenum Press, New York, 1967.
26. **Haddock, B. A. and Jones, C. W.**, Bacterial respiration, *Bacteriol. Rev.*, 41, 47, 1977.
27. **Hatefi, Y.**, The functional complexes of the mitochrondrial electron-transfer system, in *Comprehensive Biochemistry*, Vol. 14, Florkin, M. and Stotz, E. H., Eds., Elsevier, New York. 1966, 199.
28. **Horio, T. and Taylor, G. P. S.**, The photochemical determination of an oxidase of the photohetertroph, *Rhodospirillum rubrum* and the action of the inhibition of respiration by light, *J. Biol. Chem.*, 240, 1772, 1965.
29. **Iwaski, H.**, Lactate oxidation system in *Acetobacter suboxydans* with special reference to carbon monoxide binding pigment, *Plant Cell Physiol.*, 7, L199, 1966.
30. **Jones, C. W.**, Aerobic respiratory systems in bacteria, in *27th Symp. Soc. Gen. Microbiol.*, Haddock, B. A., and Hamilton, W. A., Eds., Cambridge University Press, Cambridge, 1977, 23.
31. **Jones, C. W. and Redfearn, E. R.**, Electron transport in *Azotobacter vinelandii*, *Biochim. Biophys. Acta*, 113, 467, 1966.

32. **Jones, C. W. and Redfearn, E. R.**, Preparation of red and green electron transport particles from *Azotobacter vinelandii*, *Biochim. Biophys. Acta*, 143, 354, 1967.
33. **Jones, C. W. and Redfearn, E. R.**, The cytochrome system of *Azotobacter vinelandii*, *Biochim. Biophys. Acta*, 143, 340, 1967.
34. **Jones, C. W., Brice, J. W., and Edwards, C.**, Bacterial cytochrome oxidases and respiratory chain energy conservation, in Proc. 11th FEBS Meet., *Degn, H., Lloyd, D., and Hill, G. C.*, Eds., Pergamon Press, Oxford, 1978, 89.
35. **Jurtshuk, P., Aston, P. R., and Old, L.**, Enzymatic oxidation of tetramethyl-*p*-phenylenediamine and *p*-phenylenediamine by the electron transport particulate fraction of *Azotobacter vinelandii*, *J. Bacteriol.*, 93, 1069, 1967.
36. **Jurtshuk, P. and Harper, L.**, Oxidation of D (−) lactate by the electron transport fraction of *Azotobacter vinelandii*, *J. Bacteriol.*, 96, 678, 1968.
37. **Jurtshuk, P., Jr. and Old, L.**, Cytochrome *c* oxidation by the electron transport fraction of *Azotobacter vinelandii*, *J. Bacteriol.*, 95, 1790, 1968.
38. **Jurtshuk, P., Bednarz, A. J., Zey, P., and Denton, C. H.**, L-malate oxidation by the electron transport fraction of *Azotobacter vinelandii*, *J. Bacteriol.*, 98, 1120, 1969.
39. **Jurtshuk, P., May, A. D., Pope, L. M., and Aston, P. R.**, Comparative studies on succinate and terminal oxidase activity in microbial and mammalian electron systems, *Can. J. Microbiol.*, 15, 797, 1969.
40. **Jurtshuk, P., Jr. and Schlech, B. A.**, Phospholipids of *Azotobacter vinelandii*, *J. Bacteriol.*, 97, 1507, 1969
41. **Jurtshuk, P., Jr. and Cardini, G. E.**, The mechanism of hydrocarbon oxidation by a *Corynebacterium* species, *CRC Crit. Rev. Microbiol.*, 1, 239, 1971.
42. **Jurtshuk, P. and McManus, L.**, None-pyridine nucleotide dependent L-(+)-glutamate oxidoreductase in *Azotobacter vinelandii*, *Biochim. Biophys. Acta*, 368, 158, 1974.
43. **Jurtshuk, P., Jr., Mueller, T. J., and Acord, W. C.**, Bacterial terminal oxidases, *CRC Crit. Rev. Microbiol.*, 3, 399, 1975.
44. **Jurtshuk, P., Jr., Mueller, T. J., McQuitty, D. N., and Riley, W. H.**, The cytochrome oxidase reaction in *Azotobacter vinelandii* and other bacteria, in Proc. 11th FEBS Meet., Degn, H., Lloyd, D., Hill, G. C., Eds., Pergamon Press, Oxford. 1978, 99.
45. **Kamen, M. D. and Horio, T.**, Bacterial cytochromes. I. Structural aspects, *Annu. Rev. Biochem.*, 39, 673, 1970.
46. **Kamen, M. D.**, The status of RHP and other atypical heme proteins as bacterial oxidases, in *Oxidases and Related Redox Systems*, Vol. 1, King, T. E., Mason, H. S., and Morrison, M., Eds., John Wiley & Sons, New York, 1965, 529.
47. **Kauffman, H. F. and van Gelder, B. F.**, The respiratory chain of *Azotobacter vinelandii*. I. Spectral properties of cytochrome *d*, Biochim. Biophys. Acta, 305, 260, 1973.
48. **Kauffman, H. F. and van Gelder, B. F.**, The respiratory chain of *Azotobacter vinelandii*. II. The effect of cyanide on cytochrome *d*, Biochim. Biophys. Acta, 314, 276, 1973.
49. **Kauffman, H. F. and van Gelder, B. F.**, The respiratory chain of *Azotobacter vinelandii*. III. The effect of cyanide in the presence of substrates, *Biochim. Biophys. Acta*, 333, 218, 1974.
50. **Lanyi, J. K.**, Studies of the electron transport chain of extremely halophilic bacteria. II. Salt dependence of reduced diphosphopyridine nucleotide oxidase, *J. Biol. Chem.*, 224, 2865, 1969.
51. **Lemberg, R. and Barrett, J.**, Bacterial cytochromes and cytochrome oxidases, in *Cytochromes*, Academic Press, New York, 1973, 217.
52. **Lineweaver, H.** Characteristics of oxidation by *Azotobacter*, *J. Biol. Chem.*, 99, 575, 1933.
53. **Mizushima, S. and Arima, K.**, Mechanism of cyanide resistance in *Achromobacter*. III. Nature of terminal electron transport system and its sensitivity of cyanide, *J. Biochem.*, 47, 837, 1960.
54. **Mok, T. C. K., Richard, P. A. D., and Moss, F. J.**, The carbon monoxide-reactive haemoproteins of yeast, *Biochim. Biophys. Acta*, 172, 438, 1969.
55. **Mueller, T. J. and Jurtshuk, P., Jr.**, Solubilization of cytochrome oxidase from *Azotobacter vinelandii*, *Fed. Proc. Fed. Am. Soc. Exp. Biol.*, 31, 888, 1972.
56. **Peterson, J. A.**, Cytochrome content of two psedumonads containing mixed-function systems, *J. Bacteriol.*, 103, 714, 1970.
57. **Peterson, J. A., Kusunose, M., Kusunose, E., and Coon, M. J.**, Enzymatic-oxidation. II. Function of rubredoxin as the electron carrier in hydroxylation, *J. Biol. Chem.*, 242, 4334, 1967.
58. **Porra, R. and Lascelles, J.**, Haemoproteins and haem synthesis in facultative photosynthetic and denitrifying bacteria, *Biochem. J.*, 94, 120, 1965.
59. **Repaske, R. and Josten, J. J.**, Purification and properties of reduced diphosphopyridine nucleotide oxidase from *Azotobacter*, *J. Biol. Chem.*, 233, 466, 1958.
60. **Perlish, J. S. and Eichel, H. J.**, A succinate- and DPNH-reducible *o*-type cytochrome in mitochondrial preparations from *Tetrahymena pyriformis*, *Biochem. Biophys. Res. Commun.*, 44, 973, 1971.

61. **Revsin, B. and Brodie, A. F.**, Carbon monoxide-binding pigment of *Mycobacterium phlei* and *Escherichia coli.*, *J. Biol. Chem.*, 244, 3101, 1969.
62. **Sasaki, T., Motokawa, Y., and Kikuchi, G.**, Occurrence of both *a*-type and *o*-type cytochromes as the functional terminal oxidase in *Rhodopseudomonas spheroides, Biochim. Biophys. Acta*, 197, 284, 1970.
63. **Scholes, P., Newton, N., and Smith, L.**, Succinate and cytochrome oxidases of *Micrococcus denitrificans, Fed. Proc. Fed. Am. Soc. Exp. Biol.*, 25, 740, 1966.
64. **Sekuzu, I., Takemori, S., Orii, Y., and Okunuki, K.**, Studies on cytochrome *a*. IV. Reaction of cytochrome *a* with cytochrome *c* and c_1, *Biochim. Biophys. Acta*, 37, 64, 1960.
65. **Shipp, W. S.**, Absorption bands of multiple *b* and *c* cytochromes in bacteria detected by numerical analysis of absorption spectra, *Arch. Biochem. Biophys.*, 150, 482, 1972.
66. **Sinclair, P. R. and White, D. C.**, Effect of nitrate, fumarate, and oxygen on the formation of the membrane-bound electron transport system of *Haemophilus parainfluenzae, J. Bacteriol.*, 101, 365, 1970.
67. **Smith, L.**, Bacterial cytochromes, *Bacteriol. Rev.*, 18, 106, 1954.
68. **Smith, L.**, The respiratory chain system of bacteria, in *Biological Oxidations*, Singer, T. P., Ed., Interscience, New York, 1968.
69. **Swank, R. T. and Burris, R. H.**, Purification and properties of cytochrome *c* of *Azotobacter vinelandii, Biochim. Biophys. Acta*, 180, 473, 1969.
70. **Taber, H. W. and Morrison, M.**, Electron transport in *Staphylocci.* Properties of a particulate preparation from exponential phase *Staphylococcus aureus, Arch. Biochem.*, 105, 367, 1964.
71. **Taniguchi, S. and Kamen, M. D.**, The oxidase system of heterotrophically-grown *Rhodospirillum rubrum, Biochim. Biophys. Acta*, 96, 395, 1965.
72. **Tochikubo, K.**, Changes in terminal respiratory pathways of *Bacillus subtilis* during germination, outgrowth, and vegetative growth, *J. Bacteriol.*, 108, 652, 1971.
73. **Webster, D. A. and Hackett, D. P.**, The purification and properties of cytochrome *o* from *Vitroeoscilla, J. Biol. Chem.*, 241, 3308, 1966.
74. **Webster, D. A. and Hackett, D. P.**, Respiratory chain of colorless algae. II. *Cyanophyta, Plant Physiol.*, 41, 599, 1966.
75. **Webster, D. A. and Liu, C. Y.**, NADH-cytochrome *o* reductase associated with cytochrome *o* purified from *Vitreoscilla:* evidence for an intermediate oxygenated form of cytochrome *o*, *J. Biol. Chem.*, 249, 4257, 1974.
76. **Webster, D. A. and Orii, Y.**, Oxygenated cytochrome *o*: an active intermediate observed in whole cells of *Vitreoscilla, J. Biol. Chem.*, 252, 1834, 1977.
77. **White, D. C.**, Cytochrome and catalase patterns during growth of *Haemophilus* parainfluenzae, *J. Bacteriol.*, 83, 851, 1962.
78. **White, D. C.**, Factors affecting the affinity for oxygen of cytochrome oxidases in *Hemophilus parainfluenzae, J. Biol. Chem.*, 238, 3757, 1963.
79. **White, D. C. and Sinclair, P. R.**, Branched electron-transport systems in bacteria, *Adv. Microb. Physiol.*, 5, 173, 1971.
80. **Yang, T. Y. and Jurtshuk, P., Jr.**, Purification and characterization of cytochrome *o* from *Azotobacter vinelandii, Biochim. Biophys. Acta*, 502, 543, 1978.
81. **Yang, T. Y. and Jurtshuk, P., Jr.**, Studies on the red oxidase (cytochrome *o*) of *Azotobacter vinelandii, Biochem. Biophys. Res. Commun.*, 81, 1032, 1978.
82. **Yates, M. G. and Jones, C. W.**, Respiration and nitrogen fixation in *Azotobacter, Adv. Microb. Physiol.*, 11, 97, 1974.
83. **Yonetani, T.**, Studies on cytochrome oxidase. III. Improved preparation and some properties, *J. Biol. Chem.*, 236, 1680, 1961.
84. **Yang, T. Y. and Jurtshuk, P., Jr.**, unpublished data.

Chapter 6

RESPIRATION IN METHANOGENIC BACTERIA

R. S. Wolfe

TABLE OF CONTENTS

I. INTRODUCTION

Although there has been a surge of interest in methanogens over the last few years, the following account of respiration must be brief, for only a few laboratories have contributed to the literature in this area. For an historical perspective of progress on methanogenesis the following review articles provide a chronological picture: 1956, Barker;[1] 1967, Stadtman;[2] 1971, Wolfe;[3] 1977, Zeikus;[4] 1977, Mah et al.;[5] 1978, Zehnder;[6] 1979, Wolfe and Higgins.[7]

Significant advances in the technology of isolation and cultivation of methanogens have been made in the last 5 years.[7-11] Methanogens now can be maintained and cultivated without the special competency required of experts in the Hungate technique.[12-15] These organisms are within reach of most microbiologists and biochemists. However, a word of caution is in order. In spite of recent advances, methanogens continually challenge the microbiologist to deliver the highest standards of aseptic competence in studying them. They wait to trap the unwary investigator. These organisms have a long generation time, and many of them have the ability to grow essentially as chemoautotrophs. These two properties may combine to mislead the investigator into believing that a pure culture is being cultivated, when in fact, the culture may represent a microbial zoo. During isolation of methanogens, there may be a tendency to use a minimal medium so that the growth of other organisms is retarded. In such a medium, an apparently pure culture or a pick from an isolated colony may be diluted and plated followed by incubation. This process may be repeated again and again, but as long as the culture is maintained in a minimal medium contaminants may not be seen readily. When the culture is grown in a fermentor for 5 days or more, the methanogen may secrete sufficient metabolic products to allow considerable growth of contaminants. Use of a richer medium in the fermentor will produce the same result. Thus, it is possible to detect a physiological property or to fractionate an enzyme from a "methanogen" which in fact has been provided by nonmethanogenic contaminants. The simplest way to detect contamination is to subculture only a fully grown culture to a variety of freshly prepared anaerobic media that contain various combinations of glucose, rumen fluid, yeast extract, casein digest, and cysteine-sulfide buffer under a nitrogen atmosphere. Nonmethanogens usually produce heavy growth within 24 to 30 hr.

In the presentation which follows, we shall consider only those aspects of methanogens which may be pertinent to an understanding of their respiration at the cellular and subcellular level. The picture is grossly incomplete.

II. HYDROGENASE

Respiration in methanogenic bacteria appears to represent a metabolic system that is closely associated with electron transport phosphorylation. Thus, the term respiration when applied to methanogens fits with the usage suggested by Morris.[58] That is, the term respiration should be reserved for systems in which a charge separation is a prerequisite for ATP synthesis. Methanogens are basically chemoautotrophs in their metabolism, and molecular hydrogen is a favorite substrate of all methanogens now in pure culture. For the anaerobic oxidation of hydrogen, carbon dioxide serves as the electron acceptor. Knowledge of the mechanism by which carbon dioxide is reduced to methane will be considered in Section IV. No substrate-level phosphorylation has been observed in methanogens, and it has been assumed[7,16,17] that during the oxidation of hydrogen and the subsequent transport of electrons and protons, a charge separation occurs across a membrane, with electron transport phosphorylation being the source of ATP. This subject will be considered in Section VI.

Methanogens use molecular hydrogen as an oxidizable substrate, and hydrogenase activity was detected early in whole cells or cell extracts of these organisms.[18,19] Several laboratories are currently studying hydrogenases from methanogens, but little information has been published to date. A hydrogenase-containing fraction from extracts of *Methanobacterium thermoautotrophicum* was studied recently by Gunsalus.[20] Cells were broken by passage through a French press. After centrifugation of broken cells for 2 hr at 150,000 *g*, 75 to 80% of the total hydrogenase activity was found in the supernatant solution. The hydrogenase from this organism was easily rendered soluble, and it was not necessary to use membrane-disrupting procedures (such as addition of detergents and EDTA) to obtain the enzyme in soluble form. These observations do not rule out the possibility that a distinct membrane-bound hydrogenase may be found in the 150,000 *g* pellet. The soluble hydrogenase exhibited a property common to most hydrogenases, i.e., that a variety of compounds served as electron acceptors. An apparent K_m for the reduction of coenzyme F_{420} of 25 μM was observed, whereas for methyl viologen the K_m was 1.5 m*M* with a maximal velocity of 140 μmol of methyl viologen being reduced per minute per milligram of protein. Addition of carbon monoxide produced only an 11% inhibition of hydrogenase activity when added at a partial pressure of 0.2 to the assay mixture, or when preincubated with the enzyme. The failure of carbon monoxide to be an effective inhibitor of this hydrogenase is surprising and may indicate that the methanogens in general possess hydrogenases with this property. This remains to be tested. Exposure of the enzyme to oxygen destroyed enzymic activity. The temperature optimum was 60°C, and the pH optimum was 7.0. By following the elution of the enzyme from a Bio-Gel® A-1.5 agarose column, a size in excess of 500,000 daltons was estimated.

The hydrogenase of *Methano-brevibacter ruminantium* strain PS was studied by Tzeng et al.[21] Cells were broken by sonication, and hydrogenase activity was found in the supernatant fluid after centrifugation at 22,000 *g* for 30 min. In the presence of hydrogen, a variety of electron acceptors could be reduced by crude extracts (FMN, FAD, methyl viologen, benzyl viologen, Fe^{3+}, and coenzyme F_{420}). The hydrogenase of *Methanobacterium* M.o.H. was studied by Eirich.[22] The K_m for reduction of coenzyme F_{420} was found to be 25 μM.

III. COENZYME F_{420}

Cheeseman[23] was the first to observe that cells of a methanogen (*Methanobacterium* M.o.H.) were highly fluorescent when exposed to oxygen, i.e., after harvest with a Sharples® centrifuge. When these cells were placed under a hydrogen atmosphere, the fluorescence gradually disappeared. Subsequently, it was found that cells stored in the fluorescent state quickly lost methanogenic activity, whereas cells in which the fluorescent compound had been reduced to the nonfluorescent state could be stored under hydrogen at −20°C for a year or more with full retention of methanogenic activity. The fluorescent compound could be used as an indicator for the safe storage of cells. Historically, this was a breakthrough, for it meant that cells of methanogens could be stockpiled in an active state for long periods. This same concept was later extended to the storage of cell extracts. Here, the effective storage time was found to vary according to the organism from which the extract was prepared. Certain enzymes were stable, e.g., the methylreductase of *M. thermoautotrophicum* was active after storage of extracts for one year,[20] whereas other enzymic activities were lost when extracts were frozen. At 4°C, storage time for retention of enzymic activity was limited to several weeks.

The fluorescent factor which served as the indicator of reducing potential was purified by Cheeseman and was found to have a strong absorbance at 420 nm. Properties of this compound were reported by Cheeseman et al.[23] under the name $Factor_{420}$ (F_{420}),

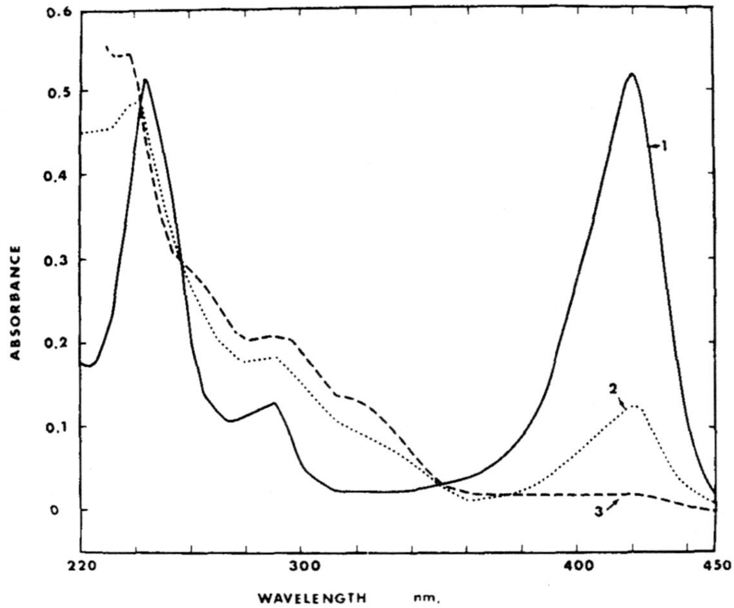

FIGURE 1. Effect of reduction of F_{420} by sodium borohydride on the absorption spectrum. Curve 1, oxidized F_{420} at pH 7.3; curve 2, reduction after 15 min; curve 3, complete reduction. (From Cheeseman, P., Toms-Wood, A., and Wolfe, R.S., *J. Bacteriol.*, 112, 529, 1972. With permission.)

since the role of this compound in the biochemistry of methanogenesis was unknown at that time. All that was known was that F_{420} was reduced by hydrogenase, but lack of specificity is a property of most hydrogenases. The UV-visible spectrum of oxidized and reduced F_{420} is shown in Figure 1.

It was not until the work of Tzeng et al.[21,24] with extracts of *M. ruminantium* PS that the role of F_{420} as a low-potential electron carrier became apparent (the name coenzyme F_{420} was applied subsequently). Crude extracts of this organism readily reduced NADP in the presence of hydrogen. Passage of these extracts through a column of Sephadex G-25® resulted in a loss of ability to reduce NADP. Similarly, when extracts were passed through a column of DEAE-cellulose, they were resolved for a component that was required for reduction of NADP. The factor removed by these procedures was not replaceable by addition of FAD, FMN, metals, yeast extract, ferredoxin, or cobalamin. Addition of coenzyme F_{420} from *Methanobacterium* M.o.H. restored the ability of the resolved extracts to reduce NADP (Figure 2). The rate of hydrogen uptake by the system was proportional to the concentration of F_{420} added (Figure 3). An apparent K_m for F_{420} was calculated to be 5×10^{-6} *M*. The system was specific for NADP. NAD was not reduced. A similar F_{420}-dependent NADP-linked hydrogenase system has been found in *Methanobacterium* M.o.H. and in *M. thermoautotrophicum.*

Although it had been known for many years that formate served as a substrate for many methanogens, it was not until the formate hydrogenylase system of *M. ruminantium* was studied that it became apparent that F_{420} was the primary electron acceptor for formate dehydrogenase.[24] F_{420} that was reduced by formate also served as an electron donor for the formation of hydrogen via hydrogenase or participated in the reduction of NADP via NADP oxidoreductase (Figure 4). Results of an experiment designed to follow the effect of formate concentration on the reduction of F_{420} revealed

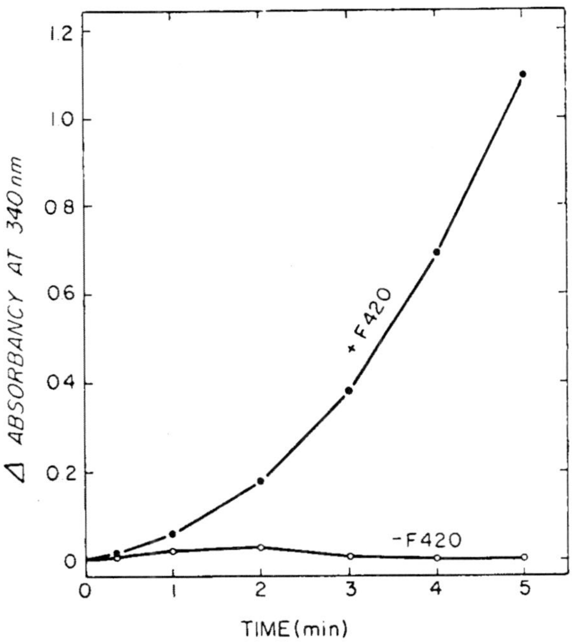

FIGURE 2. Demonstration of the requirement of F_{420} in the NADP-linked hydrogenase system of *M. ruminantium* with the spectrophotometric assay. Reaction mixtures in each Thunberg cuvette contained 5.0 μmol of NADP, 0.5 μg of F_{420} from *Methanobacterium* strain M.o.H., and 6.0 mg of protein from DEAE-G25-treated extract. (From Tzeng, S.F., Wolfe, R.S. and Bryant, M.P., *J. Bacteriol.*, 121, 188, 1972. With permission.)

a K_m of 8.3×10^{-4} *M*. A similar formate hydrogenylase system was found later in *Methanospirillum hungatii*.[25]

The structure of coenzyme F_{420} from *Methanobacterium* M.o.H. was studied by Eirich.[22] F_{420} was fractionated from whole cells, and a yield of 160 mg/kg of wet cells was obtained. The UV-visible spectra of the coenzyme at different pH values are shown in Figure 5. A proposed structure for the coenzyme has been published (Figure 6).[26] Supporting evidence for the structure was obtained by analysis of hydrolytic fragments. As shown in Figure 7, a number of fragments were generated as the coenzyme was exposed to hydrolysis in 1 *M* HCl at 110°C for increasing time periods. These fragments, P, N-2, N-1, F^+, FO, FA, and SAC were named on the basis of electrophoretic mobility. Additional fragments were generated by oxidation of F_{420} with periodate, a cationic chromophore (PA) and a strongly anionic aldehyde (ALD) being formed. By use of various analytical techniques, the position of each fragment in the side chain was established (Figure 8).

The proposed structure for the chromophore was derived by use of UV-visible spectrometry, ^1H and ^{13}C NMR spectrometry, quantitative elemental analyses of F_{420} and its fragments, and by comparison of these data with similar analytical data for 8-hydroxy-FMN, 8-hydroxy-5-deaza-FMN, and 5-deazariboflavin. The spectrum of F_{420} is identical to that of synthetic 8-hydroxy-5-deazariboflavin,[59] but is shifted 3 to 6 nm, and this shift is believed to be due to the 7-CH_3 group on the latter compound. Figure 9 presents partial spectra of F_{420} at pH 5.9 and 8.8 as compared to similar spectra (insert) of 1,5-dihydro-5-deazariboflavin.

When the stoichiometry of the reduction of F_{420} with molecular hydrogen was studied, it was found that 13.6 μmol of hydrogen was utilized during the reduction of 13.6

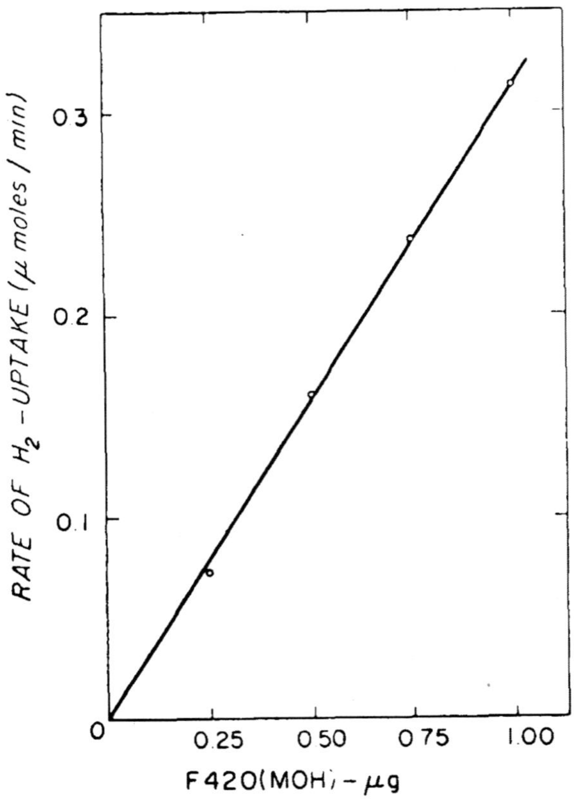

FIGURE 3. Relationship between F_{420} from *Methanobacterium* strain M.o.H. and the rate of hydrogen uptake in the NADP-linked hydrogenase system of *M. ruminantium*. Reaction mixtures contained 5.0 μmol of NADP, 6.0 mg of protein from DEAE-G25-treated extract, and various amounts of F_{420} from *Methanobacterium* strain M.o.H. as indicated. The rate of hydrogen uptake given is the mean for the first 5 min of the reaction. (From Tzeng, S.F., Wolfe, R.S., and Bryant, M.P., *J. Bacteriol.*, 121, 188, 1972. With permission.)

μmol of F_{420}, indicating a two electron transfer. In addition, reduction was found to occur at positions 1 and 5 of the ring system. No evidence could be obtained that supported the existence of a semiquinone. An ESR signal was not observed when F_{420} was examined under different states of partial reduction at pH 8. From electrochemical measurements a redox potential (E'_o = −0.373 V) was determined.[60] So, evidence indicates that coenzyme F_{420} is a low-potential electron carrier (a derivative of 8-hydroxy-7-demethyl-5-deaza-FMN) that participates in two-electron transfer reactions, and not as a semiquinone.

The role of the lactyl-diglutamyl moiety of the side chain is not defined at the present time but, presumably, the long side chain plays a role related to the active site of the enzyme during electron transfer. In extracts of *Methanobacterium* M.o.H., the apparent K_m for reduction by hydrogenase of each of the following compounds was F_{420}, 25 *μ M*; F^+, 100 *μM*; FO, 100 *μM*.[22] In the NADP-linked hydrogenase system, an apparent K_m of 27 *μM* was determined for F_{420}. The apparent K_m for F^+ was 110 *μM* and for FO was 44 *μM*. Thus, the negative charges on the phosphate group of F^+ appear to slow catalysis of the F_{420}-linked NADP oxidoreductase system, whereas FO lacks the phosphate group and functions almost as well as F_{420}. PA and FA proved to be sur-

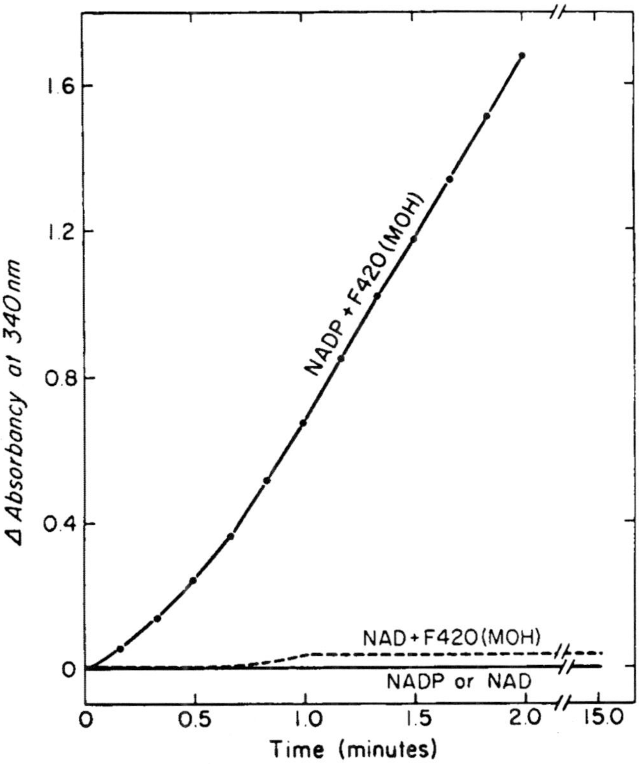

FIGURE 4. The reduction of NADP by formate via F_{420}-with DEAE-G25-treated extract of *M. ruminantium*., (From Tzeng, S.F., Wolfe, R.S., and Bryant, M.P., 121, 195, 1972. With permission.)

prising, since they were not able to serve as electron acceptors for the hydrogenase of *Methanobacterium* (and also were inactive in the F_{420}-linked NADP oxidoreductase).[22] We have no explanation for this effect at present.

The F_{420}-linked hydrogenase system of *Methanobacterium* M.o.H. was used by Eirich to assay for F_{420} in other methanogens.[22] Levels that were detected are presented below, each value having been normalized on the basis of milligrams of F_{420} per kilogram of wet cells: *M. thermoautotrophicum,* 324; *M. formicicum,* 206; *M. hungatii,* 319; Black Sea isolate JR1, 120; *Methanobacterium* AZ, 306; *M. ruminantium* M-1, 6; *M. barkeri,* 16.4; and *Methanobacterium* G, 226. F_{420} has been found only in methanogenic organisms. When F_{420} from various methanogens was subjected to electrophoresis, the electrophoretic mobility was about the same for each isolate (5.9) cm to the anode) with the exception of *M. barkeri,* which was 7.5. This is an interesting observation, since *M. barkeri* has other distinctive characteristics.

Extracts of *M. thermoautotrophicum* were examined by Zeikus et al.[27] and by Fuchs et al.[28] for enzymic activity of several oxidoreductases. Good activity was detected (nmol/min/mg of protein) for pyruvate dehydrogenase, 275; α-ketoglutarate dehydrogenase, 100; fumarate reductase, 360; malate dehydrogenase, 240; and glyceraldehyde-3-phosphate dehydrogenase, 100. Coenzyme F_{420} was found to be the electron acceptor for pyruvate dehydrogenase and for α-ketoglutarate dehydrogenase. Fumarate reductase did not couple with coenzyme F_{420}, NAD, or NADP. Menaquinone was not found to be present in this organism. NADP was the preferred electron acceptor for glyceraldehyde-3-phosphate dehydrogenase, but malate dehydrogenase was most active with NAD. These enzymes are believed to be involved in synthetic reactions.

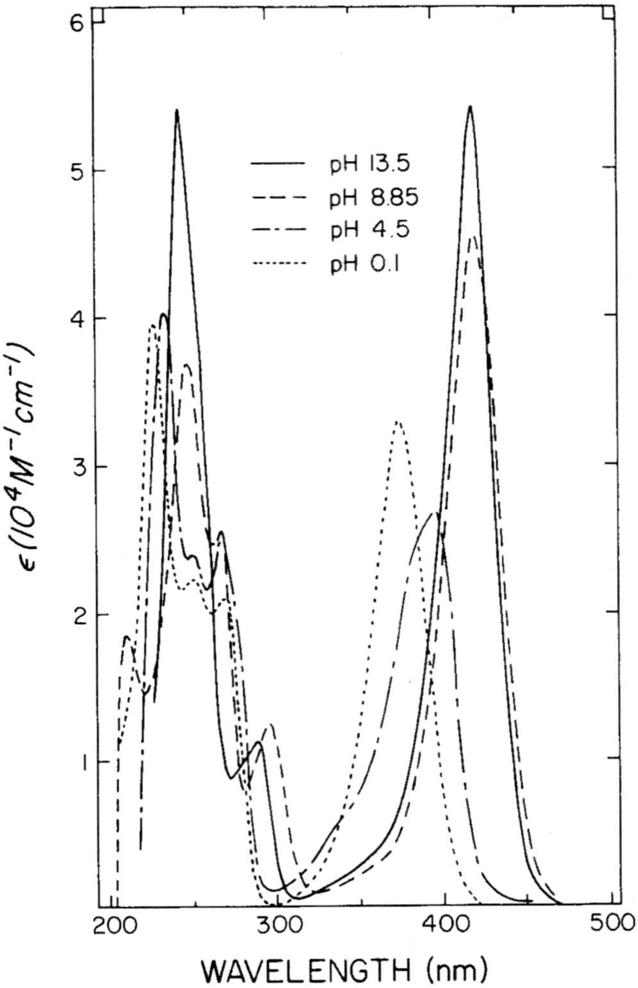

FIGURE 5. UV-visible absorption spectra of coenzyme F_{420} obtained at the indicated pH values. (From Eirich, L.D., Vogels, G.D., and Wolfe, R.S., *Biochemistry*, 17, 4589, 1978. With permission.)

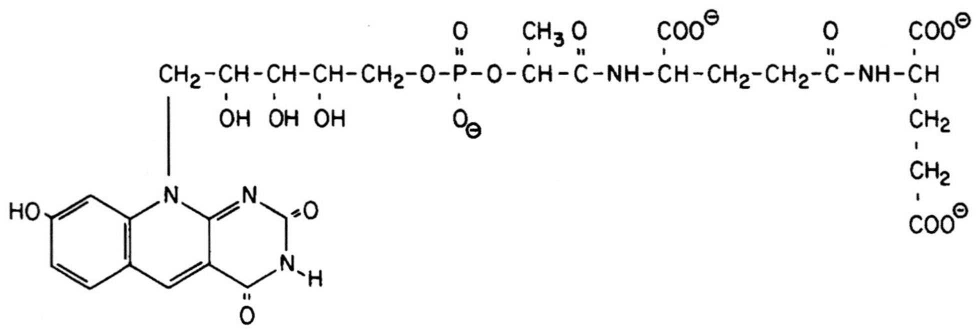

FIGURE 6. Proposed structure for coenzyme F_{420}. From Eirich, L.D., Vogels, G.D., and Wolfe, R.S., *Biochemistry*, 17, 4584, 1978. With permission.)

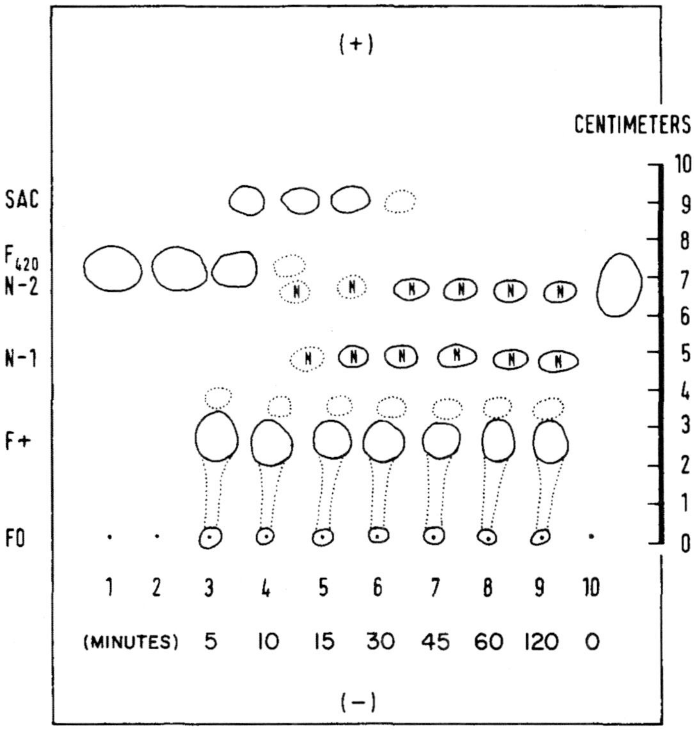

FIGURE 7. *Thin-layer electrophoresis of acid-hydrolyzed coenzyme* F_{420}. Time-dependent formation of hydrolysis products and their intermediates. Electrophoresis was accomplished on Eastman cellulose plates in buffer for 45 min at 400 V. 1, 3, and 14 are F_{420} standards; 2 is an L-glutamic acid standard; 4 to 13 represent samples of F_{420} hydrolyzed in 1.0 *N* HCl at 110°C for the indicated lengths of time. "N" indicates ninhydrin-positive spots. "P" represents inorganic phosphate spots. Dashed lines indicate spots of low intensity. Symbols for the hydrolysis products of F_{420} are presented in the left margin. Each symbol corresponds to the position of the compound it represents. (From Eirich, L.D., Vogels, G.D., and Wolfe, R.S., *Biochemistry*, 17, 4586, 1978. With permission.)

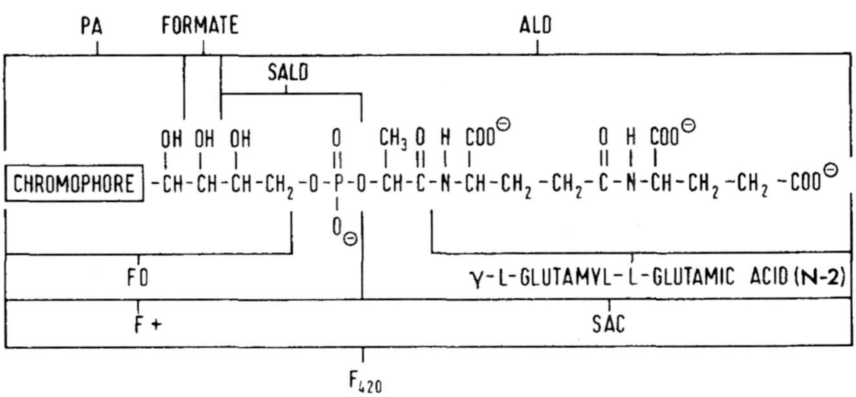

FIGURE 8. Fragments of the side chain of coenzyme F_{420}. Symbols for the products of acid hydrolysis of F_{420} are indicated below the structure. Symbols of the periodate oxidation fragments are shown above the structure. (From Eirich, L.D., Vogels, G.D., and Wolfe, R.S., *Biochemistry*, 17, 4586, 1978. With permission.)

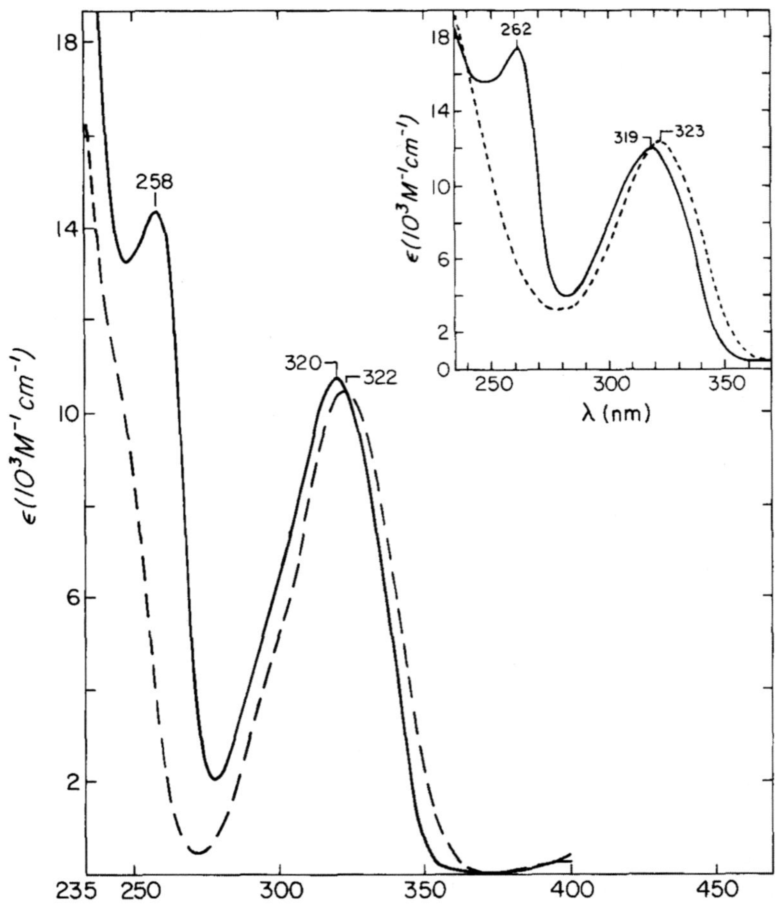

FIGURE 9. The UV-visible absorption spectra of reduced F_{420} and 1,5-dihydro-5-deazariboflavin. The spectra were obtained in 0.1 *M* phosphate buffer at pH 5.9 (broken line) and 8.8 (solid line). The spectra were measured in 1-cm anaerobic quartz cuvettes. F_{420} in a hydrogen atmosphere was reduced by the addition of 110 μg of *Methanobacterium* cell extract. A buffer solution, with enzyme, but without F_{420} added, was used as a blank. The insert shows the UV-visible spectra of 1,5-dihydro-5-deazariboflavin at pH 5.7 (broken line) and 8.6 (solid line). (From Eirich, L.D., Vogels, G.D., and Wolfe, R.S., *Biochemistry*, 17, 4592, 1978. With permission.)

IV. CO_2 REDUCTION

A. Coenzyme M

1. Illustrative Scheme

Figure 10 presents a scheme which portrays current concepts of the manner in which CO_2 serves as the electron acceptor in methanogens and is reduced to methane. This scheme is modified from Romesser.[29] Much of the scheme is speculative, but it has been drawn in this manner to account for observed phenomena. For many years, the scheme suggested by Barker[1] has served as a model for reduction of carbon dioxide to methane. It now appears that there is a close connection between the terminal reaction and CO_2 activation, and a definite cycle has been proposed.[7,29]

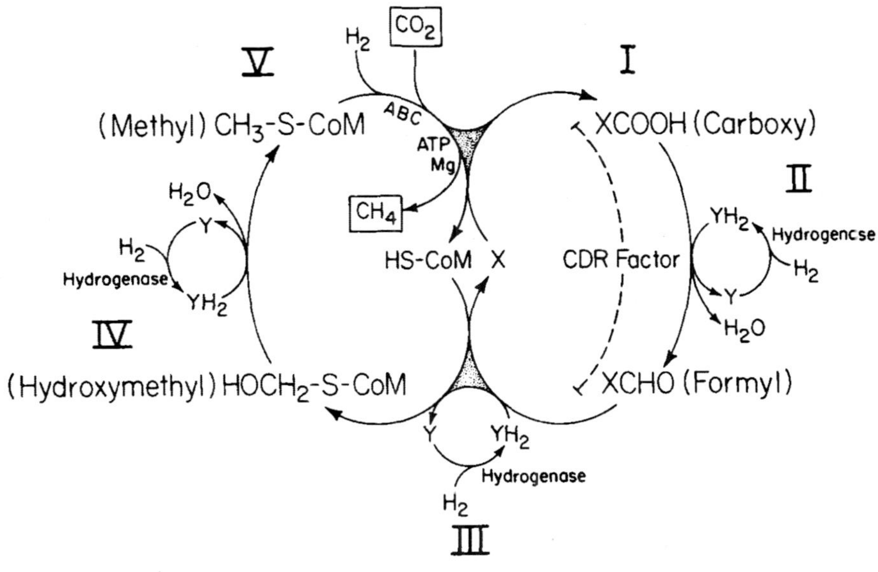

FIGURE 10. Scheme which summarizes current knowledge of the reduction of CO_2 to CH_4 by methanogens. (Modified from Romesser, J.A., Doctoral Dissertation, University of Illinois, Urbana, 1978 and from Wolfe, R.S., *Antonie van Leeuwenhoek; J. Microbiol.*, With permission.)

2. CH_3-*S*-CoM Methylreductase

a. Components

The most thoroughly studied area of this scheme is reaction V, the CH_3-*S*-CoM methylreductase. The conversion of the methyl group of 2-(methylthio)ethanesulfonic acid to methane involves a rather complex series of events. Although at one time it was considered that N^5-methyltetrahydrofolate and methylcobalamin were involved in methanogenesis at the methyl level (since these compounds could donate a methyl group for reduction), there is at present no evidence for their involvement as natural components of the methylreductase system.

The CH_3-*S*-CoM methylreductase of *M. thermoautotrophicum* was studied by Gunsalus.[20] Cell extract was fractionated into two protein components, an oxygen-sensitive protein with hydrogenase activity (A) and an oxygen-stable protein (C), as well as a heat-stable, dialyzable, oxygen-sensitive factor (component B). These components were not purified to homogeneity, but they were resolved to the extent that all three were required for methane formation from CH_3-*S*-CoM, and all components were found in the supernatant fraction after centrifugation for 1 hr at 100,000 *g*. An optimal temperature of 65°C and an optimal pH range of 5 to 6.5 were found. We have obtained no evidence that coenzyme F_{420} or other electron carriers act as an intermediate between hydrogenase and CH_3-*S*-CoM. Hydrogenase appears to be a component of the methylreductase.

b. ATP Requirement

The ATP requirement for the formation of methane by cell extracts has been known for many years.[30,31] After the discovery of coenzyme M[32] and the synthesis of CH_3-*S*-CoM[33] it was shown that ATP was required at the methylreductase level. This requirement was studied recently and was found to be catalytic.[34] Results are presented in Table 1. For each mole of ATP added, 15 mol of methane was produced from CH_3-

TABLE 1

Molar Ratio of Methane Produced Per ATP Added in Crude Cell Extract[a]

Addition (nmol ATP)	nmol CH_4 Produced	nmol CH_4-Bkg[b]	CH_4/ATP
None	191	0	—
50	924	733	14.7
100	1744	1553	15.5

[a] Reaction mixtures (0.25 m*l*) contained TES buffer pH 6, 20 μmol; $MgCl_2$, 5 μmol; CH_3-*S*-CoM, 2 μmol; cell extract (8.2 mg protein); and ATP as indicated. Reaction time was 100 min at 60°C.

[b] Bkg (background) represents the amount of methane produced (191 nmol) when no ATP was added.

From Gunsalus, R. P. and Wolfe, R. S., *J. Bacteriol.*, 135, 855, 1978. With permission.

S-CoM. The nature of the ATP requirement is not understood at the present time. An experiment was carried out to see if the interaction of ATP involved a stable modification of a protein. A reaction mixture that contained ATP, Mg, CH_3-*S*-CoM, and cell extract was incubated under hydrogen and then passed down an anaerobic Sephadex G-25® column. ATP and other small molecules were removed. When samples of the effluent from the column were tested for methanogenesis, no activity was detected. Upon the addition of 1 μmol of ATP to the reaction mixture, 465 nmol of methane was formed per hour per milligram of protein. The methylreductase could not be isolated in a stable, activated form, indicating that a covalent modification of a protein by ATP probably did not occur. The methylene analogs, α,β-ATP and β,γ-ATP had no effect on the CH_3-*S*-CoM methylreductase.

The optimal concentration of ATP was 1 m*M*, and for $MgCl_2$ it was 40 m*M*. In the presence of crude cell extract, several nucleotides were partially active when substituted for ATP.[34] The values are ATP (100%), ADP (49%), dATP (38%), CTP (61%), UTP (58%), GTP (42%), TTP, AMP, cyclic AMP, or adenosine (0.0%). The presence of an ATP-dependent CH_3-*S*-CoM methylreductase in other methanogens also was established.[34] The specific activity (nmol methane produced per hour per milligram of protein) was *M. hungatii*, 196; *M. barkeri*, 289; *M. formicicum*, 185; *M. ruminantium*, 285; *Methanobacterium* M.o.H., 157; and *M. thermoautotrophicum*, 566.

c. Inhibitors

A number of compounds have been reported over the years to inhibit methanogenesis by whole cells and cell extracts.[3,35-37] Recently a variety of these as well as other compounds have been tested for their ability to inhibit the CH_3-*S*-CoM methylreductase.[34] These results are presented in Table 2. An interesting observation was that neither sodium cyanide (10 m*M*) nor carbon monoxide (partial pressure, 0.1) caused inhibition even at rather high levels. Sulfite is a known inhibitor of methanogenesis. So, it was a surprise to find that if dithionite was carefully put into solution under strict anaerobic-conditions, no inhibition of methylreductase was observed. Although nitrous oxide inhibits methanogenesis in whole cells,[38] it had no effect on the methylreductase at a partial pressure of 0.2. Viologen dyes were found to be potent inhibitors of the methylreductase, as were 5,5′-dithiobis(2-nitrobenzoic acid), 2,4-dinitrotrophenol, and chloral hydrate. In previous studies, arsenate, selenite, and tellurite were reduced and methylated by extracts in the presence of methylcobalamin.[39] When CH_3-*S*-CoM was the methyl donor, these compounds were reduced, but not methylated.[34] This indicates that under natural conditions the methylreductase of methanogens probably is not involved in transmethylation reactions to other acceptors.

TABLE 2

Ability of Various Compounds to Inhibit CH_4 Formation from CH_3-*S*-CoM and H_2[a]

Compound	Concentration (m*M*)	% Inhibition
Chloral hydrate	0.050	100
Chloramphenicol	0.100	100
Chloroform	0.500	100
Nitromethane	0.500	100
Dichlorodifluoromethane	—[b]	100
Sodium dithionite	10.000	0
Sodium azide	10.000	100
Sodium cyanide	10.000	0
Carbon monoxide	—[b]	0
Oxygen	—[b]	100
Nitrogen	—[c]	0
Nitrous oxide	—[c]	0
Sodium nitrite	1.000	99
Sodium nitrate	4.000	0
2,4-Dinitrophenol	0.050	100
Methyl viologen	0.001	94
Benzyl viologen	0.005	100
DTNB	0.030	100
1,10-Phenanthroline	2.000	100

[a] The reaction mixtures (0.25 m*l*) contained TES-buffer, 25 μmol, pH 6.0; $MgCl_2$, 5 μmol; ATP, 1 μmol; CH_3-*S*-CoM, 2 μmol; and cell-free extract, (1.8 mg protein). Temperature, 60°C.

[b] A partial pressure of 0.1 used, balance H_2.

[c] A partial pressure of 0.2 used, balance H_2.

From Gunsalus, R. P. and Wolfe, R. S., *J. Bacteriol.*, 135, 855, 1978. With permission.

d. Coenzyme M Analogues

Gunsalus et al.[40] recently reported the results of a series of experiments that were designed to test the specificity of the CH_3-*S*-CoM methylreductase in *M. thermoautotrophicum.* A number of analogues of coenzyme M were tested for their ability to serve as substrates or to inhibit methane formation from CH_3-*S*-CoM. When the alkyl group of CH_3-*S*-CoM was lengthened (CH_3CH_2-*S*-CoM, $CH_3CH_2CH_2$-*S*-CoM) and these compounds were tested (Figure 11), ethane was produced from ethyl-*S*-CoM at a rate about 20% that of methane from CH_3-*S*-CoM. No formation of propane was detectable. Although the methylreductase was able to activate the lengthened alkyl moiety of ethyl-*S*-CoM, the enzyme system was not able to produce methane from the analogues, 3-(methylthio)propanesulfonic acid or 4-(methylthio)butanesulfonic acid, indicating the critical proportions of the ethylene bridge of the coenzyme between the sulfide and sulfonate moieties. The addition of even one additional methylene carbon completely destroyed coenzyme activity. The ability of each of the following compounds to donate a methyl group for reduction to methane by the methylreductase was tested, but none of the compounds was active: 2-(dimethylsulfonium)ethanesulfonate, *N*-methyltaurine, 2-(methylthio)ethanol, ethyl-3-(methylthio)propionate, methyl-3-(methylthio)propionate, 3-(methylthio)propylamine, methioninol, L-methionine, D-methionine, or *S*-methylcysteine. In addition, these analogues failed to serve as a substrate for the methylation of HS-CoM in a standard reaction mixture for the methylreductase.

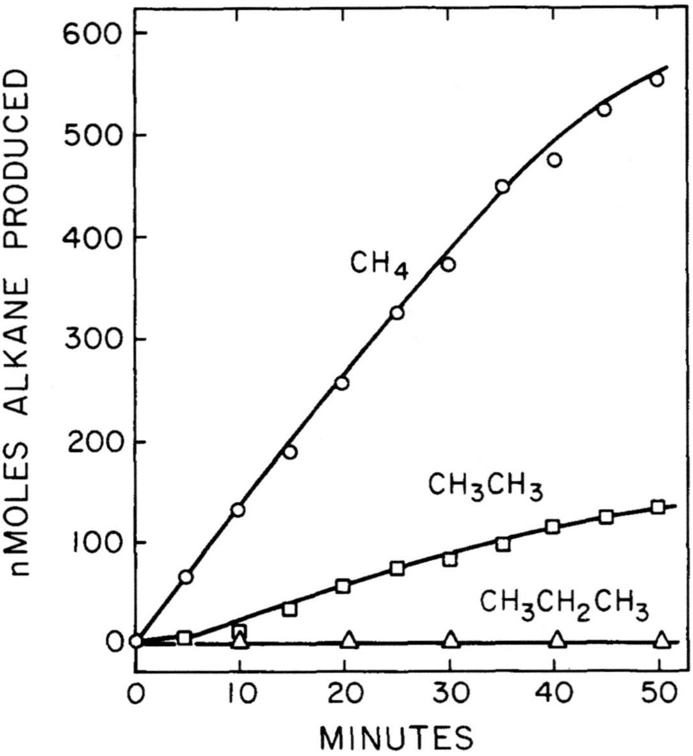

FIGURE 11. Ability of alkyl-coenzyme M analogues to be dealkylated to the corresponding alkane by cell extract. Each reaction vial contained 30 μmol of Tes buffer; 5 μmol of $MgCl_2$; 1 μmol of ATP; 30 μℓ of cell extract (1.51 mg of protein); and 2.5 μmol of methyl-, ethyl-, or propyl-*S*-coenzyme M as indicated. (From Gunsalus, R.P., Romesser, J.A., and Wolfe, R.S., *Biochemistry*, 17, 2376, 1978. With permission.)

As shown in Figure 12, mercaptan or halogenated analogues were tested at various concentrations for inhibitory properties in the standard reaction mixture that contained CH_3-*S*-CoM as substrate. Bromoethanesulfonic acid was the most potent inhibitor of methane formation and produced a 50% inhibition at 7.9×10^{-6} *M*. Chloroethanesulfonic acid also was an effective inhibitor. The other compounds tested (Figure 12) were poor inhibitors and only showed significant inhibition when added at concentrations above 10^{-2} *M*. Results of studies with additional analogues are presented in Figure 13. These analogues were not potent inhibitors. Other analogues were tested in a similar manner. At a ratio of analogue to CH_3-*S*-CoM of 17, ethyl-*S*-CoM produced a 50% inhibition. Propyl-*S*-CoM yielded a similar inhibition at a ratio of 22. At a ratio of analogue to substrate of 100, the following analogues produced no inhibition of methylreductase: $HO(CH_2)_2SO^-_3$, $H_2N(CH_2)_2SO^-_3$, $CH_3NH(CH_2)_2SO^-_3$, $^-O_3S(CH_2)_2SO^-_3$, and $CH_3S(CH_2)_2OH$.

e. RPG Effect

During study of the CH_3-*S*-CoM methylreductase, an unexpected finding was made by Gunsalus.[41] When carbon dioxide was added to the standard reaction mixture for the methylreductase assay (Table 1), the rate of methane formation was stimulated 30-fold, and the total amount of methane formed was 12-fold higher. For each mole of CH_3-*S*-CoM added, an additional 11 mol of methane was formed, suggesting that in

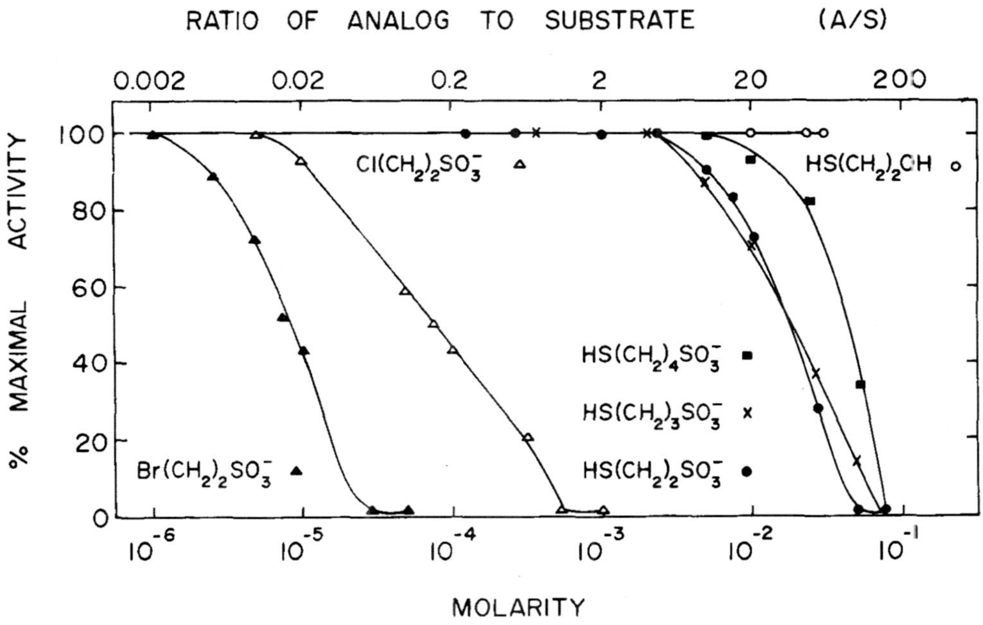

FIGURE 12. Effect of various mercaptan and halogenated analogues of coenzyme M on CH_4 formation from CH_3-S-CoM by cell extract. Each reaction vial contained 25 μmol of Tes buffer; 5 μmol of $MgCl_2$; 1 μmol of ATP; 0.125 μmol of CH_3-S-CoM; 50 μℓ of cell extract (2.3 mg of protein); and the indicated concentration of CoM analogue. (From Gunsalus, R.P., Romesser, J.A., and Wolfe, R.S., *Biochemistry*, 17, 2376, 1978. With permission.)

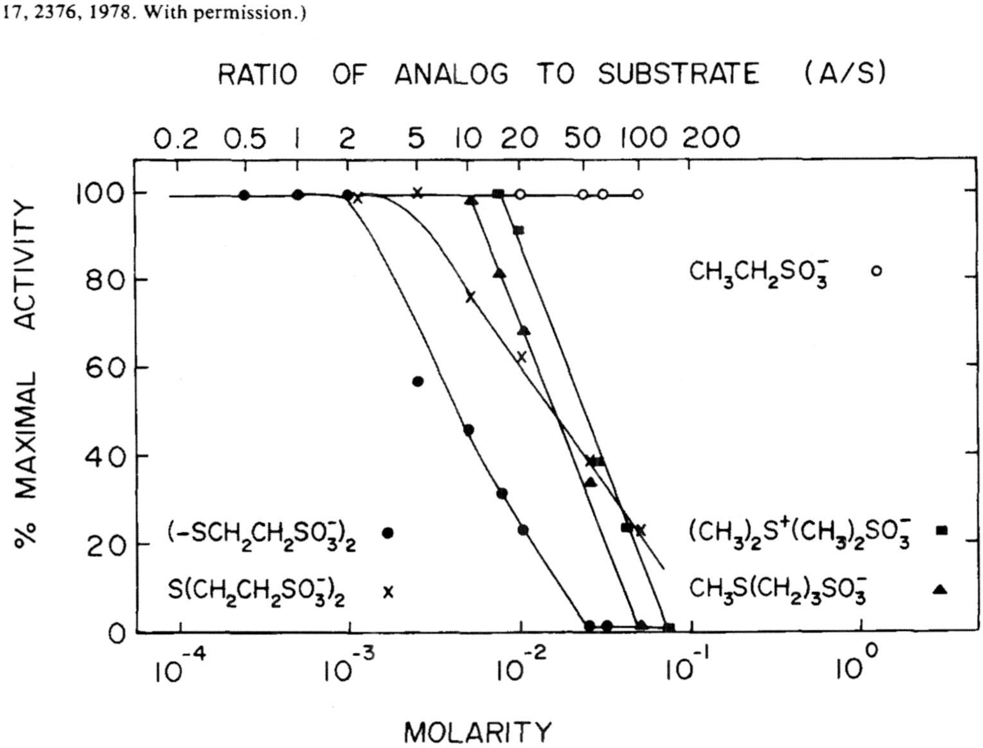

FIGURE 13. Effect of various coenzyme M analogues on methane formation from CH_3-S-CoM by cell extracts. Conditions were similar to those of Figure 12. (From Gunsalus, R.P., Romesser, J.A., and Wolfe, R.S., *Biochemistry*, 17, 2376, 1978. With permission.)

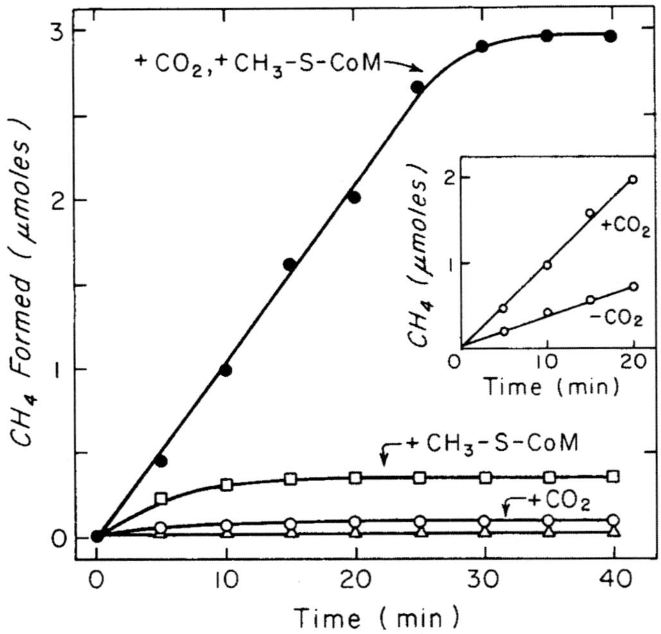

FIGURE 14. Stimulation of CO_2 reduction by CH_3-*S*-CoM. Effect of reaction components on CH_4 formation. Each reaction vial contained the standard reaction components for the methylreductase assay plus 50 *μl* cell extract (1.7 mg protein) under an H_2 atmosphere. Additions to the indicated vials were 0.2 μmol CH_3-*S*-CoM (□—□); 20% CO_2 gas phase, balance H_2 (○—○); 0.2 μmol CH_3-*S*-CoM plus 20% CO_2 gas phase (●—●); and none (△—△). *Insert* shows reaction time course in presence and absence of CO_2 gas phase and 2 μmol CH_3-*S*-CoM. (From Gunsalus, R.P. and Wolfe, R.S., *Biochem. Biophys. Res. Commun.*, 76, 792, 1977. With permission.)

some manner an active complex was formed in the presence of CH_3-*S*-CoM by which carbon dioxide was activated and subsequently reduced. Data which illustrate the effect of addition of carbon dioxide are presented in Figure 14. Reaction V (Figure 10) now takes on new significance since carbon dioxide appears to be involved. Romesser[29] was able to separate carbon dioxide stimulation of the CH_3-*S*-CoM methylreductase from net reduction of carbon dioxide to methane. Results of his experiments indicated that carbon dioxide is an effector for the methylreductase, stimulating the rate of the reaction by a mechanism that remains to be elucidated.

3. C_1 and C_2 Derivatives of Coenzyme M

To obtain information on the possible role of coenzyme M as a carrier of C_1 moieties more oxidized than a methyl group, formyl-*S*-CoM (CHO-*S*-CoM) and hydroxymethyl-*S*-CoM ($HOCH_2$-*S*-CoM) were synthesized and tested in cell extracts of *M. thermoautotrophicum.*[29] Synthesis of carboxy-*S*-CoM was not attempted because of its predicted instability in aqueous solution. In the reaction mixture, CHO-*S*-CoM was hydrolyzed rapidly to formate, and no formation of methane was detectable. (Formate is not a substrate for methanogenesis in *M. thermoautotrophicum*.) $HOCH_2$-*S*-CoM was reduced to methane by extracts, and the rate approached that for the reduction of CH_3-*S*-CoM to methane. In solution, $HOCH_2$-*S*-CoM was observed to be in equilibrium with formaldehyde and HS-CoM. When added to cell extract, formaldehyde was enzymically converted to methane at a rate equal to that from $HOCH_2$-*S*-CoM. Catalytic amounts of HS-CoM were required for the reduction of formaldehyde to methane. The hemimercaptals, thiazolidine, and thiazolidine-4-carboxylate were reduced to methane by extracts. This reduction also required catalytic amounts of HS-CoM. When a preparation of adenosyl-*S*-CoM was tested, addition of this compound neither replaced ATP nor had any detectable effect on the methylreductase.[29] An additional preparation of adenosyl-*S*-CoM as well as a preparation of adenosyl-$S^+(CH_3)$-CoM kindly provided by W. R. Cullen, University of British Columbia, also were tested. Neither compound replaced ATP nor had an inhibitory effect.[61] Adenosyl-$S^+(CH_3)$-CoM did not serve as a substrate for the methylreductase.

If we now return to Figure 10, the following observations may be made. It appears unlikely that coenzyme M is involved as a C_1 carrier at the carboxy or formyl levels. X has been used as a carrier for reactions I, II, and III to indicate the unknown nature of the carrier. We know that carbon dioxide is an effector for reaction V, and the upper stippled area is intended to indicate an interaction between the activated carbon dioxide of the methylreductase and its transfer to the carrier X. The electron carrier Y could be coenzyme F_{420} or another electron carrier. Y need not be the same carrier for each reaction, II, III, or IV. No information is yet available at these reactions, but we know (as first pointed out by Barker[1]) that the reduction of carbon dioxide must proceed through these oxidation states as it is reduced to methane. The lower stippled area suggests the interaction of the two C_1 carriers, the transfer of the C_1 moiety, and its reduction to the formaldehyde level. Free formaldehyde is a possible product of reaction III, but so far, attempts to detect it have failed. It is interesting, though, that formaldehyde is converted to methane at a good rate,[29] and that this conversion requires HS-CoM. Though hydroxymethyl-*S*-CoM has not been detected as a natural intermediate, there is the possibility that it could be one. Therefore, it is placed in the scheme.

4. Carbon Dioxide Reduction Factor (CDR Factor)

One additional aspect of the scheme presented in Figure 10 requires consideration. Romesser[29] was able to resolve the carbon dioxide activation step of reaction V from the early steps of carbon dioxide reduction. He discovered a heat-stable, dialyzable cofactor about which little is known at present. Until the site of action of this cofactor is identified, we shall refer to it as the carbon dioxide reduction factor (CDR factor). When this cofactor was removed from cell extracts, reaction V was stimulated by addition of carbon dioxide, but there was only a stoichiometric conversion of CH_3-*S*-CoM to methane. The RPG effect was abolished. However, upon addition of the factor back to the reaction mixture, the RPG effect was restored. The factor was not required for the conversion of $HOCH_2$-*S*-CoM or formaldehyde to methane. So, the CDR factor is believed to function prior to the aldehyde level of oxidation, and this is indicated by the dashed line in Figure 10.

V. ACETATE AND OTHER SUBSTRATES

Conversion of acetate to methane attracted interest when Buswell and Sollo[42] reported that carbon dioxide was not a major precursor of methane in sewage sludge and that acetate was converted to methane. These results were of interest because it had been assumed that acetate was probably degraded to carbon dioxide and hydrogen or reducing equivalents prior to methanogenesis. In a series of elegant experiments, Stadtman and Barker,[43] Pine and Barker,[44] and Pine and Vishniac[45] documented by use of isotopically labeled acetate that methane was formed from the methyl carbon of acetate, carbon dioxide being formed from the carboxyl group. In addition when

CD_3COOH was used as substrate, CD_3H was formed, indicating that the deuterium atoms remained with the carbon atom. The methyl group was reduced intact to methane. Additional confirmation of this process was obtained by carrying out the reaction with CH_3COOH in the presence of D_2O. In this instance, CH_3D was isolated, showing that only one proton was added to the methyl group. For over 20 years these experiments represented the limit of knowledge on the conversion of acetate to methane. The reason for this is apparent. Acetophilic methanogens are difficult to mass culture, and they remain very difficult to isolate in pure culture. Preparation of active, subcellular systems is a special challenge.

Recently, there has been a revival of interest in the conversion of acetate to methane. These studies have dealt largely with the ability of whole cells to grow and produce methane from acetate. So far, only *Methanosarcina* has been shown to grow on acetate and to convert it to methane in pure culture, although highly purified enrichment cultures of other acetophilic methanogens are being studied by van den Berg[46] and by Zehnder, Heuser, and Wuhrmann[62] as well as by Boone and Bryant.[63]

The adaptation of *Methanosarcina* to grow on acetate requires patience on the part of the investigator, as well as special conditions for the cells. Failure to recognize the delicate nature of the adaptation process has caused some confusion in the literature, with the concept being promoted that *Methanosarcina* could not grow on acetate as the sole source of energy and carbon in a mineral medium.[47,48] This clearly is not the case. Definitive work from the laboratory of Mah[49] has shown that *M. barkeri* strain 227 obtained energy from the conversion of acetate to methane and carbon dioxide without involvement of hydrogen gas. At first, it was thought that strain 227 was a special acetate-utilizing *Methanoscarcina* since it was the first strain to be adapted to grow well on acetate, but similar results were obtained with *M. barkeri* strain MS of Bryant. In addition, strain 227 appears to be a typical strain of *M. barkeri* on the basis of 16S rRNA catalogues.[64]

Results of studies on acetate utilization[49] showed that 1 mol of methane was produced per mole of acetate used, and the rate of methane formation was limited by acetate concentrations below 0.05 *M*. Adaptation to acetate did not occur in the presence of hydrogen or methanol, these substrates being preferred over acetate. When acetate-grown cells were placed in a medium that contained methanol and acetate, a diauxic growth occurred, with acetate not being used until methanol was exhausted. Similar results were obtained when hydrogen was present. A "mechanism resembling catabolite repression" was suggested as an explanation for these observations. Results of studies by Winter[50] in which cocultures of *M. barkeri* strain MS and *Acetobacterium woodii* were used for a complete conversion of carbohydrate to methane and carbon dioxide complement the findings of Smith and Mah.[49] An adaptation time of 21 days was required in an acetate medium under nitrogen before detectable methanogenesis began. The maximal rate of methane formation reached 9.6 mmol./ℓ/day after 50 days. As shown in Figure 15, cells that had been grown on acetate could metabolize hydrogen and carbon dioxide immediately. After growth on hydrogen and carbon dioxide for 18 days, a sample of cells was removed and added back to an acetate medium under nitrogen gas. There was no lag, and methanogenesis was detected after 3 days rather than 3 weeks. So, once cells had adapted to acetate, the ability to metabolize acetate was retained and was not lost within a 3-week period. Results presented in Figure 16 document the complete conversion of sugar to methane and carbon dioxide by a mixture of *A. woodii* and *M. barkeri* MS.

Coenzyme M was found in *M. barkeri*, when the cells were grown with hydrogen and carbon dioxide or with methanol as substrates.[9] Smith and Mah[49] obtained evidence that coenzyme M functioned in methanogenesis by acetate-grown cells. About

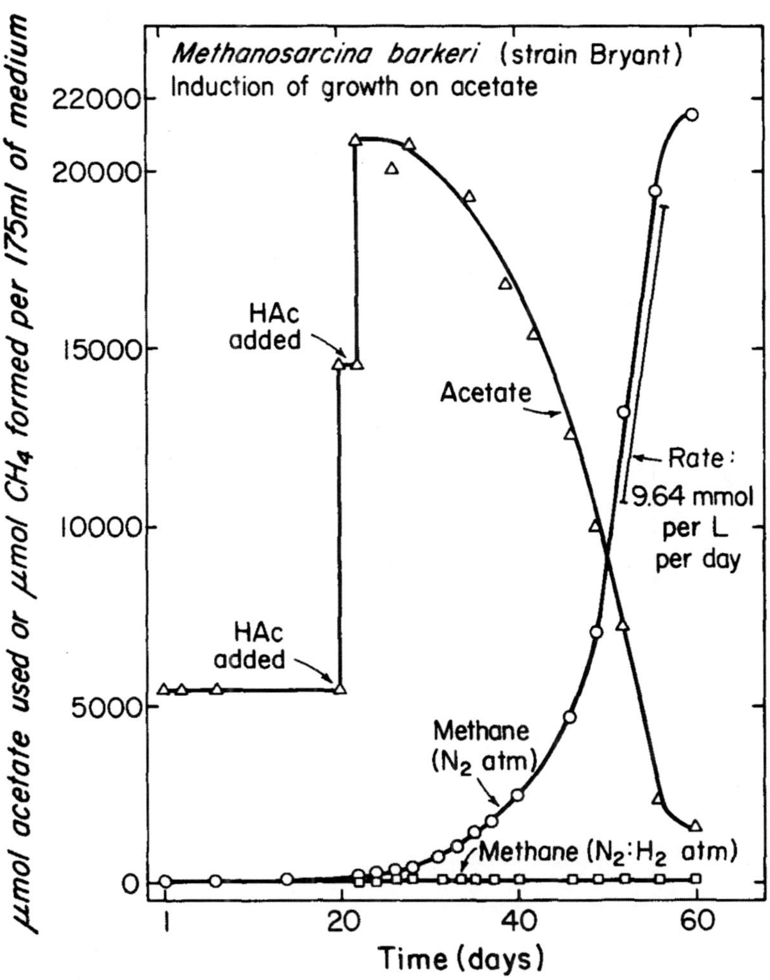

FIGURE 15. Adaptation of H_2/CO_2-grown *M. barkeri* to growth on acetate. Medium contained acetic acid to a final concentration of 0.2%. The medium, 175 mℓ under a gas phase of N_2 or N_2/H_2 (96:4), was contained in a 1.2 ℓ bottle and received a 5% inoculum of H_2/CO_2-grown cells. Incubation was stationary at 37°C. (From Winter, J. and Wolfe, R.S., *Arch. Microbiol.*, 121, 99, 1979. With permission.)

75% of methanogenesis was inhibited by 2-bromoethanesulfonic acid, and this inhibition was relieved by addition of HS-CoM.

Utilization of *N*-methyl compounds by *M. barkeri* has recently been studied in detail.[51] Methylamine, dimethylamine, trimethylamine, and ethyldimethylamine served as substrates for growth and methane formation by *M. barkeri*, a methanogen that probably has the most versatile substrate capacity. Results of the fermentations were consistent with the stoichiometry of the following equations:

$$4\,CH_3NH_3Cl + 2H_2O \longrightarrow 3CH_4 + CO_2 + 4NH_4Cl$$

$$2\,(CH_3)_2NH_2Cl + 2H_2O \longrightarrow 3CH_4 + CO_2 + 2NH_4Cl$$

$$4\,(CH_3)_3NHCl + 6H_2O \longrightarrow 9CH_4 + 3CO_2 + 4NH_4Cl$$

In addition, creatine, sarcosine, choline, and betaine were fermented to methane by

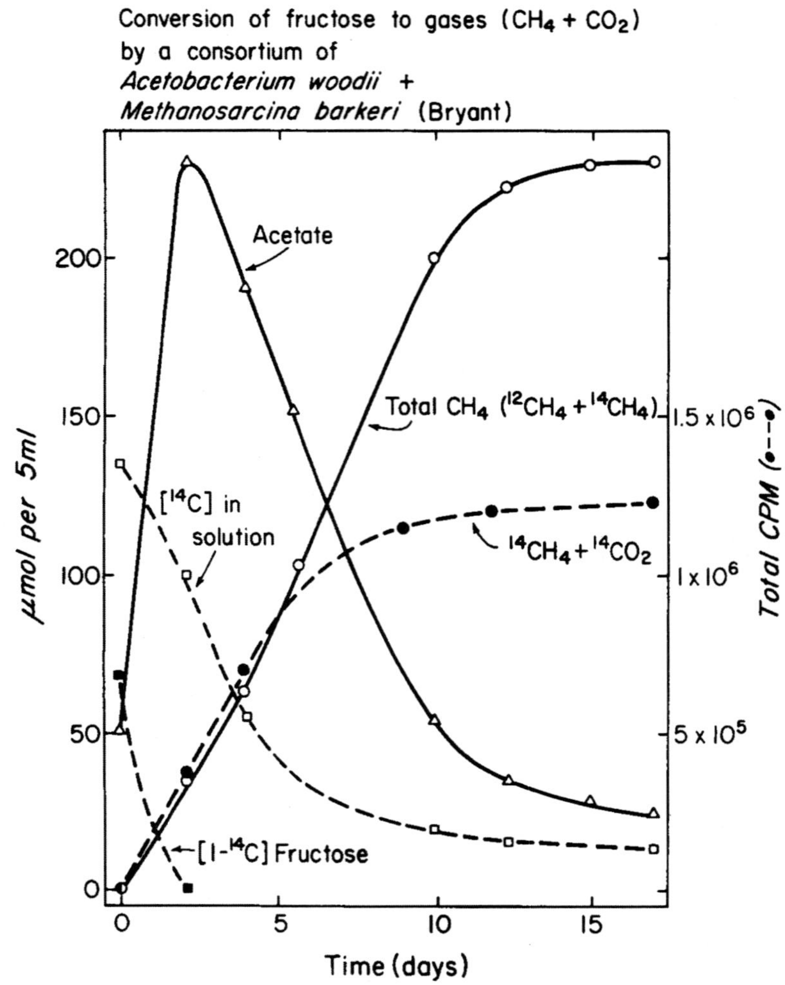

FIGURE 16. Kinetics of fructose fermentation to gases by a coculture of *A. woodii* and *M. barkeri.* Acetate-N_2 medium was used to grow *M. barkeri* on acetate under a N_2/CO_2 (80:20) gas phase. When acetate became limiting, 5 m*l* of the culture was transferred to a sealed tube which contained an atmosphere of N_2/CO_2 (80:20). A culture of *A. woodii* was added to form a 2% inoculum, and 70 μmol of [1-^{14}C]-fructose was added. Incubation temperature was 32°C. (From Winter, J. and Wolfe, R.S., *Arch. Microbiol.*, 121, 100, 1979. With permission.)

mixed cultures. These results suggest that methanogens may have a more direct role in the anaerobic microbial degradation of nitrogen compounds than previously appreciated.

VI. ATP SYNTHESIS

Since no substrate-level ATP synthesis has been found so far in extracts of methanogens, and since anaerobic hydrogen oxidation is a common property of methanogens in pure culture, it has been assumed for many years that ATP synthesis must take place via electron transport phosphorylation. In early studies with cell-extracts from the culture of "*Methanobacterium omelianskii*" and later with extracts from *Methanobacterium* M.o.H., an active ATPase was found to be a major problem in following the role as well as the stoichiometry of the ATP requirement for methanogenesis.[30,31,32] Roberton was the first to study ATP pool sizes in whole cells of a methanogen. Roberton demonstrated that the pool size increased as hydrogen was oxidized and carbon dioxide was reduced to methane.[16] When hydrogen was replaced with other gases or by air, the ATP pool size decreased. Addition of uncouplers of electron transport phosphorylation to the cell suspension caused a decrease in pool size.

So far, the work of Doddema et al.[52] is the most definitive to appear on ATPase and ATP synthesis in a methanogen. These investigators showed that ATPase activity in extracts of *M. thermoautotrophicum* was in the supernatant fraction after centrifugation at 10,000 *g* for 10 min, but after 3 hr at 140,000 *g*, ATPase activity was pelleted from this fraction and could be stored at −90°C for a year without loss of activity. The optimal pH for the enzyme was 8 at 65 to 70°C. The reaction was Mg^{2+}-dependent, and a ration ratio of ATP/Mg of 0.5 yielded optimal activity. Under these conditions, the K_m for ATP was 2 m*M*. Other divalent cations (Mn^{2+}, Co^{2+}, Cu^{2+}, and Zn^{2+}) replaced Mg^{2+} to a significant extent. TES [*N-tris*-(hydroxymethyl)methyl-2-aminoethanesulfonic acid], Tris, [*tris*(hydroxymethyl)aminomethane], and diethanolamine buffers were inhibitory. The enzyme complex also hydrolyzed GTP and UTP, but ADP and AMP were not hydrolyzed. Hydrolysis of ATP was inhibited by DCCD (*N,N'*-dicyclohexylcarbdiimide). ATP synthesis in whole cells was stimulated by an artificially imposed protonmotive force. The internal pH of the cells was found to be 7.6. By lowering the external pH, an increase in intracellular ATP was observed, the shift from pH 7.5 to 3.0 being most effective in generating ATP. When valinomycin at a concentration of 20 μM was added, ATP formation also was observed, and potassium ions were extruded from the cells.

The possibility that fumarate reductase might be coupled to electron transport phosphorylation in methanogens was proposed recently by Gottschalk et al.[53] This proposal suggested that a fumarate-succinate cycle could be coupled to methylreductase. In light of the work of Gunsalus,[20] this system appears highly unlikely to operate at the methylreductase level. When this proposal was subjected to a series of definitive experiments by Fuchs et al.,[54] [U-^{14}C, 2,3-^{3}H] succinate was found to be incorporated into cell material with a loss of only 30% of the tritium, and ^{3}H was not released into water in sufficient amounts. So, it appears that the function of the fumarate reductase of *M. thermoautotrophicum* is to synthesize succinate and not to catabolically oxidize succinate to fumarate.

Since the conversion of acetate to methane and carbon dioxide has a standard free-energy change of −7.4 kcal/mol (−31 kJ/mol),[55] it has been apparent for some time that either a new mechanism for ATP synthesis exists in acetophilic methanogens, or that during methanogenesis, a charge separation occurs and ATP synthesis from acetate is driven by generation of a protonmotive force. Various models for such a system where a charge separation occurs during acetate cleavage are available. One of these recently has been suggested.[7]

VII. ADDITIONAL FACTORS

We have discussed the chemical structure and role in enzymic catalysis of two new coenzymes, HS-CoM and coenzyme F_{420}. Two other cofactors have been resolved, and a biochemical role has been assigned. Component B participates in the CH_3-S-CoM methylreductase reaction, and the CDR factor plays a role in the reduction of carbon dioxide, a role that remains to be precisely defined. In addition, three other factors have been found. The spectrum of factor F_{430}[56] is shown in Figure 17. This factor is a yellow compound that is present in rather high amounts in methanogens, but little is

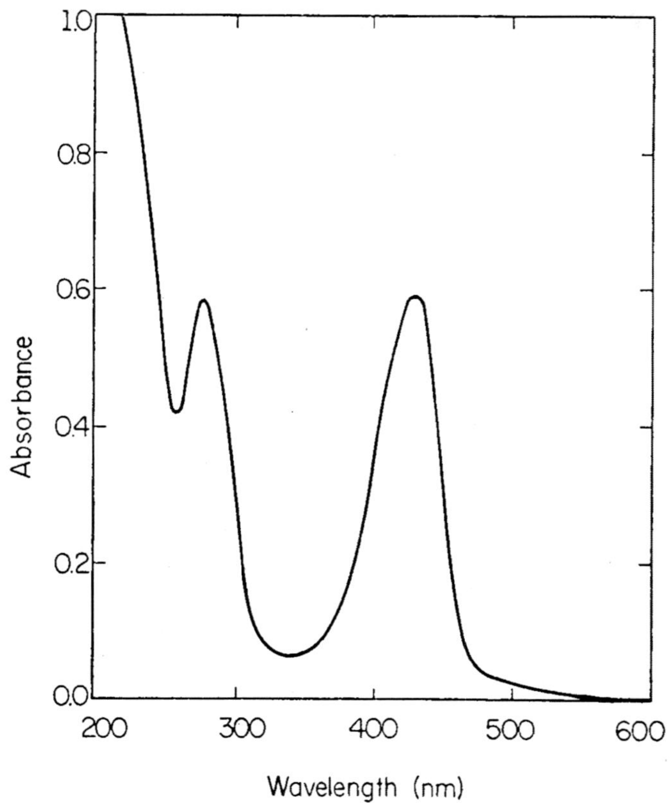

FIGURE 17. Absorption spectrum of F_{430} isolated from *M. thermoautotrophicum*. The UV-visible spectrum was obtained with a Cary 14 spectrophotometer. Sample and reference buffer was 50 m*M* Tris, pH 7.5. (From Gunsalus, R.P. and Wolfe, R.S., *FEMS Microbiol. Lett.*, 3, 193, 1978. With permission.)

known about its properties. No evidence has been obtained that it is an electron carrier or that it participates in reduction of carbon dioxide. Another factor that has been purified from methanogens is factor F_{342}.[56] The UV-visible spectrum for this factor is shown in Figure 18. The compound exhibits an intense blue fluorescence. The excitation and emission spectra are shown in Figure 19. A role for this factor has not been defined, but it is being studied in the laboratory of Vogels.[65] Daniels and Zeikus[57] have reported a yellow fluorescent factor, YFC, which becomes labeled when $^{14}CO_2$ is added to extracts. At present, little is known about its structure or whether it plays a role in synthesis of cell carbon compounds or in methanogenesis. Yet another factor (found in rumen fluid) is required as a growth factor for *Methanobacterium mobile.* None of the above factors substitute for this growth factor.[66] All of these factors are water soluble and appear to be unique to methanogens, supporting the concept that methanogens are only distantly related to typical procaryotes.[11]

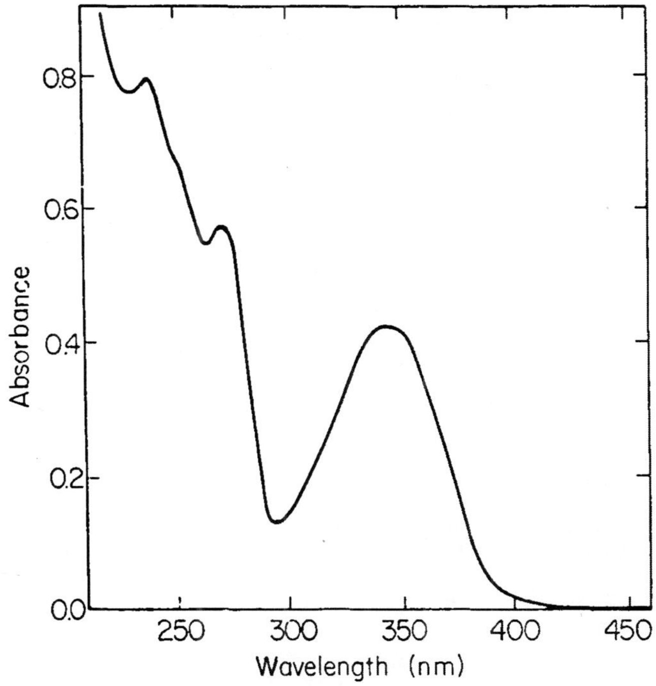

FIGURE 18. Absorption spectrum of the oxygen-stable form of F_{342} isolated from *M. thermoautotrophicum*. The UV-visible spectrum was obtained with a Cary 14 recording spectrophotometer. Sample and reference buffer was 50 m*M* Tris, pH 7.5. (From Gunsalus, R.P. and Wolfe, R.S., *FEMS Microbiol. Lett.*, 3, 191, 1978. With permission.)

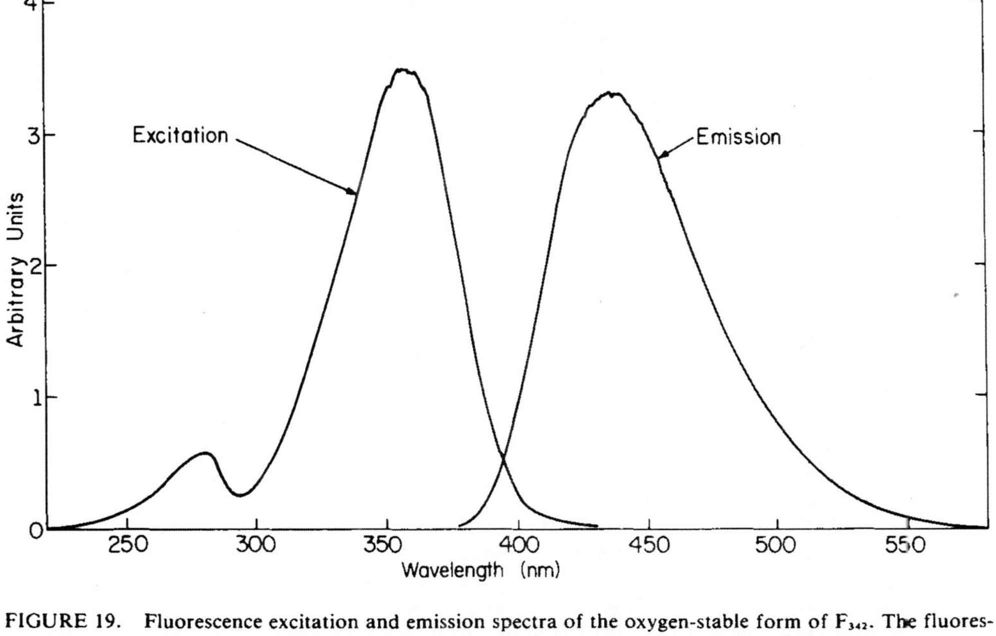

FIGURE 19. Fluorescence excitation and emission spectra of the oxygen-stable form of F_{342}. The fluorescence excitation and emission spectra were obtained with an Hitachi Perkin-Elmer MPF-2A Fluorescence Spectrophotometer. Sample buffer was 50 m*M* Tris, pH 7.5. Excitation wavelength for emission spectrum was 340 nm and emission wavelength for excitation spectrum was 435 nm. Fluorescence spectrum was uncorrected. (From Gunsalus, R.P. and Wolfe, R.S., *FEMS Microbiol. Lett.*, 3, 192, 1978. With permission.)

ACKNOWLEDGMENTS

Results of original work reported here were supported by grants from the National Science Foundation PCM 76-02652 and from the U.S. Public Health Service AI 12277. It is a pleasure to thank my students and colleagues for their exciting contributions to this field.

REFERENCES

1. **Barker, H. A.**, *Bacterial Fermentations*, John Wiley & Sons, New York, 1956.
2. **Stadtman, T. C.**, Methane fermentation, *Annu. Rev. Microbiol.*, 25, 121, 1967.
3. **Wolfe, R. S.**, Microbial formation of methane, in *Advances in Microbial Physiology*, Vol. 6, Rose, A. H. and Wilkinson, J. R., Eds., Academic Press, New York, 1971, 107.
4. **Zeikus, J. G.**, Biology of methanogenic bacteria, *Bacteriol. Rev.*, 41, 514, 1977.
5. **Mah, R. A., Ward, D. M., Baresi, L., and Glass, T. L.**, Biogenesis of methane, *Annu. Rev. Microbiol.*, 31, 309, 1977.
6. **Zehnder, A. J. B.**, Ecology of methane formation, in *Water Pollution Microbiology*, Vol. 2, Mitchell, R., Ed. John Wiley & Sons, New York, 1978, 349.
7. **Wolfe, R. S. and Higgins, I. J.**, Biochemistry of methane — a study in contrasts, in *Microbial Biochemistry*, Vol. 21, Quayle, J. R., Ed., University Park Press, Baltimore, 1979, 267.
8. **Balch, W. E. and Wolfe, R. S.**, New approach to the cultivation of methanogenic bacteria: 2-mercaptoethanesulfonic acid (HS-CoM)-dependent growth of *Methoanobacterium ruminantium* in a pressurized atmosphere, *Appl. Environ. Microbiol.*, 32, 781, 1976.
9. **Balch, W. E. and Wolfe, R. S.**, Specificity and biological distribution of coenzyme M (2-mercaptoethanesulfonic acid), *J. Bacteriol.*, 137, 256, 1979.
10. **Balch, W. E. and Wolfe, R. S.**, Transport of coenzyme M (2-mercaptoethanesulfonic acid) in *Methanobacterium ruminantium*, *J. Bacteriol.*, 137, 264, 1979.
11. **Balch, W. E., Magrum, L. J., Fox, G. E., Woese, C. R., and Wolfe, R. S.**, Methanogens, re-evaluation of a unique microbial group. *Microbiol. Rev.*, 43, 260, 1979.
12. **Bryant, M. P.**, Commentary on Hungate technique for culture of anaerobic bacteria, *Am. J. Clin. Nutr.*, 25, 1324, 1972.
13. **Hungate, R. E.**, Roll tube method for cultivation of strict anaerobes, in *Methods in Microbiology*, Vol. 3B, Norris, J. R. and Ribbons, D. W., Eds., Academic Press, New York, 1969, 117.
14. **Holdeman, L. V. and Moore, W. E. C.**, Roll-tube techniques for anaerobic bacteria, *Am. J. Clin. Nutr.*, 25, 1314, 1972.
15. **Bryant, M. P., and Robinson, I. M.**, An improved nonselective culture medium for rumen bacteria and its use in determining diurnal variation in numbers of bacteria in the rumen, *J. Dairy Sci.*, 44, 1446, 1961.
16. **Roberton, A. M. and Wolfe, R. S.**, Adenosine triphosphate pools in *Methanobacterium*, *J. Bacteriol.*, 102, 43, 1970.
17. **Roberton, A. M. and Wolfe, R. S.**, ATP requirement for methanogenesis in cell extracts of *Methanobacterium* strain M.o.H., *Biochim. Biophys. Acta*, 192, 420, 1969.
18. **Wolin, E. A., Wolin, M. J., and Wolfe, R. S.**, Formation of methane by bacterial extracts, *J. Biol. Chem.*, 238, 2882, 1963.
19. **Bryant, M. P., Wolin, E. A., Wolin, M. J., and Wolfe, R. S.**, *Methanobacillus omelianskii*, a symbiotic association of two species of bacteria, *Arch. Mikrobiol.*, 59, 20, 1967.
20. **Gunsalus, R. P.**, The Methyl Coenzyme M Reductase System in *Methanobacterium thermoautotrophicum*, Ph.D. thesis, University of Illinois, Urbana, 1977.
21. **Tzeng, S. F., Wolfe, R. S., and Bryant, M. P.**, Factor 420-dependent pyridine nucleotide-linked hydrogenase system of *Methanobacterium ruminantium*, *J. Bacteriol.*, 121, 184, 1975.
22. **Eirich, L. D.**, The Structure of Coenzyme F_{420}, a Novel Electron Carrier Isolated from *Methanobacterium* Strain M.o.H., Ph.D. thesis, University of Illinois, Urbana, 1978.
23. **Cheeseman, P., Toms-Wood, A., and Wolfe, R. S.**, Isolation and properties of a fluorescent compound, $Factor_{420}$, from *Methanobacterium* strain M.o.H., *J. Bacteriol.*, 112, 527, 1972.
24. **Tzeng, S. F., Bryant, M. P., and Wolfe, R. S.**, Factor 420-dependent pyridine nucleotide-linked formate metabolism of *Methanobacterium ruminantium*, *J. Bacteriol.*, 121, 192, 1975.

25. **Ferry, J. G. and Wolfe, R. S.**, Nutritional and biochemical characterization of *Methanospirillum hungatii*, *Appl. Environ. Microbiol.*, 34, 371, 1977.
26. **Eirich, L. D., Vogels, G. D., and Wolfe, R. S.**, Proposed structure for coenzyme F_{420} from *Methanobacterium*, *Biochemistry*, 17, 4583, 1978.
27. **Zeikus, J. G., Fuchs, G., Kenealy, W., and Thauer, R. K.**, Oxidoreductases involved in cell carbon synthesis of *Methanobacterium thermoautotrophicum*, *J. Bacteriol.*, 132, 604, 1977.
28. **Fuchs, G., Stupperich, E., and Thauer, R. K.**, Function of fumarate reductase in methanogenic bacteria *(Methanobacterium)*, *Arch. Microbiol.*, 119, 215, 1978.
29. **Romesser, J. A.**, The Activation and Reduction of Carbon Dioxide to Methane in *Methanobacterium thermoautotrophicum*, Ph.D. thesis, University of Illinois, Urbana, 1978.
30. **Wolin, M. J., Wolin, E. A., and Wolfe, R. S.**, ATP-dependent formation of methane from methylcobalamin by extracts of *Methanobacillus omelianskii*, *Biochem. Biophys. Res. Commun.*, 12, 464, 1963.
31. **Wood, J. M. and Wolfe, R. S.**, Components required for the formation of CH_4 from methylcobalamin, *J. Bacteriol.*, 92, 696, 1966.
32. **McBride, B. C. and Wolfe, R. S.**, A new coenzyme of methyl transfer, coenzyme M, *Biochemistry*, 10, 2317, 1971.
33. **Taylor, C. D. and Wolfe, R. S.**, Structure and methylation of coenzyme M, *J. Biol. Chem.*, 249, 4879, 1974.
34. **Gunsalus, R. P. and Wolfe, R. S.**, ATP activation and properties of the methyl coenzyme M reductase system in *Methanobacterium thermoautotrophicum*, *J. Bacteriol.*, 135, 851, 1978.
35. **Wolin, E. A., Wolfe, R. S., and Wolin, M. J.**, Viologen dye inhibition of methane formation by *Methanobacillus omelianskii*, *J. Bacteriol.*, 87, 993, 1964.
36. **Bauchop, T.**, Inhibition of rumen methanogenesis by methane analogues, *J. Bacteriol.*, 94, 171, 1967.
37. **McBride, B. C. and Wolfe, R. S.**, Inhibition of methanogenesis by DDT, *Nature (London)*, 234, 551, 1971.
38. **Balderston, W. L. and Payne, W. J.**, Inhibition of methanogenesis in salt march sediments and whole-cell suspensions of methanogenic bacteria by nitrogen oxides, *Appl. Environ. Microbiol.*, 32, 264, 1976.
39. **McBride, B. C. and Wolfe, R. S.**, Biosynthesis of dimethylarsine by *Methanobacterium*, *Biochemistry*, 10, 4312, 1971.
40. **Gunsalus, R. P., Romesser, J. A., and Wolfe, R. S.**, Preparation of coenzyme M analogues and their activity in the methyl coenzyme M reductase system of *Methanobacterium thermoautotrophicum*, *Biochemistry*, 17, 2374, 1978.
41. **Gunsalus, R. P. and Wolfe, R. S.**, Stimulation of CO_2 reduction to methane by methyl coenzyme M in extracts of *Methanobacterium*, *Biochem. Biophys. Res. Commun.*, 76, 790, 1977.
42. **Buswell, A. M. and Sollo, F. W.**, Mechanism of the methane fermentation, *J. Am. Chem. Soc.*, 70, 1778, 1948.
43. **Stadtman, T. C. and Barker, H. A.**, Studies on the methane fermentation. IX. The origin of methane in the acetate and methanol fermentations by *Methanosarcina*, *J. Bacteriol.*, 61, 81, 1951.
44. **Pine, M. J. and Barker, H. A.**, Studies on the methane fermentation. XII. The pathway of hydrogen in acetate fermentation, *J. Bacteriol.*, 71, 644, 1956.
45. **Pine, M. J. and Vishniac, W.**, The methane fermentation of acetate and methanol., *J. Bacteriol.*, 73, 736, 1957.
46. **van den Berg, L.**, Effect of temperature on growth and activity of a methanogenic culture utilising acetate, *Can. J. Microbiol.*, 23, 898, 1977.
47. **Weimer, P. J. and Zeikus, J. G.**, Acetate metabolism in *Methanosarcina barkeri*, *Arch. Microbiol.*, 119, 175, 1978.
48. **Zeikus, J. G., Weimer, P. J., Nelson, D. R., and Daniels, L.**, Bacterial methanogenesis: acetate as a methane precursor in pure culture, *Arch. Microbiol.*, 104, 129, 1975.
49. **Smith, M. R. and Mah, R. A.**, Growth and methanogenesis by *Methanosarcina* strain 227 on acetate and methanol, *Appl. Environ. Microbiol.*, 36, 870, 1978.
50. **Winter, J. and Wolfe, R. S.** Complete degradation of carbohydrate to carbon dioxide and methane by syntrophic cultures of *Acetobacterium woodii* and *Methanosarcina barkeri*, *Arch. Microbiol.*, 121, 97, 1979.
51. **Hippe, H., Caspari, D., Fiebig, K., and Gottschalk, G.**, Utilization of trimethylamine and other *N*-methyl compounds for growth and methane formation by *Methanosarcina barkeri*, *Proc. Natl. Acad. U.S.A.*, 76, 494, 1979.
52. **Doddema, H. J., Hutten, T. J., van der Drift, C., and Vogels, G. D.**, ATP hydrolysis and synthesis by the membrane-bound ATP synthetase complex of *Methanobacterium thermoautotrophicum*, *J. Bacteriol.*, 136, 19, 1978.

53. **Gottschalk, G., Schoberth, S., and Braun, K.,** Energy metabolism of anaerobes growing on C_1-compounds. Abstract of the symposium, *Microbial Growth on C_1 Compounds,* Scientific Center for Biological Research, Pushchino, U.S.S.R. Academy of Sciences, 1977, 157.
54. **Fuchs, G., Stupperich, E., and Thauer, R. K.,** Function of fumarate reductase in methanogenic bacteria *(Methanobacterium), Arch. Microbiol.,* 119, 215, 1978.
55. **Thauer, R. K., Jungermann, K., and Decker, K.,** Energy conservation in chemotrophic anaerobic bacteria, *Bacteriol. Rev.,* 41, 100, 1977.
56. **Gunsalus, R. P. and Wolfe, R. S.,** Chromophoric factors F_{342} and F_{430} of *Methanobacterium thermoautotrophicum, FEMS Microbiol. Lett.,* 3, 191, 1978.
57. **Daniels, L. and Zeikus, J. G.,** One-carbon metabolism in methanogenic bacteria: analysis of short-term fixation products of $^{14}CO_2$ and $^{14}CH_3OH$ incorporated into whole cells, *J. Bacteriol.,* 136, 75, 1978.
58. **Morris, G.,** personal communication.
59. **Walsh, C.,** personal communication.
60. **Doddema, H. J. and Vogels, G. D.,** personal communication.
61. **Tanner, R. S.,** personal communication.
62. **Zehnder, A., Heuser, B., and Wuhrmann K.,** personal communication.
63. **Boone, D. and Bryant, M. P.,** personal communication.
64. **Woese, C. R.,** personal communication.
65. **Vogels, G. D.,** personal communication.
66. **Balch, W. E.,** personal communication.

Chapter 7

RESPIRATION IN METHYLOTROPHIC BACTERIA

I. J. Higgins

TABLE OF CONTENTS

I. INTRODUCTION

Bacteria capable of growth on reduced carbon-one compounds (methylotrophs) have been commanding the attention of an increasing number of microbial physiologists and biochemists, especially during the last 10 years. This reflects the realization that these microorganisms play important roles in both the carbon and nitrogen cycles, and that they possess great potential for exploitation by the chemical industry as sources of single-cell protein and biocatalytic systems for petrochemical transformations. The literature prior to 1972 has been thoroughly reviewed,[1,2] and a valuable short article was published by Anthony in 1975.[3] More recently, an extensive general discussion of bacteria that generate methane (methanogens) and those that utilize it (methanotrophs)[4] has been published, and there have been a number of recent examinations of the current and potential commercial exploitation of methylotrophs.[5-7]

Five years ago, our knowledge of the bioenergetics of these bacteria was restricted to an examination of the primary oxidative enzymes in a few species and a detailed understanding of only one or two of these enzymes, most notably, methanol dehydrogenase.[8] Since then, this situation has changed dramatically, and although the data is far from complete for any one species, we now have some understanding of the total respiratory systems of a few of these bacteria. This reflects the concentrated efforts of several research groups stimulated by the importance of theoretical growth yields on methane and methanol to the single-cell protein industry.[9-11]

The methylotrophs are a large group of bacteria and are classified on nutritional grounds in Table 1. There are three types of methane utilizers, two groups being obligate methylotrophs. A considerable proportion of carbon turnover in the carbon cycle passes via methane produced by methanogenic bacteria, and it is, therefore, not surprising that methanotrophs are widely distributed in the environment.[14] They contain characteristic intracytoplasmic membranes which differ in arrangement between Type-I and Type-II organisms.[14] These two major groups also employ different carbon incorporation pathways.[33] Type-I methanotrophs use the ribulose monophosphate pathway,[34] Type-II species use the serine pathway.[35,36] To date, there have been only two descriptions of facultative methanotrophs (also see Reference 135).[15,16] A large number of both obligate and facultative methylotrophs unable to utilize methane has also been isolated (Table 1). These comprise a rather heterogeneous group. Most isolates tabulated are obligate aerobes, although a bacterium capable of growing on methane anaerobically has recently been reported.[37]

There is rather more detailed information in the literature concerning the respiratory systems of some nonmethane-utilizing methylotrophs than for methanotrophs. The former group are, therefore, examined first.

II. RESPIRATORY SYSTEMS OF METHYLOTROPHS UNABLE TO UTILIZE METHANE

A. Primary Oxidation Pathways

In addition to the carbon-one substrates, methanol and formic acid, there are a number of naturally occurring compounds which are regarded as methylotrophic substrates since they do not contain carbon-carbon bonds (Table 1). The pathways used to oxidize these compounds are summarized in Figure 1, and the evidence accumulated prior to 1975 for their operation has been discussed in detail elsewhere.[1-4] These reactions are the primary sources of electrons for energy transduction, and also of formaldehyde (either free or bound as a tetrahydrofolate derivative) from which cell material is synthesized. Formate is reduced to the oxidation level of formaldehyde for biosynthesis.

TABLE 1

Types of Methylotrophic Bacteria

Group	Examples	Compounds serving as sole carbon and energy sources for growth	Ref.
Obligate methylotrophs (Type I methanotrophs)	*Methylobacter*	Methane, methanol, dimethylether	14
	Methylococcus		12
	Methylomonas		14
	(Pseudomonas methanica)		
Obligate methylotrophs (Type II methanotrophs)	*Methylocystis*		13
	Methylosinus		
	(Methanomonas methanooxidans)		14
Facultative methanotroph (similar to Type II obligate methylotroph)	*Methylobacterium organophilum*	Methane, methanol, dimethyl ether, a variety of heterotrophic substrates	15, 16
Obligate methylotrophs unable to utilize methane	*Bacterium* 4B6	Methylamine, di- and tri-methylamine	17, 18
	Bacterium C2A1	Methanol, methylamine, di- and tri-methylamine	18
	Methylomonas M15	Methanol	19
	Methylomonas 2B36P11 MutII	Methanol	20
	Methylmonas methylovora	Methanol	21
	Methylophilus methylotrophus (formerly *Pseudomonas methylotropha*)	Methanol	22, 23
	Organism WI	Methanol, methylamine	24
Facultative methylotrophs, including Types M and L restrictive facultative methylotrophs (see Section II.A.2). (All grow on some heterotrophic substrates in addition to compounds indicated).	*Bacillus* PM6	Methylamine, di- and tri-methylamine, tetramethylammonium, and trimethylsulphonium compounds	25 26
	Bacterium 5H2	Methylamine, di- and tri-methylamine, tetramethylammonium and trimethylsulphonium compounds	17, 18
	Bacterium 5BI	Methanol, formate methylamine, di- and tri-methylamine, and trimethylamine *N*-oxide.	
	Hyphomicrobium sp.	Methanol, formate, methylamine, trimethylamine	27
	Pseudomonas AMI	Methanol, formate, methylamine, di- and tri-meth-	28
	Pseudomonas aminovornas	ylamine	29
	Pseudomonas MA	Formate, methylamine, di- and tri-methylamine, tri-	30
	Pseudomonas MS	methylamine *N*-oxide	31
	Pseudomonas M27	Methylamine	17, 18

TABLE I (continued)

Types of Methylotrophic Bacteria

Group	Examples	Compounds serving as sole carbon and energy sources for growth	Ref.
	Pseudomonas 3A2	Methylamine, di- and tri-methylamine, trimethylsulphonium compounds	17, 18
		Methanol, formate, methylamine	
		Methanol, formate, methylamine, di- and tri-methylamine, trimethylamine *N*-oxide	

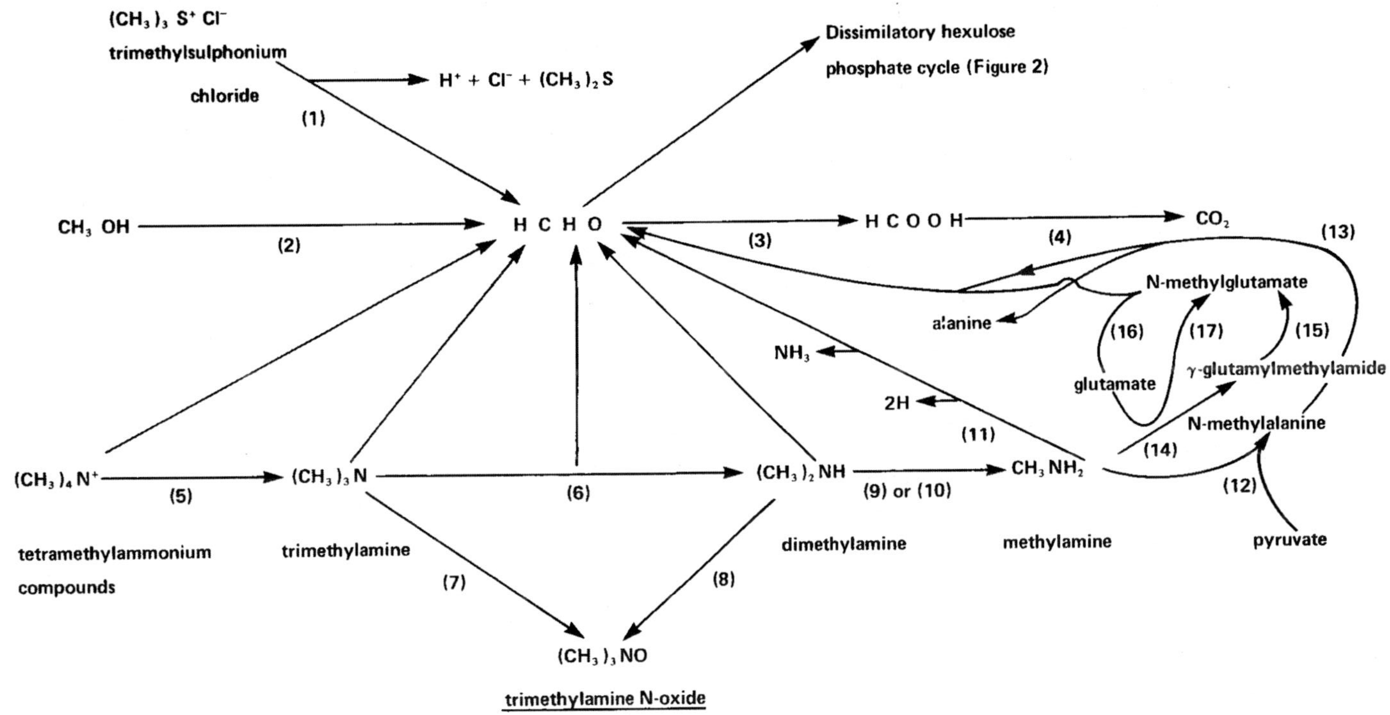

FIGURE 1. Primary oxidation pathways for methylotrophic subtrates. The enzymes involved are as follows: (1) trimethylsulphonium tetrahydrofolate methyl transferase/N^5-methyltetrahydrofolate oxidoreductase,[31,61] (2) methanol dehydrogenase,[8,32,38-48] (3) formaldehyde dehydrogenase,[3,8,18,31,38-49] (4) formate dehydrogenase,[41,56-60] (5) tetramethylammonium mono-oxygenase,[26] (6) trimethylamine dehydrogenase,[18,53,65] (7) trimethylamine mono-oxygenase,[18,62] (8) trimethylamine *N*-oxide demethylase,[18,26,54,63,64] (9) dimethylamine mono-oxygenase,[18,29,63,67,68] (10) dimethylamine dehydrogenase,[65] (11) primary amine (methylamine) dehydrogenase,[18,24,72-77] (12) *N*-methylalanine synthase,[87,88] (13) *N*-methylalanine dehydrogenase,[87-89] (14) γ-glutamylmethylamide synthase,[65,86] (15) unidentified enzyme,[65,86] (16) *N*-methylglutamate dehydrogenase,[79-81,83,85] and (17) *N*-methylglutamate synthase.[79-82,84]

1. Methanol Oxidation

In all aerobic methanol-utilizing bacteria examined, methanol is oxidized by an oxidoreductase (methanol dehydrogenase) closely similar to the one first studied by Anthony and Zatman in *Pseudomonas* M27.[8,32,38-40] This enzyme is not NAD-linked, but it can be measured in vitro by coupling via phenazine methosulphate to oxygen or by using 2,6-dichlorophenol-indophenol, or cytochrome *c* as electron acceptors. It requires ammonia or methylamine as activators, has a wide substrate specificity, oxidizing many primary alcohols. It has a high isoelectric point, a molecular weight of 120,000 to 146,000, contains two subunits of molecular weight about 60,000, and probably a pteridine prosthetic group. The mechanism of the enzyme is still unknown. The pteridine component may act simply as an electron carrier, or an $N^{5,10}$-methylene derivative of the pteridine may be formed.[2,40] At least in *Pseudomonas* M27, the in vivo electron acceptor for the enzyme is probably a flavoprotein.[38] Closely similar enzymes have since been detected in, and in many cases purified from, a number of other facultative methanol-utilizers, including *Pseudomonas* AMI, *Protaminobacter ruber, Vibrio extorquens,*[41] *Hyphomicrobium* WC, *Pseudomonas* TP-1, *Pseudomonas* WI,[42] *Pseudomonas* ZB 1 J26,[43] *Pseudomonas* C,[44] and *Pseudomonas* sp. No. 2941.[45] Methanol dehydrogenase enzymes have also recently been purified from two obligate methanol utilizers, a *Methylomonas* species[43] and *Methylophilus methylotrophus.*[46] They do not seem to differ significantly from those in facultative species.

A full understanding of the mechanism of this enzyme is awaited and is, in part, dependent upon the elucidation of the chemical structure of its prosthetic group. This deficiency in our knowledge has recently been highlighted by indications from work with pure methanol dehydrogenase from *Hyphomicrobium* X that, at least in this enzyme, the prosthetic group may in fact not be a pteridine, but a quinone containing a polyol or sugar moiety.[47] Nevertheless, there remains strong evidence for a pteridine cofactor in enzymes from other species, a 2,4-dihydroxypteridine being most likely.[48] An interesting property of these enzymes is their ability to oxidize formaldehyde at similar rates to methanol.[43-45,47,48] The physiological importance of the formaldehyde dehydrogenase activity of the enzyme is not fully clear for many species, and this is discussed in more detail in the next section. In some methanotrophs, methanol dehydrogenase may be the only enzyme capable of oxidizing formaldehyde (see Section II. 3). Sperl et al.[48] showed that the ability of the enzyme to oxidize aldehydes correlated with their ability to form stable hydrates. Because of this lack of specificity, the product of methanol oxidation by the purified enzyme is formate, and indeed, it has been reported that no free formaldehyde can be detected during the reaction.[47] Since formaldehyde is used for biosynthesis, this raises an interesting question concerning the regulation of the enzyme activity in vivo and its role in controlling the supply of formaldehyde for biosynthesis.

2. Formaldehyde Oxidation

The oxidation of formaldehyde by methanol dehydrogenase is discussed in the previous section. As well as being an intermediate in methanol oxidation, it is the oxidation product of all *N*-methyl compounds and trimethyl-sulfonium chloride (Figure 1). In addition to methanol dehydrogenase, five other mechanisms for formaldehyde oxidation have been found in methylotrophic bacteria. A glutathione-dependent, NAD-linked enzyme is present in a number of facultative methylrotophs including *Pseudomonas* 3A2, *Protaminobacter ruber, Vibrio extorquens,*[41] Bacterium 5H2[26] and *Pseudomonas aminovorans.*[3] The facultative methylotroph, *Pseudomonas* MS, when grown on amines, and the obligate organism, Bacterium 4B6, also contain NAD-linked enzymes, but they do not require glutathione.[18,31] NAD(P)-independent aldehyde dehydrogenases which are probably flavoproteins, and which can only be measured using artificial electron acceptors, have been detected in several species, but only at low specific activities.[18,31,41] Such an enzyme that can be coupled to DCPIP was first demonstrated by Johnson and Quayle in *Pseudomonas* AMI.[41] Its physiological significance is uncertain in view of very low specific activities, but these may be artifactual due to difficulties in assaying the enzyme.

It may well be that, at least in some species, free formaldehyde is not a major intermediate, and this would not be surprising in view of its toxic nature. There have been several intimations of this in the literature,[2,31,41,47] and in this context, a $N^{5,10}$-methylenetetrahydrofolate dehydrogenase present at high specific activity in *Pseudomonas* AMI[49] may be important. This enzyme could reduce $N^{5,10}$-methenyltetrahydrofolate to the methylene derivative for incorporation into cell material. Alternatively, if free formaldehyde is formed, it reacts nonenzymically with tetrahydrofolate, and the enzyme could then oxidize the methylene derivative. In *Pseudomonas* AMI, it is clear that the methanol dehydrogenase is not responsible for formaldehyde oxidation since mutants lacking methanol dehydrogenase oxidize formaldehyde at undiminished rates.[50,51]

It has recently become clear that in some species a cyclic pathway operates for the oxidation of formaldehyde. The possible involvement of the dissimilatory hexulose phosphate cycle (Figure 2) in methanotrophs was suggested in 1974 by Strom et al.[52] Colby and Zatman[53] isolated a number of restricted facultative methylotrophs. Type-M species grew only on glucose among 50 heterotrophic substrates tested while Type-L species grew on seven out of 56 such substrates. The Type-L organisms are devoid of formaldehyde and formate dehydrogenase activities, but do possess the enzymes of the dissimilatory hexulose phosphate cycle, which is, therefore, probably the only mechanism for formaldehyde oxidation in these species.[54] These authors also suggest that the cycle is probably important in the Type M and obligate species that they studied. More recently, isotopic labeling data have revealed that the cycle is the predominant mechanism for formaldehyde oxidation in *Pseudomonas* C.*[55]

3. Formate Oxidation

In methylotrophs, formate is oxidized to carbon dioxide by NAD-linked formate dehydrogenases of the type originally detected by Kaneda and Roxburgh.[56] Johnson and Quayle[41] subsequently demonstrated similar activities in several species and effected a threefold purification of the enzyme from *Pseudomonas* AMI. More recently, a soluble, inducible NAD-linked enzyme has been detected in a methylotrophic Pseudomonad,[57] and the enzyme from *Bacterium* sp. 1 has now been purified.[58,59] It is quite stable in the presence of -SH compounds, has a molecular weight of 70,000, is totally specific for formate, shows maximum activity in the pH range of 6 to 9, and has a rather high Km for formate (1.5×10^{-2} *M*). It has been suggested that this high Km may reflect a regulatory function, favoring the directing of formate into biosynthetic reactions when its concentration is low.[58,60] Substantiation of this suggestion awaits a detailed study of the regulation of oxidative metabolism in methylotrophs.

4. Trimethylsulfonium Chloride Oxidation

Pseudomonas MS grows on this compound. The initial reaction involves transfer of a methyl group to tetrahydrofolate, the resulting dimethylsulfide being released into the medium:

$$(CH_3)_3S^+Cl^- + \text{tetrahydrofolate} \rightarrow N^5\text{-methyltetrahydrofolate} + H^+ + Cl^- + (CH_3)_2S \qquad (1)$$

* Work with {^{14}C}-formaldehyde and extracts of *Pseudomonas oleovorans* suggests that the cycle operates in this species also.[171]

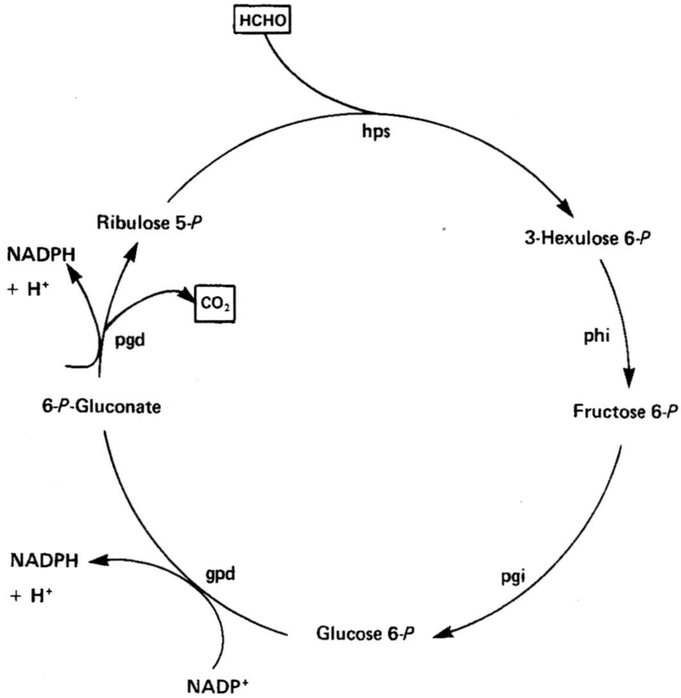

FIGURE 2. Dissimilatory hexulose phosphate cycle. gpd, glucose-6-phosphate dehydrogenase; hps, 3-hexulose phosphate synthase; pgd, 6-phosphogluconate dehydrogenase; pgi, phosphoglucoisomerase; and phi, phospho-3-hexulose isomerase. (From Strøm, T., Ferenci, T., and Quayle, G.R., *Biochem. J.*, 144, 465-476, 1974. By courtesy of the *Biochemical Journal*, London).

N^5-Methyltetrahydrofolate is then oxidized to $N^{5,10}$-methylenetetrahydrofolate.[31,61]

5. *Tetramethylammonium Oxidation*

Bacterium 5H2 which grows on tetramethylammonium chloride oxidizes this substrate via trimethylamine and trimethylamine-*N*-oxide (Figure 1). The process is initiated by a nonheme iron containing monooxygenase that catalyzes the following reaction:[26]

$$(CH_3)_4\,N^+Cl^- + O_2 + NAD(P)H + H^+ \rightarrow (CH_3)_3\,N + HCHO + H_3O^+ + NAD(P)^+ + Cl^- \qquad (2)$$

6. *Trimethylamine Oxidation*

Two routes have been identified for trimethylamine oxidation (Figure 1). Most facultative methylotrophs examined, including the type-L restricted facultative isolates of Colby and Zatman[54] possess an inducible mono-oxygenase that oxidizes the substrate to trimethylamine *N*-oxide:[18,62]

$$(CH_3)_3\,N + O_2 + NAD(P)H + H^+ \rightarrow (CH_3)_3\,NO + H_2O + NAD(P)^+ \qquad (3)$$

In these organisms and in *Bacillus* PM6 which was isolated using trimethylamine *N*-oxide, this compound is then converted to dimethylamine and formaldehyde by trimethylamine *N*-oxide demethylase.[18,26,54,63,64]

$$(CH_3)_3NO \rightarrow (CH_3)_2NH + HCHO \qquad (4)$$

The obligate methylotrophs, bacterium 4B6 and bacterium C2A1, and Type-M restricted facultative methylotrophs[53] do not use these two enzymes to oxidize trimethylamine, but contain a dehydrogenase which catalyzes an oxidative demethylation of the substrate:[18]

$$(CH_3)_3N + PMS + H_2O \rightarrow (CH_3)_2NH + PMSH_2 + H \cdot CHO \qquad (5)$$

The natural electron acceptor of this enzyme is not known, and it is assayed using phenazine methosulphate (PMS). This enzyme is also found in trimethylamine-grown *Hyphomicrobium* X which grows on this substrate both aerobically and anaerobically using nitrate as terminal electron acceptor.[65] Clearly, a monooxygenase could not be involved under anaerobic conditions. Interestingly, a similar enzyme activity had previously been detected by Large and McDougall[66] in extracts of *H. vulgare*, but was attributed to a contaminating organism.

The dehydrogenase mechanism is clearly energetically more efficient, since reducing equivalents are generated which are probably available for energy transduction, while the monooxygenase route consumes reducing equivalents in the form of NAD(P)H.

7. Dimethylamine Oxidation

An inducible secondary amine monooxygenase that contains cytochrome P_{420} and converts dimethylamine to methylamine and formaldehyde has been studied in *Pseudomonas aminovorans* by Large and his co-workers[29,67,68] who suggest that this enzyme together with those attacking trimethylamine, trimethylamine *N*-oxide, and methylamine constitute a membrane-bound association which oxidizes all these compounds to formaldehyde.[29] Dimethylamine monooxygenase catalyzes the reaction:

$$(CH_3)_2NH + O_2 + NAD(P)H + H^+ \rightarrow CH_3NH_2 + H\,CHO + NAD(P)^+ + H_2O \qquad (6)$$

A number of obligate species, as well as other facultative ones, use the same type of enzyme system.[18,29,63] Again, *Hyphomicrobium* X being capable of growth on both di- and trimethylamine under aerobic and anaerobic conditions does not use a mono-oxygenase for oxidation of dimethylamine, but rather uses a dimethylamine dehydrogenase[65,69] which is measured using artificial electron acceptors. The natural electron acceptors for this enzyme and the trimethylamine dehydrogenase are unknown, but it is unlikely that both activities are due to the same enzyme, and both have been partially purified by Meiberg and Harder.[65] Nevertheless, trimethylamine dehydrogenases do display some activity towards dimethylamine.[66,70]

A partially purified preparation of trimethylamine monooxygenase from *Pseudomonas aminovorans* oxidized dimethylamine to *N*-methylhydroxylamine, and incidently, *N,N*-dimethylhydroxylamine was found to be a substrate, as are the *N*-ethyl- and *N,N*-diethyl-analogues.[71] In addition *N*-methylhydroxylamine was slowly oxidized to formamide. The preparation, therefore, catalyzed the following sequence of reactions:

$$(CH_3)_2NH \xrightarrow[O_2]{NADPH} (CH_3)_2NOH \xrightarrow[O_2]{NADPH} HCHO + CH_3NHOH \xrightarrow[O_2]{NADPH} H\,CONH_2 \qquad (7)$$

The same enzyme was probably responsible for all these reactions, but the physiological significance is not yet clear.

8. Methylamine Oxidation

Two types of mechanism have been elucidated for the oxidation of methylamine (Figure 1). In some facultative species such as *Pseudomonas* AMI[72] and other pink pseudomonads, and in bacterium C2A1[18] and organism W1,[24] a soluble primary amine dehydrogenase catalyzes the reaction:

$$CH_3NH_2 + PMS + H_2O \rightarrow HCHO + NH_3 + PMSH_2 \qquad (8)$$

Again, the natural electron acceptor is not known, but the enzyme contains a pyridoxal-type prosthetic group which probably is converted to a pyridoxamine on reaction with methylamine and subsequently reverts to the pyridoxal form on reaction with phenazine methosulphate (PMS).[73]

A similar enzyme has recently been purified from *Methylomonas methlovora*[74] and from *Pseudomonas* sp. J.[75,76] The latter enzyme has been studied intensively. It has a molecular weight of 105,000 and is composed of subunits of mol wt 13,000 and 40,000. Cytochrome-c_{551} will act as a rather inefficient electron acceptor, and it is thought that the immediate acceptor in vivo is a flavoprotein. The methylamine dehydrogenases of both *Pseudomonas* sp. J and *Pseudomonas* AMI are $\alpha_2\beta_2$-type subunit proteins with geometrically analogous subunit structure.[76,77]

Some species, including *Pseudomonas aminovorans,*[78,79] *Pseudomonas* MA,[80,81] *Bacterium* 4B6,[18] and *Hyphomicrobium* X,[65] do not synthesise a primary amine dehydrogenase, but methylamine is first condensed with L-glutamate to form *N*-methylglutamate either directly or via γ-glutamylmethylamide. Subsequently, formaldehyde is released, and L-glutamate is regenerated. Evidence for this mechanism was first obtained by Hersh and co-workers[80,81] who demonstrated the presence in methylamine-grown *Pseudomonas* MA of *N*-methylglutamate synthase and *N*-methylglutamate dehydrogenase. It was suggested that these enzymes could catalyze the cyclic oxidation of methylamine to formaldehyde and ammonia (Figure 1):

$$CH_3NH_2 + \text{L-glutamate} \rightleftharpoons N\text{-methylglutamate} + NH_3 \cdot N\text{-}$$
$$\text{methylglutamate} + H_2O \rightarrow \text{L-glutamate} + HCHO + 2H^+ + 2e$$
$$\text{Sum: } CH_3NH_2 + H_2O \rightarrow HCHO + NH_3 + 2H^+ + 2e \qquad (9)$$

The same stoichiometry would pertain as for direct oxidation by a primary amine dehydrogenase. The synthase has been purified from this bacterium,[82] and the dehydrogenase which is membrane bound and linked to electron transport components has been solubilized and partially purified.[83] These two enzymes have since been studied in methylamine-grown *Pseudomonas aminovorans* and *Pseudomonas methylica.*[84] The dehydrogenase is probably a flavoprotein which is thought to transfer electrons via cytochrome *b* to cytochrome *c* and, thence, to cytochrome oxidase. In the latter organism, a soluble, NAD$^+$-linked *N*-methylglutamate dehydrogenase has also been detected.[85] In *Hyphomicrobium,* two enzymes are thought to be required for the synthesis of *N*-methylglutamate from methylamine and L-glutamate, namely γ-glutamylmethylamide synthase and an enzyme which converts this compound to *N*-methylglutamate.[65,86]

It has been suggested that *N*-methylalanine may in some species be an intermediate in methylamine oxidation by an analogous cyclic process to the one involving *N*-methylglutamate.[87,88] Cell-free synthesis of *N*-methylalanine was demonstrated by Kung and Wagner,[87] and an enzyme catalyzing the reaction was subsequently purified by Lin and Wagner.[88] The enzyme has a very high Km for methylamine (75m*M*), which suggests that its physiological role may be one of detoxification rather than of energy metabolism. A kinetic study of enzyme activities during adaptation of *Pseudomonas aminovorans* to growth on trimethylamine suggested that a *N*-methylalanine dehydrogenase detected in this species is not required for growth on trymethylamine,[89] and hence, that *N*-methylalanine is not an important intermediate in this case.

9. Oxidation of Reduced Carbon One Compounds by Autotrophs

Rhodopseudomonas acidophila grows anaerobically in the light on methanol[90] and oxidizes the substrate to carbon dioxide, which it incorprates using the ribulose diphosphate cycle. Under these growth conditions, it contains a methanol and formaldehyde dehydrogenase which can be coupled to phenazine methosulphate, a NAD-linked formaldehyde dehydrogenase requiring glutathione for activity, and a NAD-linked formate dehydrogenase. Although apparently similar to the methanol dehydrogenase of aerobic bacteria, the Km of the methanol dehydrogenase for methanol was increased 10-fold if the extract was prepared, and the enzyme assayed, in the absence of oxygen.[91] *Rhodopseudomonas palustris* grows autotrophically on formate, and both NAD-dependent and NAD-independent formate dehydrogenases have been implicated.[92-94] A NAD-linked enzyme is involved in autotrophic growth of *Pseudomonas oxalaticus* on formate.*[2]

B. Electron Transport Systems and Energy Transduction

The first report of the presence of cytochromes *a, b,* and *c* in a methylotroph is due to Hirsch and his co-workers[95] who were studying *Hyphomicrobium vulgare.* Subsequently, these cytochromes were also reported in *Pseudomonas* AMI,[96] *Pseudomonas* MA,[80] and another facultative methylotroph, *Pseudomonas* sp. *strain 2.*[97] In the last named organism, the overall cytochrome concentration was found to be highest after growth on carbon-one substrates (formate or methanol). Higgins and co-workers[98,99] compared the cytochrome compositions of three facultative methylotrophs (*Pseudomonas extorquens, Pseudomonas* AMI, and *Hyphomicrobium* sp.) and two obligate methanotrophs [*Pseudomonas methanica* (Type I) and *Methylosinus trichosporium* (Type II)]. After growth on methane or methanol under carefully controlled and similar conditions, the cytochrome patterns were both closely comparable and highly unusual. Disruption of these microorganisms followed by differential centrifugation revealed an extremely high concentration of a high redox potential, carbon monoxide-binding cytochrome-*c* in supernatant fractions and relatively low concentrations of cytochromes aa_3, *b* and *c* in particulate preparations. The effect of growth substrate on the cytochrome complement of *Pseudomonas extorquens* was examined, and massive differences were found between methylotrophically and heterotrophically grown bacteria. The latter contained approximately an order of magnitude more cytochromes aa_3 and membrane-bound *c* than the former. There was also a three- to fourfold increase in cytochrome *b* concentration and a small reduction in soluble cytochrome *c* concentration after heterotrophic growth. These findings suggested a radically different requirement for respiration and energy transduction between heterotrophically grown and methylotrophically grown methylotrophs. Further studies on *Pseudomonas extorquens* measuring cell-free oxidase activities and the effect of electron transport inhibitors suggested that methanol and NADH are oxidized by the same terminal oxidase in methanol-grown organisms, and that cytochrome *b* is not involved in methanol oxidation. After heterotrophic growth, a different terminal oxidase (probably cytochrome aa_3 is required for NADH oxidation and the oxidation of flavoprotein-linked

* Chemoautotrophs which grow on methanol but incorporate carbon as CO_2 have been isolated and found to contain both NAD^+- and nonNAD^+-linked formaldehyde and formate dehydrogenase enzymes. [72]

substrates such as succinate. This is present during methylotrophic growth, but its role under these conditions is uncertain. After disruption of the bacteria, methanol oxidase activity was present in supernatant fractions where the CO-binding cytochrome *c* was the only detectable cytochrome present. It was suggested that this auto-oxidizable cytochrome which was partially purified probably functions as an oxidase and may well be the only functional oxidase during growth on carbon-one substrates.[100-102] Two mutants of *Pseudomonas extorquens* were isolated, one lacking cytochrome *c*, the other lacking methanol dehydrogenase.[100,101] Extracts of neither mutant would oxidize methanol, but addition of a partially purified fraction of the CO-binding cytochrome *c* containing another (probably flavoprotein) component to extracts of the cytochrome-*c*-less mutant restored methanol oxidase activity. This indicated the involvement of other electron carriers between methanol dehydrogenase and the cytochrome *c*. Anthony had previously implicated a flavoprotein as the electron acceptor for methanol dehydrogenase in *Pseudomonas* M27.[38] A scheme representing the electron transport system of *Pseudomonas extorquens* which is consistent with the data so far obtained is depicted in Figure 3.

Anthony and his co-workers have recently extensively studied the electron transport system of *Pseudomonas* AMI. Careful removal of cytochrome *c* from membrane preparations and use of a mutant lacking cytochrome *c* permitted studies of the membrane-bound cytochrome *b*, which was shown to bind carbon monoxide, therefore having at least some of the properties of a cytochrome *o*. Both *Pseudomonas* MS and *Hyphomicrobium* X were also shown to contain a *b*-type cytochrome which binds carbon monoxide.[103,104] Failure to observe this cytochrome in earlier work[97] was due to the CO-binding spectrum of the *c*-type cytochrome obscuring that of the cytochrome-*b*. In *Pseudomonas* AMI, the cytochrome *c* was reduced in whole bacteria by all oxidizable substrates at rates determined by the primary dehydrogenases, and a mutant lacking this cytochrome oxidized all substrates except methanol, ethanol, and methylamine.[103] A similar mutant of *Pseudomonas extorquens* would not grow on or oxidize methanol.[100] As might be expected, the cytochrome *c* of *Pseudomonas* AMI was not reduced by methanol or ethanol in a mutant lacking methanol dehydrogenase.[103] Anthony[103] purified the only cytochrome *c* present in this bacterium and found it not to be autooxidizable. It was reducible by methanol dehydrogenase purified from the same organism. The pure cytochrome had a molecular weight of 20,000 and a redox potential of +260mV. Similar cytochromes purified from *Pseudomonas extorquens*[100] and *Methylosinus trichosporium* (Section II.1) are autooxidizable,* have redox potentials of +295 and +310 mV, respectively, and molecular weights of 13,000. Potassium cya-
A similar mutant of *Pseudomonas extorquens* would not grow or oxidize methanol.[100] As might be expected, the cytochrome *c* of *Pseudomonas* AMI was not reduced by methanol or ethanol in a mutant lacking methanol dehydrogenase.[103] Anthony[103] purified the only cytochrome *c* present in this bacterium and found it not to be autooxidizable. It was reducible by methanol dehydrogenase purified from the same organism. The pure cytochrome had a molecular weight of 20,000 and a redox potential of +260mV. Similar cytochromes purified from *Pseudomonas extorquens*[100] and *Methylosinus trichosporium* (Section II.1) are autooxidizable,* have redox potentials of +295 and +310OmV, respectively, and molecular weights of 13,000. Potassium cya-
nide (lm*M*) when added to suspensions of *Pseudomonas* AMI totally inhibited respiration with a number of substrates tested, but only caused 50% inhibition of endogenous respiration.[103] This suggests that there may be more than one oxidase functioning, although on the basis of further work, it was concluded that there is only one physiologically significant oxidase in this species and that the electron transport chain is fairly

* There are now thought to be two *c*-type cytochromes in this species, with molecular weights and midpoint redox potentials of 20,000 and 250mv and 10,000 and 285mv respectively.[173]

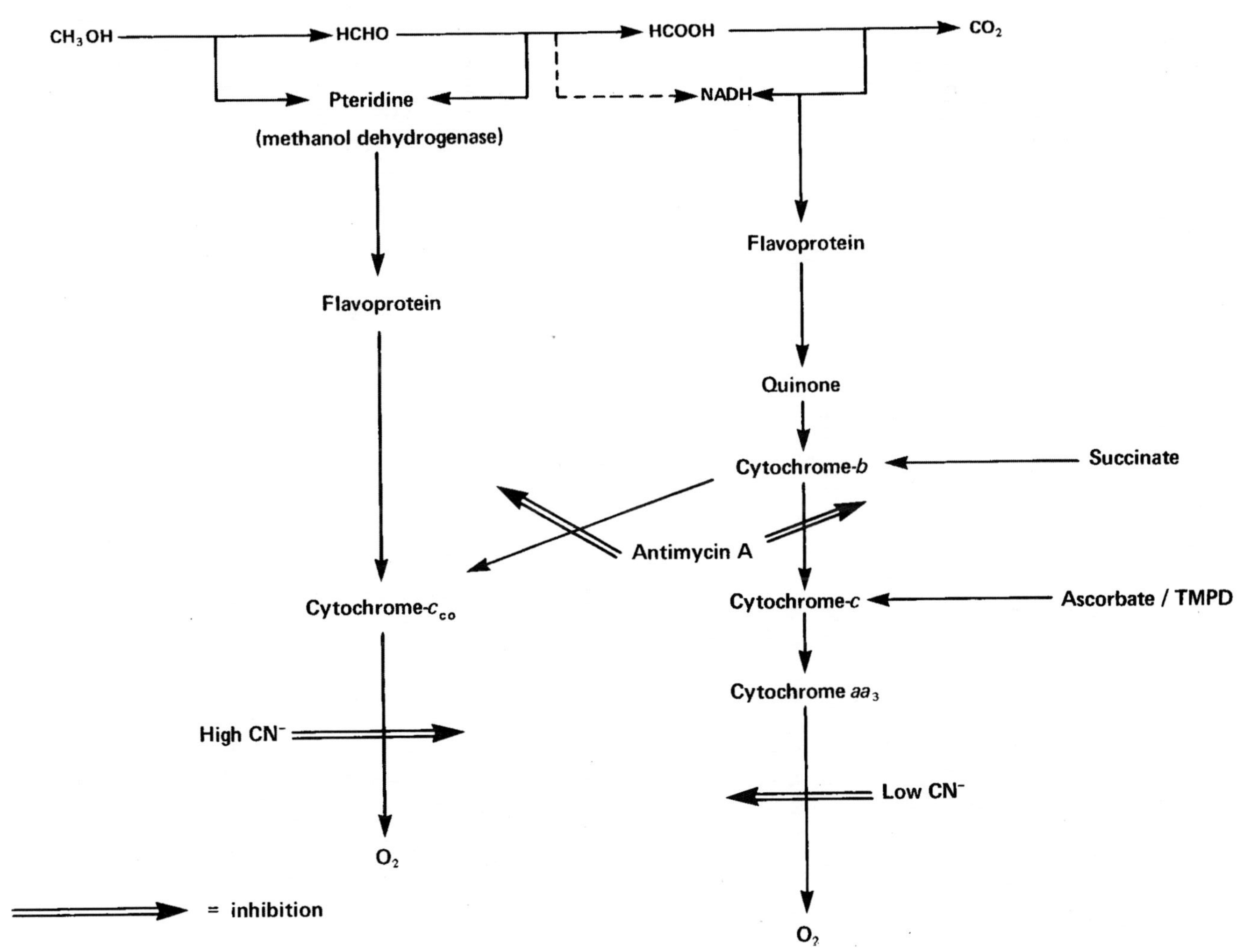

FIGURE 3. Tentative scheme for electron transport in *Pseudomonas extorquens*. (Modified from Tonge[101]).

conventional (Figure 4).[104,107] There was a large difference in cytochrome composition between heterotrophically and methylotrophically grown *Pseudomonas* AMI similar to the one previously reported in *Pseudomonas extroquens*.[98] The significance of these changes is not entirely clear in the context of a conventional electron transport chain. Results with the mutant of *Pseudomonas* AMI lacking cytochrome *c* suggested that either the cytochrome is involved only in the oxidation of methanol (and perhaps methylamine),* or that there is an alternative route for electron transport between cytochrome *b* and aa_3 in wild-type bacteria which becomes the sole route in the mutant.[103,104] In order to clarify the nature of the electron transport system of *Pseudomonas* AMI, Anthony and O'Keeffe have examined proton ranslocation in this species.[105-107] These authors found it impossible to abolish endogenous respiration by starvation, and so the values obtained for proton translocation in the presence of exogenous substrates may be composite values. The maximum → H^+/O ratio (protons translocated out of the bacteria per atom of oxygen consumed during respiration) was about four both in the wild-type bacterium and in the mutant lacking cytochrome *c*. The ratios were unaltered even in the mutant when cytochrome a_3 was inhibited by cyanide, and under these conditions, it was concluded that the single cytochrome *b* was functioning as an alternative oxidase. While two proton-translocating segments The ratios were unaltered even in the mutant when cytochrome a_3 was inhibited by cyanide, and under these conditions, it was concluded that the single cytochrome *b* was functioning as an alternative oxidase. Whilse two proton-translocating segments appear to operate during NADH oxidation, only one operates during the oxidation of formaldehyde or methylamine. It was not possible to obtain values for methanol as its oxidation is inhibited by thiocyanate. Since methylotrophs contain cytochrome c, it is a reasonable prediction that they would give a P/O ratio of three for NAD-linked substrates.[108] Since the P/O ratio is usually half the measured →H^+/O ratio,[108,109] and the maximum →H^+/O ratio is four, the maximum P/O ratio appeared to be two rather than three in this species. However, since the organism was not starved, it is possible that the true maximum →H^+/O ratio was not measured. Recently, Keevil and Anthony have shown that the endogenous →H^+/O ratio increases to six when the organism is grown carbon limited, suggesting that such conditions may lead to development of a third coupling site.[110] Molar growth yield measurements support this hypothesis. As a result of this work and other very recent measurements of oxidative phosphorylation in membrane vesicles from *Pseudomonas* AMI, it has been concluded that the scheme in Figure 4 operates under carbon excess conditions, but when carbon limited, the cytochrome *c* resides in the main electron transport chain in a conventional position between cytochromes *b* and aa_3, resulting in operation of a third coupling site.[111] Under all growth conditions, the first step in methanol oxidation is coupled to ATP synthesis. The significance of the marked differences between the cytochrome compositions of methylotrophically and heterotrophically grown organisms is still not fully understood in this species.

At present, the evidence suggests that the electron transport systems of *Pseudomonas extorquens* and *Pseudomonas* AMI are not identical, the main difference being the possibility that the carbon monoxide binding cytochrome *c* of the former species may function as an oxidase (compare Figures 3 and 4). However, more detailed studies are required to be certain that there is a real difference. Hammond and Higgins[112,113] measured →H^+/O ratios in both species, usually after successfully starving washed suspensions. However, in the case of *Pseudomonas extorquens* grown on succinate, it was not possible to starve away endogenous respiration. Evidence was obtained that proton extrusion on adding exogenous substrates in this case was composite and due

* However, in *Hyphomicrobuin* x, it has recently been shown that the cytochrome-c_{co} is not the primary acceptor for the amine dehydrogenase.[174]

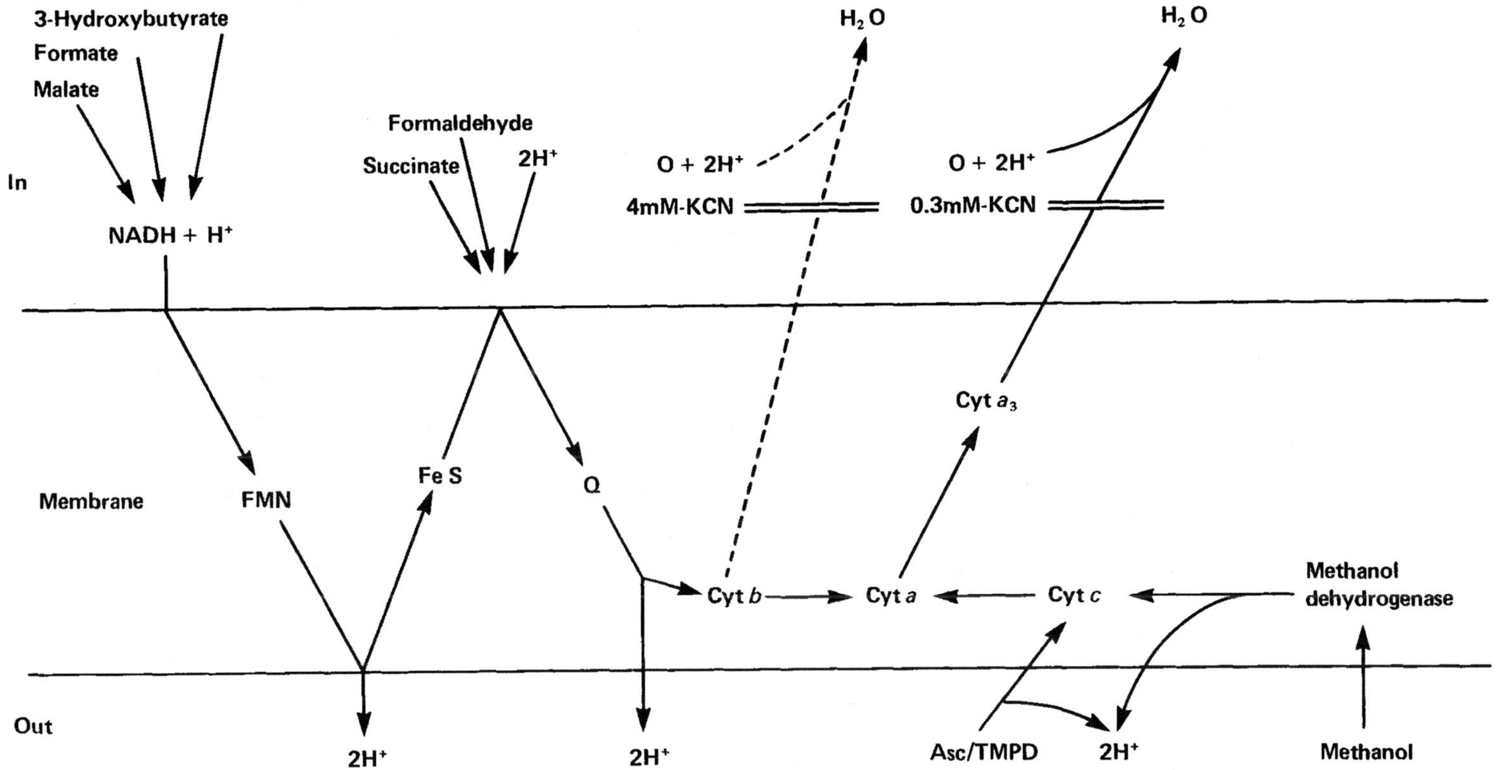

FIGURE 4. Proposed scheme for electron transport and proton translocation in *Pseudomonas* AMI. The arrows indicate flow of electrons or protons, but do not necessarily imply a direct reaction between components. Reduction of cytochromes b and c occurs on the outer side of the membrane, and reduction of molecular oxygen always occurs on the inner side. The oxidation of methanol, formaldehyde, or succinate does not necessarily occur on the side of the membrane indicated. The broken line indicates the flow of electrons to oxygen from cytochrome b that occurs when 0.3m*M* KCN inhibits cytochrome a_3. FMN, flavoprotein of NADH dehydrogenase; Fe/S, iron-sulphur proteins of the NADH-ubiquinone oxidoreductase complex; Q, ubiquinone/ubiquinol; cyt, cytochrome; and ASC/TMPD, ascorbate/tetramethylenephenylenediamine. (From O'Keeffe, D.T., and Anthony, C., *Biochem. J.*, 170, 561, 1978. By courtesy of The *Biochemical Journal*, London).

to respiration of both endogenous and exogenous substrates. In addition to thiocyanate-inhibiting methanol oxidation, the concentration used had a marked effect on $\rightarrow H^+/O$ ratios. Despite some variability in values, the results suggested only one site of energy conservation for each organism after either heterotrophic or methylotrophic growth when oxidizing methanol, three for formaldehyde, and two to three for formate. Succinate-grown organisms generally showed three conservation sites with succinate as test substrate after growth on succinate, but only two sites after growth on methanol. These values are higher than those obtained by O'Keeffe and Anthony[105-107] which may reflect successful starvation. However, $\rightarrow H^+/O$ ratios of 10 were often observed in succinate-growth *Pseudomonas extorquens*, which prompts the question of whether a site o is operating or whether the number of protons extruded per electron pair passing each conservation site is always two, sometimes greater, or even a nonintegral number. Clearly, proton extrusion data must be interpreted with caution whether it be derived from unstarved organisms where the contribution of endogenous substrate respiration to the measured pulse is unknown, or from starved organisms whose physiological state may be questionable. It is at best an indication of possible P/O ratios and is most useful in comparative studies.

If, in spite of the above reservations, the scheme for electron transport in *Pseudomonas* AMI shown in Figure 4 proves to be correct, then during growth on methanol the P/O ratio would be two for formate oxidation, one for formaldehyde oxidation, and probably one for methanol oxidation, i.e., a total of four for complete oxidation of methanol to CO_2.[107] Netrusov and co-workers[114,115] have obtained cell-free membrane preparations capable of oxidative phosphorylation from two methylotrophs, *Pseudomonas* sp. 2 and *Pseudomonas methylica*. In the former bacterium, P/O ratios of about two were obtained for NADH and formate, and 0.54 for methanol. This data might be consistent with one coupling site for methanol and two or three for formate and NADH.[114] The latter organism is thought to possess three coupling sites for formate and NADH and one for methanol.[115]

Proton translocation studies with the obligate methanol utilizer, *Pseudomonas* EN suggest a P/O ratio of 1.5 associated with methanol, formaldehyde and formate oxidation.[10] The obligate methylotroph, *Methylomonas* P11, has recently been shown to contain ubiquinone-8, cytochrome *b*, and two *c*-type cytochromes (one of which is found partly in the soluble fraction after cell disruption).[116] In this case, formaldehyde and NADH are oxidized by dehydrogenases that join the respiratory chain at the level of ubiquinone, and both cytochromes *b* and *c* participate in the oxidation of methanol, formaldehyde, and NADH. The cytochrome *c* concentration was very high, which seems to be a common finding in methylotrophs.[98] However, the methylamine-utilizer, *Pseudomonas* MS, presents an interesting contrast in that it contains only about 10% of the amount of cytochrome *c* found in other species examined, and after disruption of this bacterium, it is all found in the soluble fraction.[104] A more detailed study of electron transport in this species would be worthwhile.

Methylophilus methylotrophus, an obligate methylotroph being used for single-cell protein production, shows marked differences in the nature of its cytochromes from the facultative species discussed above.[117] While the concentrations of cytochrome *b* and *c* are fairly constant, the concentration of cytochrome oxidase (aa_3) varied greatly with growth conditions and was sometimes undetectable. The cytochrome *c* did not have the high affinity for carbon monoxide observed in other methylotrophs,[98] but some was easily removed from the membranes. Two easily distinguishable cytochromes *b* were present. One of them combines with carbon monoxide and may be the alternative oxidase (cytochrome *o*) when cytochrome aa_3 is absent. Proton translocation measurements suggest that the presence of cytochrome aa_3 is associated with a proton translocating segment which is absent when this cytochrome is absent. This difference from the electron transport pathways in *Pseudomonas extorquens* and *Pseudomonas* AMI may reflect the different pathways used for formaldehyde oxidation. In the two facultative organisms, the process involves formaldehyde and formate dehydrogenase. The former enzyme is entirely absent from extracts of *Methylophilus methylotrophus*, and the latter is present with low activity. This organism uses the dissimilatory hexulose phosphate cycle for formaldehyde oxidation (Figure 2).[23,52] Therefore, in this species, about 44% of electron transport to oxygen must be from NAD(P)H. In contrast, for organisms which do not use this cycle, the corresponding figure is probably less than 6%.[117,118]

III. RESPIRATORY SYSTEMS OF METHANOTROPHS

There is far less detailed information concerning methanotroph respiratory systems than for the nonmethane-utilizing methylotrophs. This no doubt reflects, in part, reluctance to work with organisms which are grown on inflammable gas mixtures.

A. Primary Oxidation Pathways

Most evidence favors the following route for methane oxidation,

$$\underset{(1)}{CH_4 \rightarrow} \underset{(2)}{CH_3OH \rightarrow} \underset{(3)}{HCHO \rightarrow} \underset{(4)}{HCOOH \rightarrow} CO_2 \qquad (10)$$

Ultimately, it is from this sequence of reactions that energy is derived. A summary of the overall energetics of the sequence, the enzymes involved, and the methanotrophs in which they have been studied is shown in Table 2. At one time, it was thought that dimethyl ether might be an intermediate between methane and methanol, but this is now doubtful.[4] In some species, the dissimilatory hexulose phosphate cycle may operate for formaldehyde oxidation (Figure 2),[52] although proof of this awaits the appropriate ^{14}C-labeling experiments. Evidence for the operation of the oxidative pathway above has been extensively reviewed.[1-4]

1. *Methane Oxidation*

Much effort has been expended during the last 10 years in attempts to determine the mechanism of methane oxidation,[1,2,4] and it is now clear that a monooxygenase is involved at least in two strains of *Methylococcus capsulatus* (Texas and Bath strains), *Pseudomonas methanica* (Type-I organisms), and in *Methylosinus trichosporium* (a Type-II species).[4] The oxygen-18 experiments of Higgins and Quayle[119] using whole bacteria originally suggested involvement of an oxygenase, and Ribbons and Michalover were the first to demonstrate cell-free methane oxygenase activity.[120] Particulate preparations of *Methylococcus capsulatus* (Texas) showed activities consistent with involvement of such an enzyme, but they were both unstable and unpredictable. More detailed studies by Ribbons[121] revealed that these particulate preparations oxidize methane, methanol, and formaldehyde to formate. For methane oxidation, addition of NADH as electron donor is required. In this species, the enzymes that oxidize methane to formate are localized in membranes which can account for 40 to 60% of the total mass of the bacterium.[121]

In *Methylococcus capsulatus* (Bath), a NAD(P)H-linked methane monooxygenase is located entirely in the soluble fraction of disrupted organisms[122] after using essentially the same breakage procedures as those used by Ribbons.[121] The reaction catalyzed is:

$$CH_4 + O_2 + NAD(P)H + H^+ + O_2 \rightarrow CH_3OH + NAD(P)^+ + H_2O \qquad (11)$$

This enzyme shows a quite extraordinary lack of substrate specificity, oxidizing a wide range of hydrocarbons and related compounds.[123] It may possibly have originally evolved as an oxygen-scavenging enzyme in a primitive anaerobic methanotroph. The enzyme has been partially purified and is composed of three proteins. Component C has been most extensively characterized, being an iron flavoprotein of mol wt 44,600 and containing acid-labile sulfide. This component is thought to transfer electrons from NADH to the hydroxylase, probably protein A, which has a molecular weight of about 190,000, contains two iron atoms per molecule, is composed of four subunits of two types, and is thought to bind the substrate. Little is known about the third component, protein B, which has a molecular weight of about 20,000.

Ferenci[127] reported the isolation of cell-free particulate preparations which showed NADH-dependent methane monooxygenase activity from another Type-I methanotroph, *Pseudomonas methanica.* The enzyme also oxidises carbon monoxide which permitted the demonstration of monooxygenase stoichiometry,[128] in spite of the instability of these preparations. They can in fact be stored at −70°C[129] but are not sufficiently stable unfrozen to permit purification.

Methylosinus trichosporium is the only Type-II methanotroph in which methane monooxygenase has been studied, and the enzyme was first purified from this species in the author's laboratory.[99,130-133] A particulate, NADH-linked enzyme has recently been detected in a new facultative methanotroph which has many of the properties of a Type-II obligate methanotroph.[134] The *Methylosinus trichosporium* enzyme is a three component system. One of the components is a readily solubilized, carbon monoxide-binding high-redox-potential (+310mV) cytochrome *c* (mol wt 13,000) similar to the ones found in nonmethane-utilizing methylotrophs and discussed previously. Inhibition data suggested that this cytochrome may bind the substrate,[131] and subsequent component system. One of the components is a readily solubilized, carbon-monoxide-binding high-redox-potential (+310mV) cytochrome *c* (mol wt 13,000) similar to the ones found in nonmethane-utilizing methylotrophs and discussed previously. Inhibition data suggested that this cytochrome may bind the substrate.[131] and subsequent protein NMR and substrate-binding spectra studies support this hypothesis.[132] The other two components of the system are membrane bound, but can be solubilized quite readily. One is a copper-containing protein (mol wt 47,000), the other a small protein (mol wt 9400) which may have an electron transfer role or may be required to bind the other components into an enzymically active structure. This enzyme is not directly NADH linked, and when pure, electrons can be supplied from ascorbate, or the enzyme can be coupled to methanol oxidation using partially purified methanol dehydrogenase which reduces the cytochrome component of the oxygenase (see Section III.B). Like the *Methylococcus capsulatus* (Bath) enzyme, the methane monooxygenase of *Methylosinus trichosporium* shows broad substrate specificity, oxidizing a wide range of hydrocarbons.*[135,136]

The relationship between methane monooxygenases and associated electron transport systems is discussed in Section III.B.

* While the *Methylococcus capsulatus* (Bath) and *Methylosinus trichosporium* enzymes show similar broad specificities, the one in *Pseudomaonas methanica* has a narrower substrate range.[175,176] Recent studies in the author's laboratory and that of Dalton have revealed difficulties in reproducing ascorbate-linked activity in *Methylosinus trichosporium* and have suggested that under some growth conditions, the *Methylosinus trichosporium* enzyme may be similar to the *Methylococcus capsulatus* Bath system.[177,178] However, it is now known that the nature and location of the enzyme in *Methylosinus trichosporium* depends upon growth and storage conditions.[179]

TABLE 2

Reactions Involved in Oxidation of Methane by Methanotrophs

Step in methane oxidation	Number in scheme in text	$\Delta G'_{o}pH7$ (Kcal/mol) for reaction in far left column	Enzyme catalyzing reaction	Reaction catalyzed	Cofactor requirement	Microorganisms in which enzyme has been studied	Ref.
$CH_4 + \frac{1}{2}O_2 \rightarrow CH_3OH$	1	−26.12	Methane monooxygenase (hydroxylase)	$CH_4 + XH_2 + O_2 \rightarrow CH_3OH + H_2O + X$	NADH	*Methylococcus capsulatus*(Texas and Bath strains)	120—126 127—129
						Pseudomonas methanica	134
						Facultative isolate R6	
					NADH, methanol or ascorbate	*Methylosinus trichosporium*	99, 136—133
$CH_3OH + \frac{1}{2}O_2 \rightarrow HCHO + H_2O$	2	−44.18	Methanol dehydrogenase	$CH_3OH + PMS \rightarrow HCHO + PMSH_2$	Phenazine methosulphate	*Pseudomonas methanica*	41, 128
						Methylosinus sporium	39, 42, 137, 141
						Methylosinus trichosporium	130, 131, 141 141
						Methylocystic parvus	
						Methylococcus capsulatus(Texas)	137—141
$HCHO + \frac{1}{2}O_2 \rightarrow HCOO^- + H^+$	3	−57.15	Formaldehyde dehydrogenase	$HCHO + PMS + H_2O \rightarrow HCOOH + PMSH_2$	Phenazine methosulphate	*Methylococcus capsulatus*(Texas)	138, 140
						Methylosinus sporium	141
				$HCHO + NAD^+ + H_2O \xrightarrow{GSH} HCOOH + NADH + H^+$	NAD^+, reduced glutathione	*Pseudomonas methanica*	145
				$HCHO + NAD^+ + H_2O \rightarrow HCOOH + NADH + H^+$	$NAD(P)^+$	*Methylococcus capsulatus*(Bath)	146

TABLE 2 (continued)

Reactions Involved in Oxidation of Methane by Methanotrophs

Step in methane oxidation	Number in scheme in text	ΔG′opH7 (Kcal/mol) for reaction in far left column	Enzyme catalyzing reaction	Reaction catalyzed	Cofactor requirement	Microorganisms in which enzyme has been studied	Ref.
				$HCHO + DCPIP + H_2O \rightarrow HCOOH + DCPIPH_2$	Dichlorophenolindophenol	*Pseudomonas methanica*	41
$HCOO^- + H^+ + \frac{1}{2}O_2 \rightarrow CO_2 + H_2O$	4	−58.25	Formate dehydrogenase	$HCOOH + NAD^+ \rightarrow CO_2 + NADH + H^+$	NAD^+	*Pseudomonas methanica*	41
						Methylococcus capsulatus (Texas)	137
						Facultative isolate R6	134

2. Methanol Oxidation

Methanotrophs possess NAD-independent methanol dehydrogenase enzymes similar to those studied in other methylotrophs (Section II). The enzyme from *Methylococcus capsulatus* (Texas) has been extensively purified.[137-139] It catalyzes the oxidation of methanol and formaldehyde to formate and is composed of two subunits, each of mol wt 62,000, into which it dissociates at acid pH. In this bacterium, no other mechanism has been demonstrated for oxidation of formaldehyde to formate, and so presumably, the enzyme fulfills a dual role. It is closely similar in amino acid composition to the enzyme from the facultative methylotroph *Pseudomonas* M27. Wadzinski and Ribbons[140] have also purified the *Methylococcus* enzyme and have shown that it is probably membrane bound in vivo. About 60% of the enzyme was found in particle fractions after cell breakage, and methanol- and formaldehyde-oxidase activities were present showing connection of the enzyme to electron transport components.

Patel and Felix[141] examined several species of Type-I and Type-II methanotrophs for distribution of methanol dehydrogenase and methanol oxidase activities. Type-I organisms showed a distribution of methanol dehydrogenase between soluble and particulate fractions. Methanol oxidase activity resided entirely in particulate fractions. Type-II species showed only soluble methanol dehydrogenase and no oxidase activity. However, Tonge and co-workers[130,131] found methanol oxidase activity distributed between particulate and supernatant fractions obtained from sonicated *Methylosinus trichosporium*, but Weaver and Dugan[142] found all methanol dehydrogenase activity in soluble fractions from this species. These discrepancies are probably due to the different breakage methods used. Methanol dehydrogenase has now been purified and crystallized from the closely related species, *Methylosinus sporium*.[141] Again, it is similar to the enzymes studied in other methylotrophs, and formate is the reaction product from methanol, but there are some minor differences in properties compared with enzymes from other species. The *Methylosinus sporium* enzyme oxidizes various primary alcohols at similar rates, whereas in enzymes purified from other species, the rate decreases as the alkyl chain length of the alcohol increases.[39,42,138] In addition, this enzyme gave only weak precipitin bands with antibodies to enzymes from Type-I organisms, but much stronger bands with those to enzymes from Type-II species. Immunochemically, the methanol dehydrogenases from Type-II methanotrophs are more closely related to those from facultative methanol utilizers than to those from Type-I methanotrophs.[138]

The methanol dehydrogenase of the facultative methanotroph, *Methylobacterium organophilum* has recently been purified.[143] Again, it oxidizes methanol to formate, but unlike the enzymes from other species, it is not dependent on ammonia. Its fluorescence spectrum suggests that it may have a different prosthetic group. It is serologically related to alcohol dehydrogenases from *Hyphomicrobium, Pseudomonas methanica*, and *Methylosinus trichosporium*, but not to ones from *Rhodopseudomonas acidophila* and *Methylococcus* sp. An enzyme with ammonia dependence and slightly different substrate specificity is found in the recently isolated facultative methanotroph, isolate R6.[134]

3. Formaldehyde Oxidation

As discussed in Section III.A.2, all methanol dehydrogenases also oxidize formaldehyde. In the case of *Methylococcus capsulatus* (Texas), this may be the only mechanism for formaldehyde oxidation. In methanotrophs containing other formaldehyde dehydrogenase activities, we have no knowledge of their importance relative to the methanol dehydrogenase in the overall process of formaldehyde oxidation. The enzymes necessary for the oxidation of formaldehyde by the dissimilatory hexulose phosphate cycle (Figure 2) are present in a number of methanotrophs,[52,144] but the isotopic labeling experiments required to test whether it operates in vivo have not been reported.

A NAD-linked formaldehyde dehydrogenase which requires reduced glutathione has been reported in *Pseudomonas methanica.*[145] It catalyzes the reaction:

$$HCHO + NAD^+ + H_2O \xrightarrow[\text{glutathione}]{\text{reduced}} HCOOH + NADH + H^+ \quad (12)$$

An enzyme which oxidises formaldehyde in the presence of the artificial electron acceptor, dichlorophenol-indophenol has been described in another strain of this species:[41]

$$HCHO + DCPIP + H_2O \rightarrow HCOOH + DCPIPH_2 \quad (13)$$

In the case of *Pseudomonas methanica*, therefore, there may be three, or even four, mechanisms for oxidizing formaldehyde.

A high specific activity, glutathione-independent, $NAD(P)^+$-dependent formaldehyde dehydrogenase has recently been purified from *Methylococus capsulatus* (Bath).[146] A similar enzyme, but of much lower specific activity is present in extracts of *Methylosinus trichosporium*. However, in this case, the Km for formaldehyde is high (65 m*M*). It is, therefore, of doubtful physiological significance in this species.[147] A phenazine methosulphate-linked, ammonium ion dependent formaldehyde dehydrogenase capable of oxidizing a variety of aldehydes has been detected recently in the facultative methanotroph, isolate R6.[134] In most methylotrophs, and especially in methanotrophs, our understanding of formaldehyde oxidation is incomplete, especially with regard to the relative importance of the various enzymes found to oxidize formaldehyde and the mechanisms by which they are regulated.

4. Formate Oxidation

NAD-linked formate dehydrogenases appear to be responsible in methanotrophs for the oxidation of formate to carbon dioxide as is the case in other methylotrophs. However, formate dehydrogenase activity has only been examined in a few methanotrophs It is present in *Pseudomonas methanica,*[41] *Methylococcus capsulatus* (Texas),[137] and the facultative organism, isolate R6.[134]

B. Electron Transport Systems and Energy Transduction

There is much less published data concerning electron transport in methanotrophs than in other methylotrophs, but among the methanotrophs, *Methylosinus trichosporium* is the best understood in this respect. Davey and Mitton[148] examined whole organisms of the Type I species, *Methylomonas albus*, and *Methylosinus trichosporium*, a Type II organism, and concluded that these bacteria had qualitatively similar cytochrome patterns. They contained cytochromes *a* and *c*, but there was no clear indication of cytochrome *b* from different spectra. However, there was strong absorbance at 416 nm in carbon monoxide binding spectra in addition to the cytochrome a_3 band. This resembled an earlier finding in *Methylococcus capsulatus* (Texas),[1] and it was concluded that the band was due to a cytochrome *o*.[148] The presence of cytochromes *c* and *a* in *Methylosinus trichosporium* was confirmed by Weaver and Dugan,[142] who also demonstrated by using gentle disruption techniques that these cytochromes are located in the membrane fractions together with the ATPase activity. Monosov and

Netrusov[149] demonstrated the presence of cytochrome *b* in addition to cytochromes *a* and *c* in both *Methylosinus trichosporium* and *Methylomonas agile*. The terminal oxidase and ATPase activities are present in the intracytoplasmic membranes.[149]

Methylosinus trichosporium and *Pseudomonas methanica* were included in a comparative study of methylotroph cytochrome systems by Tonge and co-workers.[98,99] After growth on methane or methanol, the cytochrome complements were closely similar and somewhat unusual in that after sonication, the concentrations of particle-bound cytochromes *a*, *b*, and *c* were low, but soluble fractions contained a high concentration of an unusual carbon monoxide-binding cytochrome *c*. In the case of *Methylosinus trichosporium*, this was subsequently shown to be a component of the methane mono-oxygenase enzyme system.[99,101,130,131,133] It is likely that it was the CO-binding spectrum of this cytochrome that was thought by Davey and Mitton[143] to be due to a cytochrome *o*. If any of the relatively small amount of cytochrome *b* in the particle fractions after cell disruption bound CO, the spectrum would have been obscured by absorbance due to the CO-binding spectrum of the large amounts of cytochrome *c* present. Whether cytochrome *o* is present in methanotrophs is, therefore, still uncertain. Cytochromes c_{co} with similar properties to the one studied in *Methylosinus trichosporium* have been purified from *Methylococcus capsulatus*[150] and the facultative species, *Methylobacterium organophilum*.[143]

There is little information concerning other electron transport components in methanotrophs, but ubiquinone-10 and flavoproteins are present in *Methylosinus trichosporium*.[101,113] Inhibitor studies with *Methylosinus trichosporium* and *Pseudomonas methanica* suggest that there are two terminal oxidases in both bacteria, probably cytochrome aa_3 and cytochrome c_{co}.[98,99,131,133,151]

Methylosinus trichosporium is the only methanotroph in which measurements of respiration-induced proton extrusion have been successfully made.[152] Normal pulses could not be measured in *Methylococcus capsulatus*,[153] and this may reflect differences in the arrangement of the intracytoplasmic membranes (and hence energy coupling sites) between the two species. Indeed, there may have been few (if any) membranes present in the *Methylosinus trichosporium* used (see Section IV). Data obtained with this bacterium suggests a P/O ratio of one for methane, methanol, formaldehyde, and formate. A value of one for formate is unexpected and suggests that two of the three proton translocating loops normally associated with NADH oxidation may be missing. However, in this species, the formate dehydrogenase reaction appears to be the only NADH-generating step during the oxidation of methane to carbon dioxide. Since most of this reducing power is probably required for CO_2 fixation and biosynthesis, there will be little NADH available for energy transduction and, hence, little selective advantage in developing more than one coupling site. The apparent P/O ratio of one for methane is interesting since during the oxidation of methane to methanol, half the oxygen consumed is incorporated into the product rather than acting as an electron acceptor. This suggests that the true P/O ratio might be two, implying an extra coupling site for electrons derived from the further oxidation of methanol and used in the mono-oxygenase reaction. However, recent growth-yield data for *Methylococcus capsulatus* strongly suggest that the methane to methanol step is energetically "neutral," i.e., it does not require ATP or consume reducing power that would otherwise be used for energy transduction, neither is any extra ATP generated over and above that obtained when methanol is substrate.[154,155] While this may be so for *Methylococcus capsulatus*, other species may differ, especially in the case of *Methylosinus trichosporium* where the methane oxygenase may not be NADH linked in vivo.

A possible scheme for electron transport and energy transduction in *Methylosinus trichosporium*, proposed largely on the basis of studies with electron transport inhibitors, is shown in Figure 5.[4,99,130,133] The scheme shows a recycling mechanism for sup-

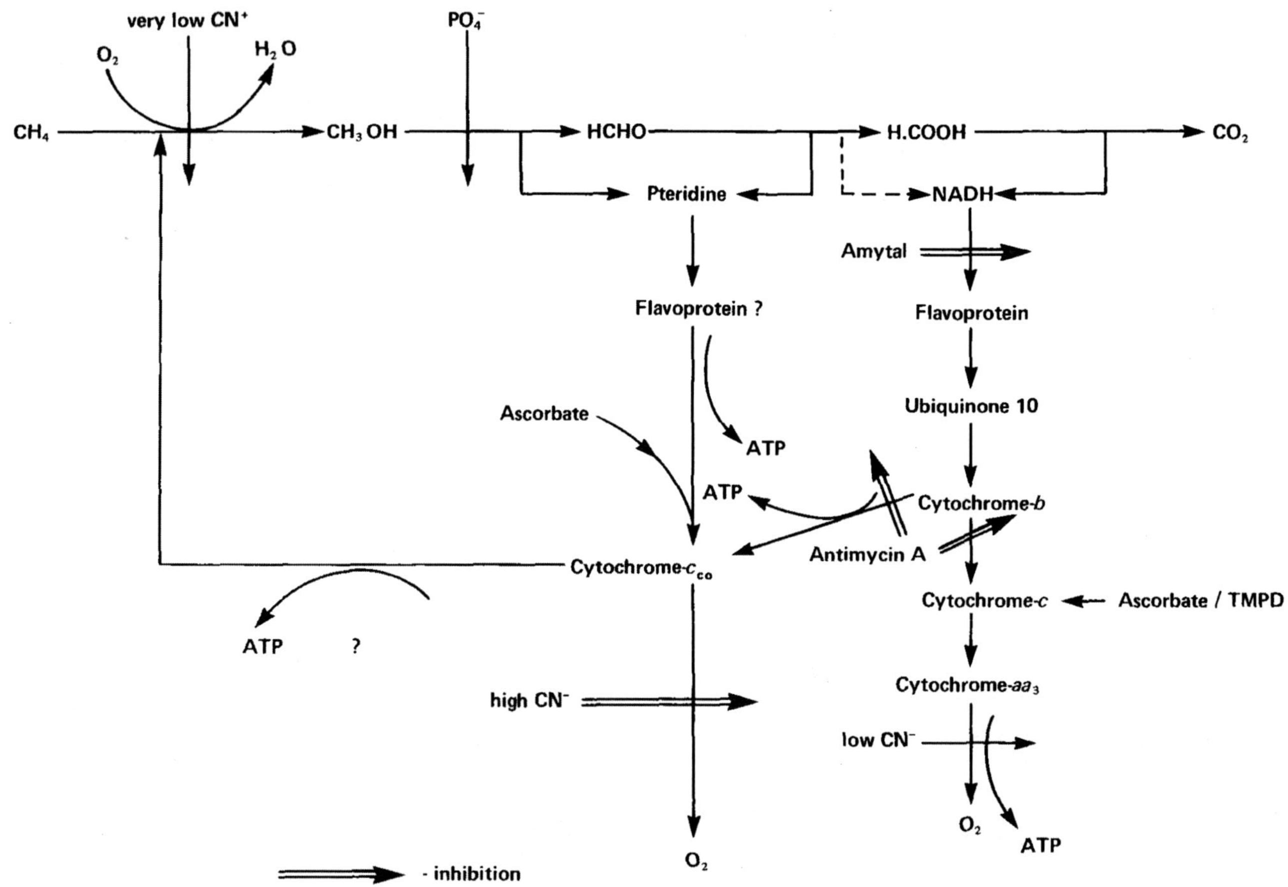

FIGURE 5. Tentative scheme for electron transport and energy transduction in *Methylosinus trichosporium*. (From Wolfe, R.S. and Higgins, I.J., in *MTP International Review of Biochemistry, series II: Microbial Biochemistry, Quayle, J.R., Ed.*, 1979. Courtesy of University Park Press, Baltimore.)

plying electrons from the further oxidation of methanol and formaldehyde to the methane monooxygenase.* Methanol addition causes methane disappearance in crude extracts,[99,130] methanol in the presence of partially purified methanol dehydrogenase will supply the reducing power for purified methane monooxygenase,[131] and the cytochrome c_{co}, while being a component of the monooxygenase, is an electron acceptor for reducing power from methanol dehydrogenase. Therefore, this proposal seems likely.[101] Clearly, NADH must be capable of supplying reducing power to the oxygenase, otherwise growth on methane could not be initiated. In crude extracts, NADH does function as an electron donor. Without the recycling mechanism, there is some difficulty in this species in accounting for an adequate supply of NAD(P)H to drive the monooxygenase reaction since it appears that only on NADH molecule is generated per molecule of methane oxidized to carbon dioxide. Since carbon is incorporated at the level of formaldehyde, substantially less than one NADH can be generated per methane molecule oxidized. Therefore, in this species, an obligatorily NAD(P)H-linked monooxygenase would necessitate reversed electron transport as pointed out by van Dijken and Harder.[9] Such a process would probably be energetically expensive and perhaps, inconsistent with molar growth yield data (but see Section IV). This argument would not be valid if a dissimilatory hexulose phosphate cycle was present, but there is no evidence for this in Type II methanotrophs. The scheme (Figure 5) includes two c-type cytochromes, although it is not yet clear whether there are two present. In the absence of methane, it is thought that the cytochrome c_{co} can function as an oxidase.

There is evidence from whole-organism studies that the *Pseudomonas methanica* methane monooxygenase may not be obligatorily NAD(P)H-linked since ethanol stimulates carbon monoxide oxidation by the methane monooxygenase, and this alcohol is oxidized to acetate by NAD(P)$^+$-independent enzymes.[127,128] This led Ferenci et al.[128] to suggest two possible mechanisms of electron supply to the enzyme (Figure 6). Route (a) involves reverse electron transport, while route (b) is analogous to the one proposed for *Methylosinus trichosporium* (Figure 5). However, ethanol would not serve as the reductant in cell-free systems, perhaps due to partial disruption of the electron transport chain. The reversed electron transport scheme (route (a), Figure 6) does not necessarily require ATP. It depends on the half reduction potential of the X-XH_2 couple.[128] The elucidation of the nature of in vivo electron supply to the methane monooxygenase of *Pseudomonas methanica* awaits further studies of electron transport in this species.

There is no evidence for an electron recycling system, or even for alternative electron donors, for the NAD(P)H-linked methane monooxygenases of *Methylococcus capsulatus* (strains Texas and Bath).[121,122] There are close similarities between the properties of the particulate methane monooxygenases of *Methylosinus trichosporium, Pseudomonas methanica,* and *Methylococcus capsulatus* (Texas),[4] and it seems possible, therefore, that in vivo there may be an electron recycling system in all three species. However, the soluble *Methylococcus capsulatus* (Bath) enzyme seems to be quite different, sharing only the property of broad substrate specificity with the *Methylosinus* enzyme. The demonstration of a highly active NAD(P)$^+$-linked formaldehyde dehydrogenase in *Methylococcus capsulatus* (Bath)[146] suggests that there is sufficient NADH generated from the further oxidation of methanol to drive an NADH-linked methane monooxygenase. In this species, therefore, reversed electron transport or electron re-

* In view of recent uncertainties that have arisen concerning the nature of the *Methylosinus trichosporium* methane monooxygenase, the evidence for the recycling mechanism is in question and hence so also are the implications of such a mechanism. This scheme and its implications should therefore be regarded as highly speculative.

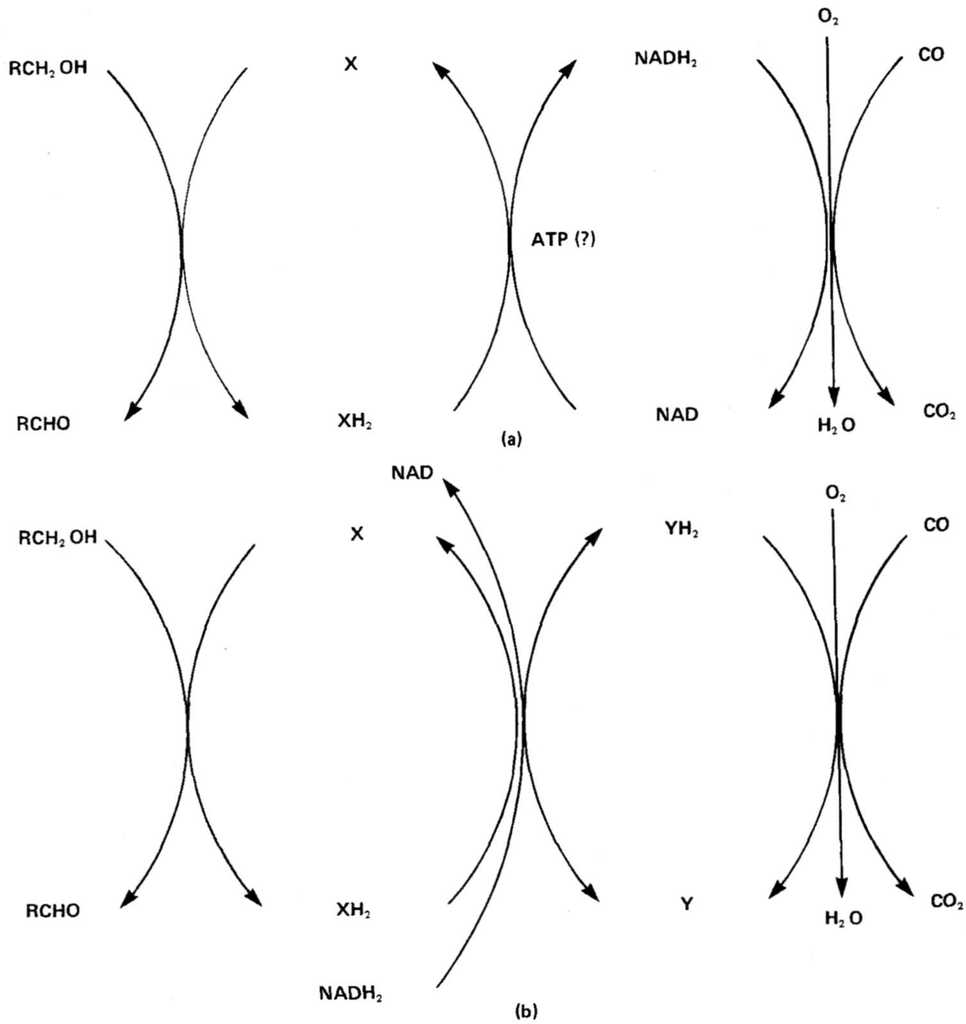

FIGURE 6. Possible routes for the channeling of reductant from alcohol oxidation to the CO-mono-oxygenase of *Pseudomonas methanica*. X represents the unknown physiological acceptor for the reactions catalyzed by the primary alcohol dehydrogenase, and Y represents a postulated carrier or set of electron carriers between NADH and carbon monoxide (methane) mono-oxygenase. (From Ferenci, T., Strφm, T., and Quayle, J.R., *J. Gen. Microbiol.*, 91, 79, 1975. By courtesy of the *Journal of General Microbiology*, Reading.)

cycling are unnecessary. If there is no recycling system, the molar growth yield on methane would probably be lower than that on methanol, but yield determinations are not available for this strain.

IV. PHYSIOLOGICAL ASPECTS AND GROWTH YIELDS

This discussion would not be complete without reference to studies of respiration in whole bacteria. Although there have been a number of publications containing figures for rates of substrate oxidation (e.g., Higgins and Quayle[119]), there is little information concerning substrate affinity. However, Harrison[156] found that *Pseudomonas extorquens* has an extremely low Km for methanol (20 μM). A methane-utilizing pseudomonad had a similarly low Km for methane (26 μM). Of possible relevance to affinity for methane are the intracytoplasmic membranes present in methanotrophs. It is generally accepted that all methane-utilizing bacteria contain these membrane systems,[1,2,4] and there is much evidence that the enzymes involved in oxidizing methane to formate, the electron transport components, and the energy transducing systems are located in membranes. However, there are few reports which distinguish between activities found in the pericytoplasmic membrane and those in the intracytoplasmic membranes. Weaver and Dugan,[142] using a gentle, low-efficiency disruption technique, probably succeeded in separating the two types of membrane from *Methylosinus trichosporium*, and their data showed the presence of the cytochrome system, NADH-cytochrome *c* reductase, and ATPase activities in the intracytoplasmic membranes. However, it was not shown whether or not these systems were present also in the periplasmic membrane. Intracytoplasmic membranes are clearly not usually necessary for growth on methanol since most facultative methylotrophs are devoid of them with the exception of *Hyphomicrobium.*[157] The facultative methanotroph, *Methylobacterium organophilum*, contains membranes when grown on methane, but not when grown on glucose or methanol.[15,158] In addition, there is an increase in the amount of membranes present in response to low oxygen tension during growth on methane.[158] The obligate methane utilizer, *Methylococcus capsulatus,* when grown on methanol retains the membranes characteristic of growth on methane.[155] Yet in *Methylosinus trichosporium*, membranes are not present in continuous cultures growing on methane, but are present in organisms growing in batch culture. They are synthesized apparently in response to oxygen limitation.[159] Intracytoplasmic membranes are not, of course, unique to methane utilizers. They are found in photosynthetic bacteria,[160] ammonia and nitrite oxidizers,[161,162] blue-green algae,[163] and in some higher hydrocarbon utilizers.[164,165] There has been much speculation about the role of these membranes, and there is good evidence that they are involved in oxidation processes, energy transduction, and perhaps, in some biosynthetic processes.[4] At least in *Methylosinus trichosporium*, they do not seem to be obligatory for growth on methane. Samples of this organism rich in intracytoplasmic membranes show slightly higher respiration rates than those without,[159] but this effect is small. It is possible that they are synthesized in order to decrease the Km for substrate, oxygen, or both, or that they increase chemiosmotic coupling efficiency, which might be expected if protons are translocated into an internal cavity rather than across the pericytoplasmic membrane.

There have recently been several detailed analyses of theoretical cell yields in methylotrophs,[9-11,154,166,167] and reviews and recent determinations of actual yields.[9,10,155,167-169] Although accurate yields on methane are relatively difficult to obtain, most recent data is sound, and there is a fair level of agreement between predicted and actual yields for both methane and methanol. A detailed analysis of cell yields is not within the scope of this review. However, since the mechanisms of respiration and energy transduction are major factors in determining yields, the subject is discussed briefly here. Since carbon incorporation pathways in methylotrophs are now well documented,[4] it is relatively easy to calculate requirements for ATP and reducing power for biosynthesis and how the different pathways and their variations will affect relative yields. However, it should be clear from the rest of this review that our understanding of the bioenergetics of methylotrophs is far from complete. We have a reasonable idea of the likely ATP and NAD(P)H yields during the oxidation of carbon-one substrates for a few species, but for most there is insufficient data. Anthony[11] has recently published a thorough analysis of growth-yield prediction in methylotrophs and draws an important conclusion, namely that ATP yield from substrate oxidation is not, in many methylotrophs, as important a factor in determining yields as is often assumed. Predictions of P/O ratios from cell-yield measurements are often not possible in these organisms because these values are not directly related. This is due to the fact that many methylotrophs are similar to autotrophs in that growth yield is more dependent on the potential yield of reducing equivalents than on the ATP supply. This is particularly true of Type II methanotrophs, facultative methylotrophs that use the serine pathway for biosynthesis, and for methanotrophs that have a methane monooxygenase obligatorily dependent on NAD(P)H. In all these organisms, the concept of Y_{ATP} loses the significance it has in a "normal heterotroph" and estimates of Y_{ATP} from growth yields and P/O ratios are invalid.[11] This is because growth is limited by NAD(P)H supply and not by ATP yield.

Recent reliable measurements of molar growth yields on methane for *Methylococcus capsulatus* show values of about 0.8 gm/mol with either methane or methanol as substrate,[155,166] which is closely similar to values obtained with facultative methylotrophs growing on methanol.[170] Our current knowledge of the biochemistry of *Methylococcus capsulatus* might lead us to expect a higher molar growth yield on methanol than on methane, although it is difficult to predict the possible toxic effects of methanol on the yield and the effect of oxygenation of methanol by methane monooxygenase.[123] The reason for this unexpected similarity in yields on methane and methanol may well be found in the mechanisms of regulation of methylotrophic energy metabolism of which we know very little.

ACKNOWLEDGMENT

I would like to thank Dr. R. C. Hammond for valuable discussions on proton translocation in methylotrophs.

REFERENCES

1. **Ribbons, D. W., Harrison, J. E., and Wadzinski, A. M.,** Metabolism of single carbon compounds, *Annu. Rev. Microbiol.*, 24, 135, 1970.
2. **Quayle, J. R.,** The metabolism of one-carbon compounds by microorganisms, *Adv. Microbiol Physiol.*, 7, 119, 1972.
3. **Anthony, C.,** The biochemistry of methylotrophic micro-organisms, *Sci. Prog. Oxf.*, 62, 167, 1975.
4. **Wolfe, R. S. and Higgins, I. J.,** The microbial production and utilization of methane — a study in contrasts, in *MTP International Review of Biochemistry, Series II, Microbial Biochemistry*, 21, 267, 1979.
5. **Tannenbaum, S. R.,** Single cell protein, in *Food Proteins*, Whitaker, J. R. and Tannenbaum, S. R., Eds., AVI Publishing, Westport, Conn., 1977, 315.
6. **Mateles, R. I.,** The physiology of biomass production, in *Symp. Soc. Gen. Microbiol.*, 29, 29, 1979.
7. **Higgins, I. J. and Hill, H. A. O.,** Microbial generation and interconversion of energy sources, *Symp. Soc. Gen. Microbiol.*, 29, 359, 1979.
8. **Anthony, C. and Zatman, L. J.,** The microbial oxidation of methanol. II. The methanol-oxidising enzyme of *Pseudomonas sp.* M27, *Biochem. J.*, 92, 614, 1964.
9. **van Dijken, J. P. and Harder, W.,** Growth yields of microorganisms on methanol and methane. A theoretical study, *Biotechnol. Bioeng.*, 17, 15, 1975.
10. **Barnes, L. J., Drozd, J. W., Harrison, D. E. F., and Hamer, G.,** Process considerations and techniques specific to protein production from natural gas, in *Microbial Production and Utilization of Gases*, Schlegel, H. G., Gottschalk, G., and Pfennig, N., Eds., E. Goltze, Gottingen, Federal Republic of Germany, 1975, 301.
11. **Anthony, C.,** The prediction of growth yields in methylotrophs, *J. Gen. Microbiol.*, 104, 91, 1978.
12. **Dworkin, M., and Foster, J. W.,** Studies on *Pseudomonas methanica* (Sohngen). Nov. Comb., *J. Bacteriol.*, 72, 646, 1956.

13. **Brown, L. R., Strawinski, R. J., and McCleskey, C. S.**, The isolation and characterization of *Methanomonas methanooxidans*, Brown and Strawinski, *Can. J. Microbiol.*, 10, 791, 1964.
14. **Whittenbury, R., Phillips, K. C., and Wilkinson, J. F.**, Enrichment, isolation and some properties of methane-utilizing bacteria, *J. Gen. Microbiol.*, 61, 205, 1970.
15. **Patt, T. E., Cole, G. C., Bland, J., and Hanson, R. S.**, Isolation and characterization of bacteria that grow on methane and organic compounds as sole sources of carbon and energy, *J. Bacteriol.*, 120, 955, 1974.
16. **Patt, T. E., Cole, G. C., and Hanson, R. S.**, *Methylobacterium*, a new genus of facultatively methylotrophic bacteria, *Int. J. Syst. Bacteriol.*, 26, 226, 1976.
17. **Colby, J. and Zatman, L. J.**, Hexose phosphate synthase and tricarboxylic acid-cycle enzymes in Bacterium 4B6, an obligate methylotroph, *Biochem. J.*, 128, 1373, 1972.
18. **Colby, J. and Zatman, L. J.**, Trimethylamine metabolism in obligate and facultative methylotrophs, *Biochem. J.*, 132, 101, 1973.
19. **Sahm, H. and Wagner, F.**, Isolation and characterisation of an obligate methanol-utilising bacterium *Methylomonas* M-15, *Eur. J. Microbiol.*, 2, 147, 1975.
20. **Raczynska-Bojanowska, K., Lukaszkiewicz, Z., Michalik, J., and Kurzatkowski, W.**, *Eur. J. Appl. Microbiol.*, in press, 1980.
21. **Kouno, K., Oki, T., Nomura, H., and Ozaki, A.**, Isolation of new methanol-utilising bacteria and its thiamine requirement for growth, *J. Gen. Appl. Microbiol.*, 19, 11, 1973.
22. **Byrom, D. and Ousby, J. C.**, Identification of a methanol oxidising Pseudomonad, in *Microbial Growth on* C_1 *Compounds*, Terui, G., Ed., Society of Fermentation Technology, Tokyo, 1975, 23.
23. **Taylor, I. J.**, Carbon assimilation and oxidation by *Methylophilus methylotrophus* — the ICI SCP organism, in *Microbial Growth on* C_1 *Compounds*, Skyrabin, G. K., Ivanov, M. V., Kondratjeva, E. N., Zavarzin, G. A., Trotsenko, Yu.A., and Nesterov, A. I., Eds., U.S.S.R. Academy of Sciences, Pushchino, 1977, 52.
24. **Dahl, J. S., Mehta, R. J., and Hoare, D. S.**, New obligate methylotroph, *J. Bacteriol.*, 109, 916, 1972.
25. **Myers, P. A. and Zatman, L. J.**, The metabolism of trimethylamine *N*-oxide by *Bacillus* PM6, *Biochem. J.*, 121, 10P, 1971.
26. **Hampton, D., and Zatman, L. J.**, The metabolism of tetramethylammonium chloride by Bacterium 5H2, *Biochem. Soc. Trans.* 1, 667, 1973.
27. **Harder, W., Attwood, M. M., and Quayle, J. R.**, Methanol assimilation by *Hypomicrobium sp.*, *J. Gen. Microbiol.*, 78, 155, 1973.
28. **Peel, D. and Quayle, J. R.**, Microbial growth on C_1 compounds. I. Isolation and characterization of *Pseudomonas* AM1, *Biochem. J.*, 81, 465, 1961.
29. **Eady, R. R., Jarman, J. R., and Large, P. J.**, Microbial oxidation of amines. Partial purification of a mixed function secondary-amine oxidase system from *Pseudomonas aminovorans* that contains enzymically active cytochrome-P-420 type haemoprotein, *Biochem. J.*, 125, 449, 1971.
30. **Shaw, W. V., Tsai, L., and Stadtman, E. R.**, The enzymatic synthesis of *N*-methylglutamic acid, *J. Biol. Chem.*, 241, 935, 1966.
31. **Kung, H. F. and Wagner, C.**, Oxidation of C_1 Compounds by *Pseudomonas sp.* MS, *Biochem. J.*, 116, 357, 1970.
32. **Anthony, C. and Zatman, L. J.**, The microbial oxidation of methanol. I. Isolation and properties of *Pseudomonas sp.* M27, *Biochem. J.*, 92, 609, 1964.
33. **Lawrence, A. J. and Quayle, J. R.**, Alternative carbon assimilation pathways in methane-utilizing bacteria, *J. Gen. Microbiol.*, 63, 371, 1970.
34. **Kemp, M. B. and Quayle, J. R.**, Microbial growth on C_1 compounds. Uptake of (^{14}C) formaldehyde and (^{14}C) formate by methane grown *Pseudomonas methanica* and determination of the hexose labelling pattern after brief incubation with (^{14}C) methanol, *Biochem. J.*, 102, 94, 1967.
35. **Large, P. J., Peel, D., and Quayle, J. R.**, Microbial growth on C_1 compounds. II. Synthesis of cell constituents by methanol — and formate — grown *Pseudomonas* AM1, and methanol-grown *Hyphomicrobium vulgare*, *Biochem. J.*, 81, 470, 1961.
36. **Lawrence, A. J., Kemp, M. B., and Quayle, J. R.**, Synthesis of cell constituents by methane-grown *Methylococcus capsulatus* and *Methanomonas methanooxidans*, *Biochem. J.*, 116, 631, 1970.
37. **Panganiban, Jr., A. T., Patt, T. E., Hart, W., and Hanson, R. S.**, Oxidation of methane in the absence of oxygen in lakewater samples, *Appl. Environ. Microbiol.*, 37, 303, 1979.
38. **Anthony, C. and Zatman, L. J.**, The microbial oxidation of methanol. The alcohol dehydrogenase of *Pseudomonas sp.* M27, *Biochem. J.*, 96, 808, 1965.
39. **Anthony, C. and Zatman, L. J.**, The microbial oxidation of methanol. Purification and properties of the alcohol dehydrogenase of *Pseudomonas sp.* M27, *Biochem. J.*, 104, 953, 1967.

40. **Anthony, C. and Zatman, L. J.**, The Microbial oxidation of methanol. The prosthetic group of the alcohol dehydrogenase of *Pseudomonas sp.* M.27: a new oxidoreductase prosthetic group, *Biochem. J.*, 104, 960, 1967.
41. **Johnson, P. A. and Quayle, J. R.**, Microbial growth on C_1 compounds. VI. Oxidation of methanol, formaldehyde and formate by methanol-grown *Pseudomonas* AMI, *Biochem. J.*, 93, 281, 1964.
42. **Sperl, G. T., Forrest, H. S., and Gibson, D. T.**, Substrate specificity of the purified primary alcohol dehydrogenase from methanol-oxidising bacteria, *J. Bacteriol.*, 118, 541, 1974.
43. **Michalik, J. and Raczynska-Bojanowska, K.**, Oxidation of methanol by facultative and obligate methylotrophs, *Acta Biochim. Pol.*, 23, 375, 1976.
44. **Goldberg, I.**, Purification and properties of a methanol-oxidising enzyme in *Pseudomonas* C., *Eur. J. Biochem.*, 63, 233, 1976.
45. **Yamanaka, K. and Matsumoto, K.**, Purification, crystallisation and properties of primary alcohol dehydrogenase from a methanol-oxidising *Pseudomonas sp.* No. 2941, *Agric. Biol. Chem.*, 41, 467, 1977.
46. **Ghosh, R. and Quayle, J. R.**, Purification and physical properties of the methanol dehydrogenase from *Methylophilus methylotrophus*, *Proc. Soc. Gen. Microbiol.*, 5, 42, 1978.
47. **Duine, J. A., Frank, J., and Westerling, J.**, Purification and properties of methanol dehydrogenase from *Hyphomicrobium X. Biochim. Biophys. Acta*, 524, 277, 1978.
48. **Sperl, G. T., Forrest, H. S., and Gibson, D. T.**, Substrate specificity of the purified primary alcohol dehydrogenases from methanol-oxidising bacteria, *J. Bacteriol.*, 118, 541, 1974.
49. **Large, P. J. and Quayle, J. R.**, Microbial growth on C_1 compounds. V. Enzyme activities in extracts of *Pseudomonas* AMI, *Biochem. J.*, 87, 386, 1963.
50. **Heptinstall, J. and Quayle, J. R.**, Pathways leading to and from serine during growth of *Pseudomonas* AMI on C_1 compounds or succinate, *Biochem. J.*, 117, 563, 1970.
51. **Dunstan, P. M., Anthony, C., and Drabble, W. T.**, Microbial metabolism of C_1 and C_2 compounds. The involvement of glycollate in the metabolism of ethanol and of acetate by *Pseudomonas* AMI, *Biochem. J.*, 128, 99, 1972.
52. **StrØm, T., Ferenci, T., and Quayle, J. R.**, The carbon assimilation pathways of *Methylococcus capsulatus, Pseudomonas methanica* and *Methylosinus trichosporium* (OB3b) during growth on methane, *Biochem. J.*, 144, 465, 1974.
53. **Colby, J. and Zatman, L. J.**, Tricarboxylic acid cycle and related enzymes in restricted facultative methylotrophs, *Biochem. J.*, 148, 505, 1975.
54. **Colby, J. and Zatman, L. J.**, Enzymological aspects of the pathways of trimethylamine oxidation and C_1 assimilation in obligate methylotrophs and restricted facultative methylotrophs, *Biochem. J.*, 148, 513, 1975.
55. **Ben-Bassat, A. and Goldberg, I.**, Oxidation of C_1-compounds by *Pseudomonas C*, *Biochim. Biophys. Acta*, 497, 586, 1977.
56. **Kaneda, T. and Roxburgh, J. M.**, A methanol-utilising bacterium. II. Studies on the pathway of methanol assimilation, *Can. J. Microbiol.*, 5, 187, 1959.
57. **Netrusov, A. I.**, Formate dehydrogenase and NADH-oxidase in Pseudomonas species oxidising methanol, *Biol. Nauki (Moscow)*, 16, 99, 1973.
58. **Rodionov, Yu.V., Avilova, T. V., Zakharova, E. V., Platonenkova, L. S., Egorov, A. M., and Berezin, I. V.**, Purification and basic properties of NAD-dependent formate dehydrogenase from methylotrophic bacteria, *Biokhimiya*, 42, 1497, 1977.
59. **Rodionov, Yu. V., Avilova, T. V., and Popov, V.**, NAD-dependent formate dehydrogenase from methylotrophic bacteria. Study of the properties and stability of soluble and immobilised enzymes, *Biokhimiya*, 42, 2020, 1977.
60. **Kato, N., Kano, M., Tani, K., and Ogata, K.**, Purification and characterization of formate dehydrogenase in a methanol-utilizing yeast, *Kloeckera sp.* No. 2201, *Agric. Biol. Chem.*, 38, 111, 1974.
61. **Wagner, C., Lusty, S. M., Kung, H-F., and Rogers, N. L.**, Preparation and properties of trimethylsulfonium-tetrahydrofolate methyltransferase, *J. Biol. Chem.*, 242, 1287, 1967.
62. **Large, P. J., Boulton, C. A., and Crabbe, M. G. C.**, The reduced nicotinamide-adenine dinucleotide phosphate- and oxygen-dependent *N*-oxygenation of trimethylamine by *Pseudomonas aminovorans*, *Biochem. J.*, 128, 137P, 1972.
63. **Myers, P. A. and Zatman, L. J.**, The metabolism of trimethylamine *N*-oxide by *Bacillus* PM6, *Biochem. J.*, 121, 10P, 1971.
64. **Large, P. J.**, Non-oxidative demethylation of trimethylamine *N*-oxide by *Pseudomonas aminovorans*, *FEBS Lett.*, 18, 297, 1971.
65. **Meiberg, J. B. M. and Harder, W.**, Synthesis of certain assimilatory and dissimilatory enzymes during bacterial adaptation to growth on trimethylamine, *J. Gen. Microbiol.*, 101, 151, 1977.
66. **Large, P. J. and McDougall, H.**, An enzymic method for the microestimation of trimethylamine, *Anal. Biochem.*, 64, 304, 1975.

67. **Eady, R. R. and Large, P. J.**, Bacterial oxidation of dimethylamine, a new mono-oxygenase reaction, *Biochem. J.*, 111, 37P, 1969.

68. **Jarman, T. R., Eady, R. R., and Large, P. J.**, An enzymically active P-420-type cytochrome involved in the mixed-function diemthylamine oxidase system of *Pseudomonas aminovorans*, *Biochem. J.*, 119, 55P, 1970.

69. **Meiberg, J. B. M. and Harder, W.**, Aerobic and anaerobic degradation of trimethylamine and dimethylamine by *Hyphomicrobium X*, *Proc. Soc. Gen. Microbiol.*, 4, 45, 1976.

70. **Colby, J. and Zatman, L. J.**, Purification and properties of the trimethylamine dehydrogenase of Bacterium 4B6, *Biochem. J.*, 143, 555, 1974.

71. **Boulton, C. A. and Large, P. J.**, Oxidation of *N*-alkyl and *NN*-dialkylhydroxylamines by partially purified preparations of trimethylamine mono-oxygenase from *Pseudomonas aminovorans*, *FEBS Lett.*, 55, 286, 1975.

72. **Eady, R. R. and Large, P. J.**, Purification and properties of an amine dehydrogenase from *Pseudomonas* AMI and its role in growth on methylamine, *Biochem. J.*, 106, 245, 1968.

73. **Eady, R. R. and Large, P. J.**, Microbial oxidation of amines. Spectral and kinetic properties of the primary amine dehydrogenase of *Pseudomonas* AMI, *Biochem. J.*, 123, 757, 1971.

74. **Mehta, R. J.**, Methylamine dehydrogenase from the obligate methylotroph, *Methylomonas methylovora*, *Can. J. Microbiol.*, 23, 402, 1977.

75. **Matsumoto, J.**, Methylamine dehydrogenase of *Pseudomonas sp.* J., purification and properties, *Biochim. Biophys. Acta*, 522, 291, 1978.

76. **Matsumoto, J.**, Methylamine dehydrogenase of *Pseudomonas sp.* J., isolation and properties of the subunits, *Biochim. Biophys. Acta*, 522, 303, 1978.

77. **Matsumoto, T. and Tobari., J.**, Methylamine dehydrogenase of *Pseudomonas sp.* J and *Pseudomonas* AMI, *J. Biochem.*, 84, 461, 1978.

78. **Large, P. J. and Carter, R. H.**, Specific activities of enzymes of the serine pathway of carbon assimilation in *Pseudomonas aminovorans* and *Pseudomonas* MS grown on methylamine, *Biochem. Soc. Trans.*, 1, 1291, 1973.

79. **Bamforth, C. W., and Large, P. J.**, Solubilization, partial purification and properties of *N*-methylglutamate dehydrogenase from *Pseudomonas aminovorans*, *Biochem. J.*, 161, 357, 1977.

80. **Hersh, L. B., Peterson, J. A., and Thompson, A. A.**, An *N*-methyl glutamate dehydrogenase from *Pseudomonas* MA., *Arch. Biochem. Biophys.*, 145, 115, 1971.

81. **Bellion, E. and Hersh, L. B.**, Methylamine metabolism in a *Pseudomonas* species, *Arch. Biochem. Biophys.*, 153, 368, 1972.

82. **Pollock, R. J. and Hersh, L. B.**, *N*-Methylglutamate synthetase. Purification and properties of the enzyme, *J. Biol. Chem.*, 246, 4737, 1971.

83. **Hersh, L. B., Stark, M. J., Worthen, S., and Fiero, M. K.**, *N*-Methyl glutamate dehydrogenase: kinetic studies on the solubilized enzyme, *Arch. Biochem. Biophys.*, 150, 219, 1972.

84. **Loginova, N. V. and Trotsenko, Yu. A.**, Enzymes involved in the metabolism of methanol and methylamine by *Pseudomonas methylica*, *Mikrobiologiya*, 43, 979, 1974.

85. **Netrusov, A. I.**, NAD-dependent *N*-methylglutamate dehydrogenase — a new enzyme metabolising methylamine in methylotrophs, *Mikrobiologiya*, 44, 552, 1975.

86. **Loginova, N. V., Shishkina, V. N., and Trotsenko, Yu. A.**, Primary metabolic pathways of methylated amines in *Hyphomicrobium vulgare*, *Mikrobiologiya*, 45, 34, 1976.

87. **Kung, H. K. and Wagner, C.**, The enzymatic synthesis of *N*-methylalanine, *Biochim. Biophys. Acta*, 201, 513, 1970.

88. **Lin, M. C-M. and Wagner, C.**, Purification and characterization of *N*-methylalanine dehydrogenase, *J. Biol. Chem.*, 250, 3746, 1975.

89. **Boulton, C. A. and Large, P. J.**, Synthesis of certain assimilatory and dissimilatory enzymes during bacterial adaptation to growth on trimethylamine, *J. Gen. Microbiol.*, 101, 151, 1977.

90. **Quayle, J. R. and Pfennig, N.**, Utilisation of methanol by *Rhodospirallaceae*, *Arch. Microbiol.*, 102, 193, 1975.

91. **Sahm, H., Cox, R. B., and Quayle, J. R.**, Metabolism of methanol by *Rhodopseudomonas acidophila*, *J. Gen. Microbiol.*, 94, 313, 1976.

92. **Stokes, J. E. and Hoare, D. S.**, Reductive pentose cycle and formate assimilation in *Rhodopseudomonas palustris*, *J. Bacteriol.*, 100, 890, 1969.

93. **Yoch, D. C. and Lindstrom, E. S.**, Photosynthetic conversion of formate and CO_2 to glutamate by *Rhodopseudomonas palustris*, *Biochem. Biophys. Res. Commun.*, 28, 65, 1967.

94. **Yoch, D. C. and Lindstrom, E. S.**, Nicotinamide adenine dinucleotide-dependent formate dehydrogenase from *Rhodopseudomonas palustris*, *Arch. Microbiol.*, 67, 182, 1969.

95. **Hirsch, P., Morita, S., and Conti, S. F.**, Cytochromes and CO-binding hemoproteins of *Hyphomicrobium vulgare*, *Bacteriol. Proc.*, 97, 1963.

96. **Anthony, C.**, Cytochrome *c* and the oxidation of C_1 compounds in *Pseudomonas* AMI, *Biochem. J.*, 119, 54P, 1970.
97. **Netrusov, A. I., Verkhoturov, V. N., Kirikova, N. N., and Kondratjeva, E. N.**, Study of the respiratory system in *Pseudomonas sp.* assimilating one-carbon compounds, *Mikrobiologiya*, 40, 200, 1971.
98. **Tonge, G. M., Knowles, C. J., Harrison, D. E. F., and Higgins, I. J.**, Metabolism of one-carbon compounds. Cytochromes of methane — and methanol-utilising bacteria, *FEBS Lett.*, 44, 106, 1974.
99. **Higgins, I. J., Knowles, C. J., and Tonge, G. M.**, Enzymic mechanisms of methane and methanol oxidation in relation to electron transport systems in methylotrophs; purification and properties of methane oxygenase, in *Microbial Production and Utilization of Gases*, Schlegel, H. G., Gottschalk, G., and Pfennig, N., Eds., E. Goltze, Göttingen, Federal Republic of Germany, 1976, 389.
100. **Higgins, I. J., Taylor, S. C., and Tonge, G. M.**, The respiratory system of *Pseudomonas extorquens*, *Proc. Soc. Gen. Microbiol.*, 3, 179, 1976.
101. **Tonge, G. M.**, The Bioenergetics of Methylotrophic Bacteria and the Methane Oxidising Enzyme System of *Methylosinus trichosporium*, Ph.D. thesis, University of Kent, U.K., 1977.
102. **Tonge, G. M., Taylor, F., and Higgins, I. J.**, The respiratory system of *Pseudomonas extorquens*, in *Microbial Growth on C_1-Compounds*, Skryabin, G. K., Ivanov, M. V., Kondratjeva, E. N., Zavarzin, G. A., Trotsenko, Yu.A., and Netrusov, A. I., Eds., U.S.S.R. Academy of Sciences, Pushchino, 1977, 75.
103. **Anthony, C.**, The microbial metabolism of C_1 compounds; the cytochromes of *Pseudomonas* AMI, *Biochem. J.*, 146, 289, 1975.
104. **Widdowson, D. and Anthony, C.**, The microbial metabolism of C_1 compounds; the electron-transport chain of *Pseudomonas* AMI, *Biochem. J.*, 152, 349, 1975.
105. **O'Keeffe D. T. and Anthony, C.**, Proton translocation in the facultative methylotroph *Pseudomonas* AMI, *Proc. Soc. Gen. Microbiol.*, 4, 67, 1977.
106. **Anthony, C. and O'Keeffe, D. T.**, Proton translocation in a mutant of *Pseudomonas* AMI lacking cytochrome-*c*, *Proc. Soc. Gen. Microbiol.*, 4, 68, 1977.
107. **O'Keeffe, D. T. and Anthony, C.**, The microbial metabolism of C_1 compounds. The Stoicheometry of respiration-driven proton translocation in *Pseudomonas* AMI, and in a mutant lacking cytochrome-*c*, *Biochem. J.*, 170, 561, 1978.
108. **West, I. C. and Mitchell, P.**, The proton-translocating adenosine triphosphatase of *Escherichia coli*, *FEBS Lett.*, 40, 1, 1974.
109. **Jones, C. W.**, Aerobic respiratory systems in bacteria, *Symp. Soc. Gen. Microbiol.*, 27, 23, 1977.
110. **Keevil, C. W. and Anthony, C.**, The relationship between proton translocation and cell yields in the facultative methylotroph *Pseudomonas* AMI and a mutant lacking cytochrome-*c*, *Biochem. Soc. Trans.*, 7, 179, 1979.
111. **Netrusov, A. I. and Anthony, C.**, The microbial metabolism of C_1 compounds. Oxidative phosphorylation in membrane preparations of *Pseudomonas* AMI, *Biochem. J.*, 178, 353, 1979.
112. **Hammond, R. C. and Higgins, I. J.**, Respiration-driven proton translocation in *Pseudomonas extorquens* and *Pseudomonas* AMI, *Proc. Soc. Gen. Microbiol.*, 5, 43, 1978.
113. **Hammond, R. C. and Higgins, I. J.**, unpublished observations.
114. **Netrusov, A. I., Rodionov, Y. V., and Kondratjeva, E. N.**, ATP-generation coupled with C_1-compound oxidation by methylotrophic bacterium *Pseudomonas* sp. 2, *FEBS Lett.*, 76, 56, 1977.
115. **Netrusov, A. I.**, ATP generation by methylotrophic bacteria, in *Microbial Growth on C_1-Compounds*, Skryabin, G. K., Ivanov, M. V., Kondratjeva, E. N., Zavarzin, G. A., Trotsenko, Yu. A., and Netrusov, A. I., Eds., U.S.S.R. Academy of Sciences, Pushchino, 1977, 78.
116. **Drabikowska, A. K.**, The respiratory chain of a newly isolated *Methylomonas* P11, *Biochem. J.*, 168, 171, 1977.
117. **Cross, A. R., and Anthony, C.**, The cytochromes of *Methylophilus methylotrophus*, *Proc. Soc. Gen. Microbiol.*, 5, 42, 1978.
118. **Anthony, C.**, The oxidation of NAD(P)H by methylotrophs, *Proc. Soc. Gen. Microbiol.*, 5, 67, 1978.
119. **Higgins, I. J. and Quayle, J. R.**, Oxygenation of methane by methane-grown *Pseudomonas methanica* and *Methanomonas methanooxidans*, *Biochem. J.*, 118, 201, 1970.
120. **Ribbons, D. W. and Michalover, J. L.**, Methane oxidation by cell-free extracts of *Methylococcus capsulatus*, *FEBS Lett.*, 11, 41, 1970.
121. **Ribbons, D. W.**, Oxidation of C_1 compounds by particulate fractions from *Methylococcus capsulatus*: distribution and properties of methane-dependent reduced nicotinamide adenine dinucleotide oxidase (methane hydroxylase), *J. Bacteriol.*, 122, 1351, 1975.
122. **Colby, J. and Dalton, H.**, Some properties of a soluble methane mono-oxygenase from *Methylococcus capsulatus* strain Bath, *Biochem. J.*, 157, 495, 1976.

123. **Colby, J., Stirling, D. I., and Dalton, H.**, The soluble methane monooxygenase of *Methylococcus capsulatus* (Bath). Its ability to oxygenate *n*-alkanes, *n*-alkenes, ethers and alicyclic, aromatic and heterocyclic compounds, *Biochem. J.*, 165, 395, 1977.
124. **Colby, J. and Dalton, H.**, Resolution of the methane mono-oxygenase of *Methylococcus capsulatus* (Bath) into three components, *Biochem. J.*, 171, 461, 1978.
125. **Colby, J. and Dalton, H.**, Structure and function of component C (NADH-acceptor reductase) of the methane mono-oxygenase of *Methylococcus capsulatus* (Bath), *Proc. Soc. Gen. Microbiol.*, 5, 101, 1978.
126. **Dalton, H.**, Personal communication.
127. **Ferenci, T.**, Carbon monoxide-stimulated respiration in methane-utilizing bacteria, *FEBS Lett.*, 41, 94, 1974.
128. **Ferenci, T., Strøm, T., and Quayle, J. R.**, Oxidation of carbon monoxide and methane by *Pseudomonas methanica*, *J. Gen. Microbiol.*, 91, 79, 1975.
129. **Colby, J., Dalton, H., and Whittenbury, R.**, An improved assay for bacterial methane mono-oxygenase: some properties of the enzyme from *Methylomonas methanica*, *Biochem. J.*, 151, 459, 1975.
130. **Tonge, G. M., Harrison, D. E. F., Knowles, C. J., and Higgins, I. J.**, Properties and partial purification of the methane-oxidising enzyme system from *Methylosinus trichosporium*, *FEBS Lett.*, 58, 293, 1975.
131. **Tonge, G. M., Harrison, D. E. F., and Higgins, I. J.**, Purification and properties of the methane mono-oxygenase enzyme system from *Methylosinus trichosporium* OB3b, *Biochem. J.*, 161, 333, 1977.
132. **Higgins, I. J.**, Microbial oxidation of methane, *Abstr. 12th Int. Congr. Microbiol.*, Munich, 1978, 17.
133. **Higgins, I. J., Tonge, G. M., and Hammond, R. C.**, Methane mono-oxygenase and respiration in *Methylosinus trichosporium*, in *Microbial Growth on C_1-Compounds*, Skryabin, G. K., Ivanov, M. V., Kondratjeva, E. N., Zavarzin, G. A., Trotsenko, Yu. A., and Netrusov, A. T., Eds., U.S.S.R. Academy of Sciences, Pushchino, 1977, 65.
134. **Patel, R. N., Hou, C. T., and Felix, A.**, Microbial oxidation of methane and methanol: isolation of methane-utilising bacteria and characterisation of a facultative methane-utilising isolate, *J. Bacteriol.*, 136, 352, 1978.
135. **Higgins, I. J.**, British provisional patent 35123/78, 1978.
136. **Stirling, D. I., Colby, J., and Dalton, H.**, Oxidation of various carbon compounds by methane-utilising bacteria, *Proc. Soc. Gen. Microbiol.*, 5, 101, 1978.
137. **Patel, R. N. and Hoare, D. S.**, Physiological studies of methane- and methanol-oxidising bacteria: oxidation of C-1 compounds by *Methylococcus capsulatus*, *J. Bacteriol.*, 107, 187, 1971.
138. **Patel, R. N., Bose, H. R., Mandy, W. J., and Hoare, D. S.**, Physiological studies of methane- and methanol-oxidising bacteria: comparison of a primary alcohol dehydrogenase from *Methylococcus capsulatus* (Texas strain) and *Pseudomonas* species M27, *J. Bacteriol.*, 110, 570, 1972.
139. **Patel, R. N., Mandy, W. J., and Hoare, D. S.**, Physiological studies of methane and methanol-oxidising bacteria: immunological comparison of a primary alcohol dehydrogenase from *Methylococcus capsulatus* and *Pseudomonas* species M27, *J. Bacteriol.*, 113, 937, 1973.
140. **Wadzinski, A. M. and Ribbons, D. W.**, Oxidation of C_1 compounds by particulate fractions of *Methylococcus capsulatus*: properties of methanol oxidase and methanol dehydrogenase, *J. Bacteriol.*, 122, 1364, 1975.
141. **Patel, R. N. and Felix, A.**, Microbial oxidation of methane and methanol: crystallization and properties of methanol dehydrogenase from *Methylosinus sporium*, *J. Bacteriol.*, 128, 413, 1976.
142. **Weaver, T. L. and Dugan, P. R.** Methylotrophic enzyme distribution in *Methylosinus trichosporium*, *J. Bacteriol.*, 122, 433, 1975.
143. **Wolf, H. J. and Hanson, R. S.**, Alcohol dehydrogenase from *Methylobacterium organophilum*, *Appl. Environ. Microbiol.*, 36, 105, 1978.
144. **Davey, J. F., Whittenbury, R., and Wilkinson, J. F.**, The distribution in the Methylobacteria of some key enzymes concerned with intermediary metabolism, *Arch. Microbiol.*, 87, 359, 1972.
145. **Harrington, A. A. and Kallio, R. E.**, Oxidation of methanol and formaldehyde by *Pseudomonas methanica*, *Can. J. Microbiol.*, 6, 1, 1960.
146. **Stirling, D. I. and Dalton, H.**, Purification and properties of $NAD(P)^+$-linked formaldehyde dehydrogenase from *Methylococcus capsulatus* (Bath), *J. Gen. Microbiol.*, 107, 19, 1978.
147. **Stephens, G. M. and Higgins, I. J.**, unpublished observations.
148. **Davey, J. F. and Mitton, J. R.**, Cytochromes of two methane-utilising bacteria, *FEBS Lett.*, 37, 335, 1973.
149. **Monosov, E. Z. and Netrusov, A. I.**, Localization of energy generators in methane-oxidising bacteria, *Mikrobiologiya*, 45, 598, 1976.

150. **Gvozdev, R. I., Sadkov, A. P., Belova, V. S., and Pilyashenko-Novokhatny, A. I.**, The investigation of the mechanism of biological methane oxidation, in *Microbial Growth on C_1-Compounds*, Skryabin, G. K., Ivanov, M. V., Kondratjeva, E. N., Zavarzin, G. A., Trotsenko, Yu. A., and Netrusov, A. I., Eds., U.S.S.R. Academy of Sciences, Pushchino, 1977, 68.
151. **Ferenci, T.**, Oxygen metabolism in *Pseudomonas methanica, Arch. Microbiol.*, 108, 217, 1976.
152. **Tonge, G. M., Drozd, J. W., and Higgins, I. J.**, Energy coupling in *Methylosinus trichosporium, J. Gen. Microbiol.*, 99, 229, 1977.
153. **Tongu, G. M., Drozd, J. W., and Higgins, I. J.**, Respiration and energy coupling in *Methylosinus trichosporium, Proc. Soc. Gen. Microbiol.*, 3, 179, 1976.
154. **Drozd, J. W., Linton, J. D., Downs, J., Stephenson, R., Bailey, M. L., and Wren, S. J.**, Growth energetics in methylotrophic bacteria, in *Microbial Growth on C_1-Compounds*, Skyrabin, G. K., Ivanov, M. V., Kondratjeva, E. M., Zavarzin, G. A., Trotsenko, Yu. A., and Nesterov, A. I., Eds., U.S.S.R. Academy of Sciences, Pushchino, 1977, 91.
155. **Linton, J. D. and Vokes, J.**, Growth of the methane utilising bacterium *Methylococcus* NCIB 11083 in mineral salts medium with methanol as the sole source of carbon, *FEMS Microbiol. Lett.*, 4, 125, 1978.
156. **Harrison, D. E. F.**, Studies on the affinity of methanol- and methane-utilizing bacteria for their carbon substrates, *J. Appl. Bacteriol.*, 36, 301, 1973.
157. **Conti, S. F. and Hirsch, P.**, Biology of budding bacteria. III. Fine structure of *Rhodomicrobium* and *Hyphomicrobium spp., J. Bacteriol.*, 89, 503, 1965.
158. **Patt, J. E. and Hanson, R. S.**, Intracytoplasmic membrane, phospholipid and sterol content of *Methylobacterium organophilum* cells grown under different conditions, *J. Bacteriol.*, 134, 636, 1978.
159. **Brannan, J. and Higgins, I. J.**, Effect of growth conditions on the intracytoplasmic membranes of *Methylosinus trichosporium* OB3b, *Proc. Soc. Gen. Microbiol.*, 5, 69, 1978.
160. **Oelze, J. and Drews, G.**, Membranes of photosynthetic bacteria, *Biochim. Biophys. Acta*, 265, 209, 1972.
161. **Murray, R. G. E. and Watson, S. W.**, Structure of *Nitrocystis oceanus* and comparison with Nitrosomonas and *Nitrobacter, J. Bacteriol.*, 89, 1594, 1965.
162. **Pope, L. M., Hoare, D. S., and Smith, A. J.**, Ultrastructure of *Nitrobacter agilis* grown under autotrophic and heterotrophic conditions, *J. Bacteriol.*, 97, 936, 1969.
163. **Gantt, E. and Conti, S. F.**, Ultrastructure of blue-green algae, *J. Bacteriol.*, 97, 1486, 1969.
164. **Kennedy, R. S. and Finnerty, W. R.**, Microbial assimilation of hydrocarbons. II. Intracytoplasmic membrane induction in *Acinetobacter sp., Arch. Microbiol.*, 102, 85, 1975.
165. **Stirling, L. A., Watkinson, R. J., and Higgins, I. J.**, Microbial metabolism of alicyclic hydrocarbons: isolation and properties of a cyclohexane-degrading bacterium, *J. Gen. Microbiol.*, 99, 119, 1977.
166. **Drozd, J. W., Khosrovi, B., Downs, J., Bailey, M. L., Barnes, L. J., and Linton, L. J.**, Biomass production from natural gas, in *Proc. Int. Continuous Culture Symp.*, Sikyta, B., Ed., Czechoslovakia Academy of Science, Prague, in press, 1980.
167. **Harder, W. and van Dijken, J. P.**, Theoretical considerations on the relation between energy production and growth of methane-utilizing bacteria, in *Microbial Production and Utilization of Gases*, Schlegel, H. G., Gottschalk, G., and Pfennig, N., Eds., E. Goltze, Göttingen, Federal Republic of Germany, 1975, 403.
168. **Goldberg, I., Rock, J. S., Ben-Bassat, A., and Mateles, R. I.**, Bacterial yields on methanol, methylamine, formaldehyde and formate, *Biotechnol. Bioeng.*, 18, 1657, 1976.
169. **Nagai, S., Mori, T., and Aiba, S.**, Investigation of the energetics of methane-utilising bacteria in methane- and oxygen-limited chemostat cultures, *J. Appl. Chem. Biotechnol.*, 23, 549, 1973.
170. **Goldberg, I.**, Production of SCP from methanol-yield factors, *Process Biochem.*, 12 (9) 12, 1977.
171. **Sokolov, A. P. and Trotsenko, Yu. A.**, Cyclic pathways of formaldehyde oxidation in *Pseudomonas oleovorans, Mikrobiologiya*, 46, 1119, 1977.
172. **Loginova, N. V. and Trotsenko, Y. A.**, Autotrophic growth on methanol by bacteria isolated from activated sludge, *FEMS Microbiol. Lett.*, 5, 239, 1979.
173. **O'Keeffe, D. T., and Anthony, C.**, The two cytochromes *c* in the facultative methylotroph, *Pseudomonas* AM1, *Soc. Gen. Microbiol. Quart.*, 6, 70, 1979.
174. **Large, P. J., Meiberg, J. B., and Harder, N.**, Cytochrome-c_{co} is not a primary electron acceptor for the amine dehydrogenase of *Hyphomicrobium X, FEMS Microbiol. Lett.*, 5, 281, 1979.
175. **Stirling, D. I., Colby, J., and Dalton, H.**, A comparison of the substrate specificity and electron donor specificities of the methane mono-oxygenases from three strains of methane-oxidising bacteria, *Biochem. J.*, 177, 361, 1979.
176. **Higgins, I. J., Hammond, R. C., Sariaslani, F. S., Best, D., Davies, M. M., Tryhorn, S. E., and Taylor, F.**, Biotransformation of hydrocarbons and related compounds by whole organism suspensions of methane-grown *Methylosinus trichosporium* OB3b, *Biochem. Biophys. Res. Commun.*, 89, 671, 1979.

177. **Stirling, D. I. and Dalton, H.**, Properties of the methane mono-oxygenase from extracts of *Methylosinus trichosporium* OB3b and evidence for its similarity to the enzyme from *Methylococcus capsulatus* (Bath), *Eur. J. Biochem.*, 96, 205, 1979.
178. **Scott, D. and Higgins, I. J.**, unpublished work.
179. **Brannan, J., Scott, D., and Higgins, I. J.**, The effects of growth- and post-growth-conditions on intracytoplasmic membrane content and location of methane monooxygenase activities in *Methylosinus trichosporium* OB 3b, *Proceedings of the 3rd International Symposium on Microbial Growth on C_1 Compounds*, Sheffield, in press, 1980.

INDEX

A

B

C

D

E

F

G

H

I

K

L

M

O

P

Q

R

S

T

U

V

W

X

Y

Z

For Product Safety Concerns and Information please contact our EU representative GPSR@taylorandfrancis.com Taylor & Francis Verlag GmbH, Kaufingerstraße 24, 80331 München, Germany